# Advances in Environmental Science

D.C. Adriano and W. Salomons

# Acidic Precipitation

## Volume 2

## Biological and Ecological Effects

Edited by D.C. Adriano and A.H. Johnson

With 47 Illustrations

Springer-Verlag
New York Berlin Heidelberg
London Paris Tokyo Hong Kong

## Volume Editors:

D.C. Adriano
Savannah River Ecology Laboratory
University of Georgia
Aiken, SC 29801
USA

A.H. Johnson
Department of Geology
University of Pennsylvania
Philadelphia, PA 19104-6316
USA

Library of Congress Cataloging-in-Publication Data
Acidic precipitation
  (Advances in environmental science)
  Bibliography: v. 1, p.
  Includes index.
  Contents: v. 1. Case studies/volume editors,
D.C. Adriano and M. Havas—v. 2. Biological and
ecological effects/D.C. Adriano, A.H. Johnson,
editors.
  I. Series.
TD19.5.42.A25 1989
363.7'386                         88-37418

Printed on acid-free paper

Typeset by McFarland Graphics and Design, Dillsburg, Pennsylvania.

9 8 7 6 5 4 3 2 1

ISBN-13:978-1-4613-8901-9      e-ISBN-13:978-1-4613-8899-9
DOI: 10.1007/978-1-4613-8899-9

# Preface to the Series

In 1986, my colleague Prof. Dr. W. Salomons of the Institute for Soil Fertility of the Netherlands and I launched the new *Advances in Environmental Science* with Springer-Verlag New York, Inc. Immediately, we were faced with a task of what topics to cover. Our strategy was to adopt a thematic approach to address hotly debated contemporary environmental issues. After consulting with numerous colleagues from Western Europe and North America, we decided to address *Acidic Precipitation*, which we view as one of the most controversial issues today.

This is the subject of the first five volumes of the new series, which cover relationships among emissions, deposition, and biological and ecological effects of acidic constituents. International experts from Canada, the United States, Western Europe, as well as from several industrialized countries in other regions, have generously contributed to this subseries, which is grouped into the following five volumes:

**Volume 1**  *Case Studies*
   (D.C. Adriano and M. Havas, editors)

**Volume 2**  *Biological and Ecological Effects*
   (D.C. Adriano and A.H. Johnson, editors)

**Volume 3**  *Sources, Deposition, and Canopy Interactions*
   (S.E. Lindberg, A.L. Page, and S.A. Norton, editors)

**Volume 4**  *Soils, Aquatic Processes, and Lake Acidification*
   (S.A. Norton, S.E. Lindberg, and A.L. Page, editors)

**Volume 5**  *International Overview and Assessment*
   (T. Bresser and W. Salomons, editors)

From the vast amount of consequential information discussed in this series, it will become apparent that acidic deposition should be seriously addressed by many countries of the world, in as much as severe damages have already been inflicted on numerous ecosystems. Furthermore, acidic constituents have also been shown to affect the integrity of structures of great historical values in various places of the world. Thus, it is hoped that this up-to-date subseries would increase the "awareness" of the world's citizens and encourage governments to devote more attention and resources to address this issue.

The series editors thank the international panel of contributors for bringing this timely series into completion. We also wish to acknowledge the very insightful input of the following colleagues: Prof. A.L. Page of the University of California, Prof. T.C. Hutchinson of the University of Toronto, and Dr. Steve Lindberg of the Oak Ridge National Laboratory.

We also wish to thank the superb effort and cooperation of the volume editors in handling their respective volumes. The constructive criticisms of chapter reviewers also deserve much appreciation. Finally, we wish to convey our appreciation to my secretary, Ms. Brenda Rosier, and my technician, Ms. Claire Carlson, for their very able assistance in various aspects of this series.

Aiken, South Carolina                                          *Domy C. Adriano*
                                                              *Coordinating Editor*

# Preface to *Acidic Precipitation*, Volume 2 (*Advances in Environmental Science*)

As a result of pioneering research in the 1960s and because of the perceived and real environmental effects described during the ensuing years, the terms *acidic rain, acidic deposition,* or *acidic precipitation* have become commonplace in scientific and popular literature. In the last decade, governments throughout the world have responded to public pressure and to the concerns of the scientific community by establishing research programs on national and international scales. These programs have been designed to enhance our understanding of the important links between atmospheric emissions and their potential environmental effects in both industrialized and developing nations. Acidic precipitation was studied initially because of its effects on aquatic systems. However, because reports from Western Europe in the early 1980s suggested a link with forest decline, acidic precipitation is now considered a potential environmental stress in terrestrial systems as well as in aquatic systems. Most recently, scientists viewed acidic precipitation as part of a larger "global change" issue along with other issues such as warming climate, increasing carbon dioxide in the atmosphere and atmospheric ozone depletion.

As has been the case with many environmental issues of the twentieth century, acidic precipitation has its origin in emissions to the atmosphere of numerous compounds from both natural and man-made sources. There are 10 chapters in this volume covering a wide array of topics on the biological and ecological effects of acidic precipitation. A chapter on soil productivity emphasizes changes in biological and chemical characters of forest soils impacted by acidic deposition. Additional chapters discuss specific effects on soil microorganisms, trees, and crops. The importance of aluminum in this environmental issue is highlighted by a discussion on the mobility and phytotoxicity of this element in acidic soils. This chapter nicely puts into perspective the biology of Al stressed plants. Two major chapters discuss the effects of acidic precipitation on forest ecosystems; one emphasizing North America, and the other, Europe. Effects of soil acidification on key soil processes, including litter decomposition and depletion of essential plant nutrients in the soil profile, are emphasized. And finally, three major chapters comprehensively cover limnological ecosystems and their response to acidic perturbation. These chapters discuss the response of stream and lake communities, both floral and faunal, to water acidification, including reduced biodiversity in these systems.

With the National Acidic Precipitation Assessment Program of the United States nearing a 1990 completion date, and with programs in Canada and many European countries accelerating to reach a consensus on the role that atmospheric emissions and acidic precipitation play in the environment, publication of this series is timely.

The editors wish to thank the contributors to this volume for their excellent discussions of some of the most relevant biological and ecosystem studies dealing with acidic deposition. The authors also wish to thank reviewers for the volume chapters: Ms. Claire Carlson, Mr. Dan Kaplan, and Dr. Ken McLeod of the Savannah River Ecology Laboratory; Dr. James Bowers of the Westing-house Savannah River Laboratory; Dr. Mike Chimney of the Normandeau Associates, Inc.; Dr. Jodi Shann of the University of Cincinnati; Dr. D. Godbold of the Forstbotanisches Institut of the University of Gottingen; and Dr. Paul Miller of the USDA, Forest Service. And finally, we are grateful for the expert assistance of my secretary, Ms. Brenda Rosier, and my technician, Ms. Claire Carlson, that made our task bearable.

Aiken, South Carolina                                                      *Domy C. Adriano*

Philadelphia, Pennsylvania                                          *Arthur H. Johnson*

# Contents

Contents

# Contributors

*C.L. Carlson,* Biogeochemistry Division, Savannah River Ecology Laboratory, Drawer E, Aiken, NC 29801, USA

*J.W. Elwood,* Environmental Science Division, Oak Ridge National Laboratory, Oak Ridge, TN 37831-6036, USA

*L.S. Evans,* Laboratory of Plant Morphogenesis, Manhattan College, The Bronx, NY 10471, USA; and Aquatic Ecology Division, Department of Applied Science, Brookhaven National Laboratory, Upton, NY 11973, USA

*I.J. Fernandez,* Department of Plant and Soil Sciences, University of Maine, One Deering Hall, Orono, ME 04469, USA

*A.J. Francis,* Department of Applied Science, Brookhaven National Laboratory, Upton, NY 11973, USA

*B.L. Haines,* Botany Department, University of Georgia, Athens, GA 30602, USA

*H.H. Harvey,* Department of Zoology, University of Toronto, Toronto M5S 1A1, Canada

*E.T. Howell,* Department of Botany, University of Manitoba, Winnipeg, Manitoba R3T 2N2, Canada

*G. Krantzberg,* Ontario Ministry of the Environment, Water Resources Branch, 1 St. Clair Avenue West, Toronto, Ontario M4V 1P5, Canada

*P.J. Mulholland,* Environmental Sciences Division, Oak Ridge National Laboratory, Oak Ridge, TN 37831-6038, USA

*W.H. Smith,* School of Forestry and Environmental Studies, Yale University, New Haven, CT 06511, USA

*P.M. Stokes,* Institute for Environmental Studies and Department of Botany, University of Toronto, Toronto, Ontario M5S 1A4, Canada

*G.J. Taylor,* Department of Botany, University of Alberta, Edmonton, Alberta T6G 2E9, Canada

*B. Ulrich,* Institute of Soil Science and Forest Nutrition, University of Göttingen, Büsgenweg 2, D-3400 Göttingen, FRG

# Effects of Acidic Precipitation on Trees

Bruce L. Haines[*] and
Claire L. Carlson[†]

## Abstract

Acidic precipitation has the potential to affect trees both directly and indirectly. Potential direct effects could include altered mineral uptake or leaching in the tree canopy, altered metabolism, and damage to tissues. Potential indirect effects operating through changes in soil chemistry include altered mineral nutrient availability, increased solubility of toxic heavy metals, and altered mycorrhizal development. Other indirect interactions with stresses such as gaseous air pollutants including ozone, insects, pathogens, and frost shock could decrease tree vigor. Available evidence from laboratory studies suggests that pollen germination and pollen tube elongation are the most susceptible stages of the tree life cycle to acidic precipitation damage. At present, evidence of direct negative effects of acidic rain on tree growth in the field is lacking. Acidic rain may be one of many interacting environmental factors contributing to forest decline, but gaseous air pollutants appear to be a more serious problem for trees.

## I. Introduction

What are the known and potential effects of acidic deposition upon the growth and physiology of trees? A search for the answer to this question requires consideration of acidic deposition effects on the tree throughout its entire life cycle, from pollination and fertilization, through seed development, seedling establishment, and vegetative growth, to tree death.

Acidic precipitation potentially could affect trees in a variety of ways, both direct and indirect. Direct effects could include injury to aboveground tissues, foliar leaching and/or uptake of ions, and changes in metabolic processes. Indirect

*Botany Department, University of Georgia, Athens, GA 30602, USA.
†Biogeochemistry Division, Savannah River Ecology Laboratory, Drawer E, Aiken, SC 29801, USA.

effects could include altered tree growth and nutrition due to changes in soil chemistry, including pH and nutrient availability, changes in mycorrhizal development and growth, and interactions with other stress agents such as ozone, insects and plant pathogens, and frost.

With a general knowledge of plants, we can ask what structures and functions are most likely to be affected by acidic deposition? The processes of pollen germination and pollen tube growth would be the most sensitive parts of the life cycle. Leaves would be the next most sensitive because they are exposed for long periods to acidic depositions. Roots would be less affected for the short term because they exist in the soil, which is buffered against rapid pH change. Tree trunks covered with bark would be least likely to be affected.

Acidic rain research with trees has frequently evaluated changes in plant structure or in rate processes between groups of plants receiving different simulated acidic rain (SAR) treatments. Results are compared to controls of pH 5.5 or 5.6. The results of these studies have been summarized in Tables 1–1 through 1–5, which show the tree species studied, the author(s), the structure or rate processes evaluated, and the plant response observed. These tables provide a comparative overview only. For details of treatments and responses, readers can consult the original papers. Results of some studies are presented in narrative fashion because they do not fit into the comparative tables.

This review takes a life-cycle approach, beginning with the effects of SAR on pollination, seed production, seed germination, and seedling growth. Effects on leaves and on tree nutrition are explored. Potential interactions with other stress agents and the potential roles of acidic rain in the forest decline/dieback phenomenon are discussed.

Earlier reviews considering the physiological effects of acidic rain on trees, including those by Evans (1984), Morrison (1984), and those in Linthurst (1984), may be of interest to readers.

## II. Effects of Acidic Precipitation on Tree Reproduction

### A. Pollen Germination and Pollen Tube Formation

Although the effects of acidic precipitation on pollen production by trees have not yet been evaluated, there have been several studies of effects of simulated acidic rain (SAR) on pollen germination and pollen tube elongation. Pollens from 13 tree species have been evaluated (Table 1–1). Inhibition of germination in some species occurred between pH 3.5 and 3.9 and in all species at pH 2.5. Pollen tube elongation was inhibited at pH 4.0 in two species. Of the trees evaluated (Table 1–1), the deciduous genera *Acer*, *Betula*, *Cornus*, and *Populus* were negatively affected by SAR at higher pH values than were the conifers *Picea*, *Pinus*, and *Tsuga*. There was a graduated decrease in both pollen germination and pollen tube elongation with decreasing pH of SAR in the study by Van Ryn et al. (1986).

Potential effects of SAR on the ability of the angiosperm stigma and of the gymnosperm pollen droplet in the pollen cavity to promote pollen germination and pollen tube elongation have yet to be investigated (Wolters and Martens, 1987).

**Table 1-1.** Pollen germination and pollen tube enlongation responses following exposure to simulated acidic rain. For studies comparing responses at various pH values against a control, the pH of the control is shown as $C$, and treatments with statistically different responses from $C$ are indicated by *. Results of studies not making statistical comparisons but showing a negative effect are indicated by —; those showing a graduated decreasing response down to the lowest pH tested are indicated as —<——————. Treatments applied at a particular pH but for which no effect was observed are indicated by 0.

| Species, authors, and response evaluated | Simulated rainfall pH | | | | | | | |
|---|---|---|---|---|---|---|---|---|
| | 2.0 2.4 | 2.5 2.9 | 3.0 3.4 | 3.5 3.9 | 4.0 4.4 | 4.5 4.9 | 5.0 5.4 | 5.5 5.9 |
| *Acer saccharum* Marsh. Cox, 1983 | | | | | | | | |
| Germination | * | * | * | | | | | C |
| Elongation | * | * | * | * | | | | C |
| *Acer saccharum* Marsh. Van Ryn et al., 1986 | | | | | | | | |
| Germination | | | | — <———————— | | | C | |
| Elongation | | | | — <——— | | | C | |
| *Betula alleghaniensis* Britton Van Ryn et al., 1986 | | | | | | | | |
| Germination | | | | — <———————— | | | C | |
| Elongation | | | | — <——— | | | C | |
| *Betula lenta* L. Van Ryn et al., 1986 | | | | | | | | |
| Germination | | | | — <———————— | | | C | |
| Elongation | | | | — <——— | | | C | |
| *Betula papyrifera* Marsh. Cox, 1983 | | | | | | | | |
| Germination | * | * | * | | | | | C |
| Elongation | * | * | * | | | | | C |
| *Cornus florida* L. Van Ryn et al., 1986 | | | | | | | | |
| Germination | | | | — <———————— | | | C | |
| Elongation | | | | — <——— | | | C | |
| *Malus domestica* Borkh. Forsline et al., 1983a | | | | | | | | |
| Germination | * | | 0 | | | 0 | | 0 |
| *Picea mariana* (Miller) BSP Cox, 1983 | | | | | | | | |
| Germination | * | | | | | | | C |
| Elongation | * | | | | | | | C |
| *Pinus banksiana* Lamb. Cox, 1983 | | | | | | | | |
| Germination | * | | | | | | | C |
| Elongation | * | | | | | | | C |

**Table 1-1.** (*Continued*)

| Species, authors, and response evaluated | Simulated rainfall pH | | | | | | | |
|---|---|---|---|---|---|---|---|---|
| | 2.0 2.4 | 2.5 2.9 | 3.0 3.4 | 3.5 3.9 | 4.0 4.4 | 4.5 4.9 | 5.0 5.4 | 5.5 5.9 |
| *Pinus resinosa* Alt. | | | | | | | | |
| Cox, 1983 | | | | | | | | |
| Germination | | * | | | | | | C |
| Elongation | | * | | | | | | C |
| *Pinus strobus* L. | | | | | | | | |
| Cox, 1983 | | | | | | | | |
| Germination | | * | * | * | | | | C |
| Elongation | | * | * | | | | | C |
| *Populus tremuloides* Michx. | | | | | | | | |
| Cox, 1983 | | | | | | | | |
| Germination | | * | * | * | | | | C |
| Elongation | | * | * | * | * | | | C |
| *Prunus pennsylvanica* L. | | | | | | | | |
| Cox, 1983 | | | | | | | | |
| Germination | | * | * | | | | | C |
| Elongation | | 0 | | | | | | C |
| *Tsuga canadensis* (L.) Carr. | | | | | | | | |
| Cox, 1983 | | | | | | | | |
| Germination | | * | | | | | | C |
| Elongation | | * | * | | | | | C |

Studies with the angiosperm herb *Oenothera parviflora* L. (Cox, 1987) and with *Zea mays* L. (Wertheim and Craker, 1987) showed decreased pollen germination on stigmas and on silks exposed to SAR. These could serve as models for future studies with trees.

## B. Seed Production and Germination

Potential effects of acidic rain on seed production have been evaluated only for *Malus domestica* Borkh. Exposure to SAR from pH 5.5 to 2.5 did not result in changes in the seed number per fruit (Forsline et al., 1983a).

At least three studies of the effects of acidic rain on seed germination have been conducted in North America (Table 1–2). In a study of 11 species subjected to SAR pH values from 5.6 to 3.0, significant stimulation of germination was shown for five species at pH 3.0 (Lee and Weber, 1979). Inhibition of germination in *Pseudotsuga menziesii* was observed at SAR of pH 2.0 (McColl and Johnson, 1983), whereas stimulation was observed in this species at pH 3.0 (Lee and Weber, 1979). Inhibition in additional species was observed between SAR pH 2.5 and pH 4.0, whereas stimulation of germination was found in *Pinus strobus* at SAR pH values between 2.0 and 3.0 (Raynal et al., 1982a).

**Table 1-2.** Summary of seed germination, shoot growth, and root growth responses to simulated acidic rain. For studies comparing responses at various pH values against a control, the pH of the control is shown as *C*, and treatments with statistically different negative responses from *C* are indicated by —*; +* means significant positive effect; 0 means effect tested for but not significant.

| Species, authors, and response evaluated | Simulated rainfall pH | | | | | | | |
|---|---|---|---|---|---|---|---|---|
| | 2.0 / 2.4 | 2.5 / 2.9 | 3.0 / 3.4 | 3.5 / 3.9 | 4.0 / 4.4 | 4.5 / 4.9 | 5.0 / 5.4 | 5.5 / 5.9 |
| *Abies balsamea* L. Scherbatskoy et al., 1987 | | | | | | | | |
| Germination | | | +* | | 0 | | *C* | |
| *Acer rubrum* L. Raynal et al., 1982a,b | | | | | | | | |
| Germination | | | —* | | —* | | | *C* |
| *Acer saccharum* Marsh Raynal et al., 1982a | | | | | | | | |
| Germination | | | 0 | | 0 | | | *C* |
| Shoot growth (high nutrient) | 0 | | 0 | | 0 | | 0 | |
| Shoot growth (low nutrient) | —* | | — | | 0 | | 0 | |
| Root growth | —* | | —* | | 0 | | 0 | |
| *Acer saccharum* Marsh Lee and Weber, 1979 | | | | | | | | |
| Germination | | | 0 | 0 | 0 | | | *C* |
| Shoot growth | | | 0 | 0 | 0 | | | *C* |
| Root growth | | | 0 | 0 | 0 | | | *C* |
| *Betula alleghaniensis* Britt. Lee and Weber, 1979 | | | | | | | | |
| Germination | | | + | 0 | 0 | | | *C* |
| Shoot growth | | | 0 | 0 | 0 | | | *C* |
| Root growth | | | 0 | 0 | 0 | | | *C* |
| *Betula alleghaniensis* Britt. Wood and Bormann, 1974 | | | | | | | | |
| Shoot growth | —* | | | | | *C* | | |
| *betula alleghaniensis* Britt. Scherbatskoy et al., 1987 | | | | | | | | |
| Germination | | | +* | | 0 | | *C* | |
| *Betula lutea* Michx. F. Raynal et al., 1982a | | | | | | | | |
| Germination | | | —* | | 0 | | | *C* |
| *Betula papyrifera* Marsh Scherbatskoy et al., 1987 | | | | | | | | |
| Germination | | | 0 | | 0 | | *C* | |

**Table 1–2.** (*Continued*)

| Species, authors, and response evaluated | 2.0 2.4 | 2.5 2.9 | 3.0 3.4 | 3.5 3.9 | 4.0 4.4 | 4.5 4.9 | 5.0 5.4 | 5.5 5.9 |
|---|---|---|---|---|---|---|---|---|
| *Carya ovata* (Mill.) K. Koch | | | | | | | | |
| Lee and Weber, 1979 | | | | | | | | |
|   Germination | | | 0 | 0 | 0 | | | C |
|   Shoot growth | | | 0 | 0 | 0 | | | C |
|   Root growth | | | 0 | 0 | 0 | | | C |
| *Cornus florida* L. | | | | | | | | |
| Lee and Weber, 1979 | | | | | | | | |
|   Germination | | | 0 | 0 | 0 | | | C |
|   Shoot growth | | | 0 | 0 | 0 | | | C |
|   Root growth | | | 0 | 0 | 0 | | | C |
| *Fagus grandifolia* Ehrh. | | | | | | | | |
| Lee and Weber, 1979 | | | | | | | | |
|   Germination | | | 0 | 0 | 0 | | | C |
|   Shoot growth | | | 0 | 0 | 0 | | | C |
| *Juniperus virginiana* L. | | | | | | | | |
| Lee and Weber, 1979 | | | | | | | | |
|   Germination | | | +* | 0 | 0 | | | C |
|   Shoot growth | | | 0 | 0 | 0 | | | C |
|   Root growth | | | 0 | 0 | 0 | | | C |
| *Liquidambar styraciflua* L. | | | | | | | | |
| Neufeld et al., 1985 | | | | | | | | |
|   Shoot growth (height) | 0 | | 0 | | 0 | | | C |
|   Shoot + root dry weight | —* | | 0 | | 0 | | | C |
| *Liriodendron tulipifera* L. | | | | | | | | |
| Neufeld et al., 1985 | | | | | | | | |
|   Shoot growth (height) | 0 | | 0 | | 0 | | | C |
|   Shoot + root dry weight | 0 | | 0 | | 0 | | | C |
| *Liriodendron tulipifera* L. | | | | | | | | |
| Lee and Weber, 1979 | | | | | | | | |
|   Germination | | | 0 | 0 | 0 | | | C |
|   Shoot growth | | | 0 | 0 | +* | | | C |
|   Root growth | | | 0 | 0 | 0 | | | C |
| *Picea rubens* Sarg. | | | | | | | | |
| Scherbatskoy et al., 1987 | | | | | | | | |
|   Germination | | | 0 | | 0 | | C | |
| *Pinus strobus* L. | | | | | | | | |
| Wood and Bormann, 1977 | | | | | | | | |
|   Shoot growth | +* | | | | | | | C |
| *Pinus strobus* L. | | | | | | | | |
| Raynal et al., 1982a | | | | | | | | |
|   Germination | +* | | +* | | 0 | | | C |

**Table 1-2.** (*Continued*)

| Species, authors, and response evaluated | Simulated rainfall pH | | | | | | | |
|---|---|---|---|---|---|---|---|---|
| | 2.0 2.4 | 2.5 2.9 | 3.0 3.4 | 3.5 3.9 | 4.0 4.4 | 4.5 4.9 | 5.0 5.4 | 5.5 5.9 |
| *Pinus ponderosa* Laws. | | | | | | | | |
| McColl and Johnson, 1983 | | | | | | | | |
|   Shoot growth | 0 | | 0 | | 0 | | | C |
| *Pinus strobus* L. | | | | | | | | |
| Reich et al., 1987 | | | | | | | | |
|   Shoot growth | | | +* | 0 | 0 | | | C |
|   Root growth | | | +* | 0 | 0 | | | C |
| *Pinus strobus* L. | | | | | | | | |
| Lee and Weber, 1979 | | | | | | | | |
|   Germination | | | +* | 0 | 0 | | | C |
|   Shoot growth | | | 0 | 0 | 0 | | | C |
|   Root growth | | | 0 | 0 | 0 | | | C |
| *Platanus occidentalis* L. | | | | | | | | |
| Neufeld et al., 1985 | | | | | | | | |
|   Shoot growth (height) | −* | | 0 | | 0 | | | C |
|   Shoot + root dry weight | −* | | 0 | | 0 | | | C |
| *Pseudotsuga menziesii* (Mirb.) Franco | | | | | | | | |
| McColl and Johnson, 1983 | | | | | | | | |
|   Germination | −* | | 0 | | 0 | | | C |
|   Shoot growth | 0 | | 0 | | 0 | | | C |
| *Pseudotsuga menziesii* (Mirb.) Franco | | | | | | | | |
| Lee and Weber, 1979 | | | | | | | | |
|   Germination | | | +* | 0 | 0 | | | C |
|   Shoot growth | | | +* | 0 | 0 | | | C |
|   Root growth | | | 0 | 0 | 0 | | | C |
| *Robinia pseudoacacia* L. | | | | | | | | |
| Neufeld et al., 1985 | | | | | | | | |
|   Shoot growth | 0 | | 0 | | 0 | | | C |
| *Tsuga canadensis* L. Carr. | | | | | | | | |
| Raynal et al., 1982a | | | | | | | | |
|   Germination | | | 0 | | 0 | | | C |

## C. Seedling Survival and Growth

Few studies have been conducted on the effects of acidic precipitation on seedling survival. Seedlings of *Pseudotsuga menziesii* germinating at pH 2.0 died following erosion of cuticles, which permitted fungal attack (McColl and Johnson, 1983). *Pinus halepensis* Mill. subjected to SAR of pH 3.1 showed 11% increased mortality compared to controls at pH 5.1 (Matziris and Nakos, 1977). Studies by Raynal et al. (1982a, b) indicate that, at least for some species, seed germination is less sensitive to acidic precipitation than is subsequent seedling establishment. Although seed germination was unaffected by treatment with acid at pH 3.0, establishment of the seedlings was deleteriously affected by the acidic treatment.

Most studies of the effects of acidic rain on seedling growth have found no effect when the applied solution was between pH values of 5.6 and 3.0 (Table 1–2). In one study, shoot growth of *Platanus occidentalis* and *Liriodendron tulipifera* was negatively affected at pH 2.0. These plants received only foliar applications of simulated acidic rain; thus their responses are due to foliar effects and not to soil acidification (Neufeld et al., 1985). Growth was stimulated in *Pinus strobus* at pH levels of 2.5 and 3.1. *Liriodendron tulipifera* was stimulated at pH 4.1, *Pinus ponderosa* was unaffected between pH 5.0 and 2.0, and growth of *Betula alleghaniensis* was inhibited at pH 2.3 (Table 1–2). Under nutrient-limiting conditions, simulated throughfall at pH 3.0 resulted in both foliar damage and stimulated growth of *Acer saccharum* seedlings, but under higher fertility conditions growth reductions took place only at pH 2.0 (Raynal et al., 1982b). Growth of *P. strobus* seedlings receiving SAR of 5.6, 4.0, 3.5, or 3.0 was little affected when planted in relatively N-rich soil, but in relatively N-poor soil growth was accelerated in the pH 3 treatment due to an N fertilization effect (Reich et al., 1987).

## III. Effects of Acidic Precipitation on Vegetative Components

### A. Leaves

Studies of structural damage to leaf surfaces have been mostly short-term experiments (Table 1–1). Foliar damage was observed in some species at pH 3.0, and in most species tested at pH 2.0. In *Picea abies* (Table 1–2), acidic mists at pH 3.0 resulted in cracks in the wax plugs in current needles. Susceptibility to damage from acidic rain was positively related to leaf wettability (Haines et al., 1985b). *Quercus palustris* leaves were damaged by SAR at pH 2.5 (Evans and Curry, 1979). Damage has also been reported for *Populus* hybrids treated with simulated acidic rain at pH levels from 2.7 to 3.4 (Evans et al., 1978).

The acid-neutralizing capacity of leaves was investigated by Pylypec and Redmann (1984), who determined the quantity of hydrogen ions required to produce a 5 microequivalents change in homogenated leaves. The order from highest to lowest buffering capacity was *Populus tremuloides* Michx., *Ledum groenlandicum* Oeder., *Larix laricina* (Du Roi) L. Koch, *Picea glauca* (Moench) Voss, *Picea marina* (Mill.) BSP, and *Pinus banksiana* Lamb. Evergreen conifers, especially *P. glauca*, seem to be less well buffered metabolically against external acidification.

The effects of acidic precipitation on transpirational water loss and $CO_2$ uptake and loss have also been studied (Table 1–3). Photosynthesis was inhibited in *Platanus occidentalis* subjected to rain at pH 2.0 but was stimulated in *Pinus*

**Table 1-3.** Summary of responses of tree leaves to simulated acidic rain. Responses evaluated are foliar damage (FDMG), mineral element leaching (LEACH), photosynthesis (PS), and stomatal conductance (COND). For studies comparing responses at various pH values against a control, the pH of the control is shown as *C,* and treatments with statistically different responses from *C* are indicated by *. Results from studies not making statistical comparisons but showing a negative effect are indicated by —; those showing a positive effect are indicated by +. Treatments applied at a particular pH but for which no effect was observed are indicated by 0.

| Species, authors, and response evaluated | Simulated rainfall pH | | | | | | | |
|---|---|---|---|---|---|---|---|---|
| | 2.0 2.4 | 2.5 2.9 | 3.0 3.4 | 3.5 3.9 | 4.0 4.4 | 4.5 4.9 | 5.0 5.4 | 5.5 5.9 |
| *Acer rubrum* L. Haines et al., 1980, 1985 | | | | | | | | |
| FDMG | + | 0 | | | | | | |
| LEACH | | 0 | | 0 | | 0 | | 0 |
| *Acer saccharum* Marsh. Reich and Amundson, 1985 | | | | | | | | |
| PS | | | 0 | | 0 | | | C |
| *Betula alleghaniensis* Britt. Paparozzi and Tukey, 1983 | | | | | | | | |
| FDMG | | + | + | | 0 | | | 0 |
| *Betula alleghaniensis* Britt. Wood and Bormann, 1974 | | | | | | | | |
| FDMG | + | | + | | 0 | 0 | | |
| *Carya illinoiensis* (Wangenh.) K. Koch Haines et al., 1980, 1985a | | | | | | | | |
| FDMG | + | 0 | | | | | | |
| LEACH | | 0 | | 0 | | 0 | | C |
| *Cornus florida* L. Haines et al., 1980, 1985a | | | | | | | | |
| FDMG | + | 0 | | | | | | |
| LEACH | | 0 | | 0 | | 0 | | C |
| *Liquidambar styraciflua* L. Neufeld et al., 1985 | | | | | | | | |
| FDMG | + | | 0 | | 0 | | | C |
| PS | 0 | | 0 | | 0 | | | C |
| COND | 0 | | 0 | | 0 | | | C |
| *Liriodendron tulipifera* L. Haines et al., 1980, 1985a | | | | | | | | |
| FDMG | + | 0 | | | | | | |
| LEACH | | 0 | | 0 | | 0 | | C |
| Liriodendron tulipifera L. Neufeld et al., 1985 | | | | | | | | |
| FDMG | +* | | 0 | | 0 | | | C |
| PS | 0 | | 0 | | 0 | | | C |
| COND | 0 | | 0 | | 0 | | | C |

**Table 1-3.** (*Continued*)

| Species, investigators, and response evaluated | Simulated rainfall pH | | | | | | | |
|---|---|---|---|---|---|---|---|---|
| | 2.0 2.4 | 2.5 2.9 | 3.0 3.4 | 3.5 3.9 | 4.0 4.4 | 4.5 4.9 | 5.0 5.4 | 5.5 5.9 |
| *Malus domestica* Borkh. Forsline et al., 1983b | | | | | | | | |
| FDMG | | + | + | 0 | 0 | 0 | 0 | |
| *Malus domestica* Borkh. Proctor, 1983 | | | | | | | | |
| FDMG | + | | + | | 0 | | | |
| *Picea rubens* Sarg. Taylor et al., 1986 | | | | | | | | |
| PS | | | | 0 | | | C | |
| COND | | | | 0 | | | C | |
| *Picea rubens* Sarg. Seiler and Paganelli, 1987 | | | | | | | | |
| PS | | | + | | | C | | |
| Growth | | | 0 | | | 0 | | |
| *Pinus ponderosa* Laws. McColl and Johnson, 1983 | | | | | | | | |
| FDMG | + | | 0 | | 0 | | | C |
| *Pinus strobus* L. Haines et al., 1980, 1985a | | | | | | | | |
| FDMG | + | 0 | | | | | | |
| LEACH | | 0 | | 0 | | 0 | | C |
| *Pinus strobus* L. Reich et al., 1987 | | | | | | | | |
| PS | | | + | 0 | 0 | | | C |
| *Pinus strobus* L. Wood and Bormann, 1977 | | | | | | | | |
| FDMG | + | | 0 | | 0 | | | C |
| *Pinus taeda* L. Seiler and Paganelli, 1987 | | | | | | | | |
| PS | | | 0 | | | C | | |
| Growth | | | 0 | | | C | | |
| *Platanus occidentalis* L. Neufeld et al., 1985 | | | | | | | | |
| FDMG | +* | | 0 | | 0 | | | C |
| PS | —* | | 0 | | 0 | | | C |
| COND (mesophyll only) | —* | | 0 | | 0 | | | C |
| *Populus* spp. hybrids Evans et al., 1978 | | | | | | | | |
| FDMG | | + | + | | | | | C |

Table 1–3. *(Continued)*

| Species, investigators, and response evaluated | Simulated rainfall pH | | | | | | | |
|---|---|---|---|---|---|---|---|---|
| | 2.0 2.4 | 2.5 2.9 | 3.0 3.4 | 3.5 3.9 | 4.0 4.4 | 4.5 4.9 | 5.0 5.4 | 5.5 5.9 |
| *Pseudotsuga menziesii* (Mirb.) Franco | | | | | | | | |
| McColl and Johnson, 1983 | | | | | | | | |
| FDMG | + | | 0 | | 0 | | | C |
| *Quercus palustris* Muenchh. | | | | | | | | |
| Evans and Curry, 1979 | | | | | | | | |
| FDMG | | + | + | | | | | C |
| *Quercus prinus* L. | | | | | | | | |
| Haines et al., 1980, 1985a | | | | | | | | |
| FDMG | + | 0 | | | | | | |
| LEACH | | 0 | | 0 | | 0 | | C |
| *Quercus rubra* L. | | | | | | | | |
| Reich and Amundson, 1985 | | | | | | | | |
| PS | | | 0 | | 0 | | | C |
| *Robinia pseudoacacia* L. | | | | | | | | |
| Haines et al., 1980, 1985a | | | | | | | | |
| FDMG | + | 0 | | | | | | |
| LEACH | | 0 | | 0 | | 0 | | C |
| *Robinia pseudoacacia* L. | | | | | | | | |
| Neufeld et al., 1985 | | | | | | | | |
| FDMG | + | | 0 | | 0 | | | C |
| PS | 0 | | 0 | | 0 | | | C |
| COND | 0 | | 0 | | 0 | | | C |

*strobus* at pH 3.0 (Table 1–3). Inhibition of photosynthesis in *Platanus* was interpreted as a metabolic effect because stomatal conductance and transpiration were not altered by the acidic rain treatment. Stimulation of photosynthesis in *Pinus strobus* was attributed to a fertilization effect. In the *Platanus* study, simulated rains were applied only to the plant canopies, so a soil fertilization effect was not possible. Photosynthesis and transpiration in *Picea rubens* did not differ between acidic treatments at pH 5.1 and 4.1, but overall growth was greater for plants receiving the more acidic treatments.

Responses of chlorophyll a and b to SAR treatments were quantified in the plants studied by Neufeld et al. (1985) (Table 1–3). No statistically significant effect of SAR on these pigments was found.

Gas exchange processes of *Picea abies* were studied in controlled environmental chambers (Table 1–4). Acidic mists at pH 3.0 inhibited the maximum carboxylation rate, the maximum net photosynthesis rate, the quantum yield, and

**Table 1–4.** Summary of responses of *Picea abies* (L.). Karst treated with mist and mist + ozone for nonfertilized treatments and current year needles where 0 = no effect, + = stimulation, − = inhibition relative to control *C*, NT = not tested, and OC = outside control (= Freiland).*

| Process and author | mist pH 3 | mist pH 3 + ozone | mist pH 5.6 | mist pH 5.6 + ozone | OC |
|---|---|---|---|---|---|
| Leaching of Ca, Mg, Fe, Mn, Cu | + | + | *C* | + | NT |
| Foliar deficiency of Mg and Ca | (for all treatments on unfertilized soil) | | | | |
| Foliar Al | − | − | *C* | + | + |
| Shoot growth | 0 | 0 | 0 | 0 | 0 |
| Bosch et al., 1986 | | | | | |
| | | | | | |
| Fine roots (<2 mm dia) dry weight | 0 | 0 | 0 | 0 | 0 |
| Large roots (>2 mm dia) dry weight | 0 | 0 | 0 | 0 | 0 |
| Weiss and Agerer, 1986 | | | | | |
| | | | | | |
| Wax plug formation in current year needles (in both fertilized and nonfertilized) | | | | | |
| a. Fused wax rodlets | 0 | + | *C* | + | |
| b. Cracks in wax plugs | + | + | *C* | 0 | |
| Magel and Ziegler, 1986a | | | | | |
| | | | | | |
| Max. carboxylation rate ambient $CO_2$ | − | − − | *C* | − − | *C* |
| Max. net photosynthesis at saturated light and $CO_2$ | − | − − | *C* | − | *C* |
| Quantum yield | − | − − | *C* | − − | *C* |
| Dark respiration | − | − − | *C* | − | *C* |
| $CO_2$ compensation point | + | ++ | *C* | + | *C* |
| Light compensation point | 0 | + | *C* | + | *C* |
| Stomatal conductance | + | − | *C* | − | *C* |
| (most − effects overcome by fertilization) | | | | | |
| Selinger et al., 1986 | | | | | |
| | | | | | |
| Peroxidase activity | (very heterogeneous response, not useful bioindicator of air pollution effects) | | | | |
| Dohmen, 1986 | | | | | |
| | | | | | |
| Adenine-nucleotide level in needles | (very heterogeneous response) | | | | |
| Energy charge ratio | 0 | 0 | *C* | + | *C* |
| ATP/ADP | + | + | *C* | + | *C* |
| Phosphorylation potential | + | + | *C* | + | *C* |
| Water-soluble sugars | − | 0 | *C* | − | *C* |
| Starch content | 0 | 0 | *C* | 0 | *C* |
| Magel and Ziegler, 1986b | | | | | |
| | | | | | |
| Monoterpene composition | (could not be related to experimental conditions) | | | | |
| Schonwitz and Merk, 1986 | | | | | |

**Table 1-4.** (*Continued*)

| Process and author | mist pH 3 | mist pH 3 + ozone | mist pH 5.6 | mist pH 5.6 + ozone | OC |
|---|---|---|---|---|---|
| Chlorophyll | + | + | C | + | C |
| Ascorbic acid | 0 | 0 | C | − | C |
| Glutathion | + | 0 | C | + | C |
| Tocopherol | + | + | C | + | C |
| Senger et al., 1986 | | | | | |
| Abscisic acid | 0 | 0 | C | + | C |
| Indoleacetic acid | + | + | C | + | C |
| Fackler et al., 1986 | | | | | |
| *p*-Hydroxyacetophenon | 0 | 0 | C | − | C |
| Osswald et al., 1986 | | | | | |

*The experimental design is described by Rehfuess and Bosch (1986) and Payer et al. (1986). Most data sets were not statistically analyzed by authors. Subjective interpretation given by present reviewers. See original papers for details of responses to fertilizer and frost shock treatments.

dark respiration. The pH 3.0 treatment also increased the $CO_2$ compensation point and the stomatal conductance but did not affect the light compensation point. The physiological effects of acidic rain were intensified when ozone treatments were applied simultaneously. Most of the acidic rain effects could be overcome by fertilizer application.

Additional studies with *Picea abies* foliage (Table 1–4) indicate that pH 3.0 mist, when compared to pH 5.6 mist, produced a heterogeneous response on adenine-nucleotide concentrations, no effect on energy charge ratios, increased ATP-ADP ratios, increased phosphorylation potential, decreased water-soluble sugars, no effect on starch content, no effect on monoterpene composition, and increased chlorophyll content. The production of the antioxidants ascorbic acid, glutathione, and tocopherol were also investigated. Acidic rain treatment at pH 3.0 stimulated production of the latter two compounds. The pH 3.0 treatment also stimulated production of indoleacetic acid (a growth regulator) but not the production of *p*-hydroxyacetophenon (a fungitoxic compound) or abscisic acid (a growth regulator).

Acidic fog effects on *Citrus sinensis* L. cv Valencia and on *Poncirus trifoliate* L. were evaluated by Musselman and Sterrett (1988), who found that 12 2-hour exposures to pH 3.2, 2.8, 2.4, 2.0, and 1.6 resulted in a significant percentage of leaf area of necrosis at pH 1.6 in both species but no changes in height growth in either species.

Effects of acidic precipitation on mineral element processing by leaves are reviewed in the section on tree nutrition below.

## B. Stems

The most physiologically active portion of the plant stem is the cambial region, from which new cells proliferate. Protected by the bark, cambial activity seems unlikely to be affected by acidic rain. Apparently there are no reports of acidic rain damage to the bark or cambium of mature trees. Seedlings of *Pseudotsuga menziesii* treated with SAR at pH 2.0 developed fissures in the stem cuticle. These fissures were subsequently invaded by fungi that killed the seedlings (McColl and Johnson, 1983).

## C. Roots

The negative effects of acidity on element uptake by plant roots are well known (Adams, 1984). However, few studies have been performed on the specific physiological effects of acidic rain on nutrient uptake. In one study where the uptake of $^{45}Ca$ by excised *Liriodendron tulipifera* roots was measured, uptake progressively decreased as the solution pH decreased from 5.0 to 4.0 and to 3.0 (Haines and Swank, 1987).

Studies of the effects of SAR on roots are listed in Tables 1–2, 1–4, and 1–5. Root growth was not affected in tree seedlings subjected to rain at pH levels of 5.6, 4.0, 3.5, or 3.0 (Lee and Weber, 1979). Numbers of mycorrhizal and nonmycorrhizal root tips in *Picea abies* stands irrigated with simulated acidic rain at pH 2.7 to 2.8 decreased compared to an irrigated control (Table 1–5). When *Picea abies* was exposed to acidic mists in a controlled environment on unfertilized soil (Table 1–4), the accumulation of biomass in fine roots and in large roots was not affected by the treatment, but foliar deficiencies of Mg and Ca developed. These plants also showed elevated leaching of Ca, Mg, Fe, Mn, and Cu from the foliage.

# IV. Effects of Acidic Precipitation on Tree Mineral Nutrition

## A. Foliar Leaching and Foliar Uptake

Rain passing through a tree canopy may lose or gain mineral elements through some combination of the natural processes of absorption and leaching (Tukey, 1970; Parker, 1987). Modification of these absorption and leaching processes by acidic rain has been investigated for both angiosperms and gymnosperms (Tables 1–3 and 1–4). Results vary with chemical elements and tree species (McLaughlin, 1985; Lindberg et al., 1986; Parker, 1987).

Absorption of H and $NH_4$ by tree canopies has been reported by Lovett et al. (1985), Nihlgard (1985), and Lindberg et al. (1986). Both $NO_3$ and $SO_4$ in precipitation can be absorbed by tree leaves, serving as a source of these often growth-limiting elements (Mollitor and Raynal, 1983; Jacobson, 1984; Skiba et al., 1986). Absorption of $NO_3$ by deciduous trees exposed to ambient acidic rain was reported by Mollitor and Raynal (1983). In the same study, $NO_3$ was leached from conifers.

**Table 1-5.** Summary of responses of *Picea abies* L. Karst. to acidic irrigation, normal irrigation, and liming of field plots at Hoglwald, Bavarian forest district of Aichach, 16 km SE of Augsburg, West Germany (Kreutzer and Bittersohl, 1986) from various authors where $0$ = no effect, $+$ − stimulation, and $-$ = inhibition relative to control *C*.

| | Treatments* | | | | | |
|---|---|---|---|---|---|---|
| | Acidic irrigation (B1) | Normal irrigation (C1) | Control (A1) | Lime (A2) | Acidic irrigation and lime (B2) | Normal irrigation and lime (C2) |
| Fine root dynamics Ulrich and Pirouz-panah, 1986 | 0 | | C | | | |
| Uptake of Al, Fe, Mg, Ca, Mn, P | | | C | + | | |
| Uptake of Mg | − | | C | | | |
| Uptake of Fe | + | | C | | | |
| Uptake of Mg, Fe, Al Stienen, 1986 | | | C | | + | |
| Mycorrhizal and nonmycorrhizal root tip numbers Blaschke, 1986 | − | | C | | | |

*Treatments were acidic irrigation (B1) of pH 2.7 to 2.8 $H_2SO_4$, normal irrigation (C1), control (A1), liming (A2) with 4000 kg dolomite/ha, acid irrigation + liming (B2), and normal irrigation + liming (C2).

Leaching of Ca, Mg, and K from leaves was reported by Blanpied (1979), Scherbatskoy and Klein (1983), Lovett et al. (1985), Lindberg et al. (1986), Skiba et al. (1986), and Wood and Bormann (1975). Accelerated leaching of Ca, Mg, Fe, Mn, and Cu occurred in *Picea abies* subjected to mists of SAR at pH 3.0 (Table 1-4).

Elements leached from tree canopies can potentially come from both external and internal plant surfaces. Elements on plant surfaces may come from wet and dry atmospheric deposition as well as from inside the plant. When element leaching from plant canopies occurs, as evidenced by enrichment of throughfall, the proportions of those elements derived from surface washoff versus the interior of the plant are difficult to determine. In one study, where six species of tree seedlings were grown in air from which most particulates had been removed by filtration, no significant effects of acidic rain on foliar leaching were found upon application of SAR at pH 5.6, 4.5, 3.5, and 2.5 (Table 1-3; Haines et al., 1985a). In another study, where the air was not filtered, the application of SAR comprised of $H_2SO_4$ significantly increased rates of foliar leaching losses of K, Ca, and Mg from *Acer saccharum* leaves (Wood and Bormann, 1975). Field studies attributing increased rates of foliar leaching to acidic rain may in fact be measuring the washing off of dry deposition rather than the movement of ions from inside the leaf to the outside of the leaf. In a field study with *Abies balsamea* (L) Mill., Reiners et al. (1986) found more ions were derived from dry deposition on the needles between rain storms than from within the needles.

## B. Changes in Soil Nutrient Availability

### 1. Essential Elements

Acidic precipitation inputs can affect the availability to trees of essential elements in several ways: (1) Inputs of S and N can have a fertilization effect, and (2) inputs of H, $SO_4$, and $NO_3$ can result in leaching losses of macronutrient cations, particularly Mg, Ca, and K, from the soil (Abrahamsen, 1980; Johnson et al., 1982; Abrahamsen, 1984; Huetterman and Ulrich, 1984; Johnson and Reuss, 1984; Woodman, 1987). Whether or not these effects take place is dependent largely on soil properties (Johnson and Cole, 1977; Johnson et al., 1982; Abrahamsen 1984; Johnson and Reuss, 1984). Acidic inputs could increase tree growth on sites deficient in S or N (Johnson et al., 1982; Abrahamsen, 1984; Johnson and Reuss, 1984; McLaughlin, 1985). Wood and Bormann (1977) found that short-term application of a simulated acidic rain composed of $H_2SO_4$, $HNO_3$, and HCl stimulated the growth of *Pinus strobus* seedlings. Similarly, Schier (1986) reported that application of simulated rain at pH 3.0 increased biomass production in *Pinus rigida* Mill. seedlings. These effects are probably a direct result of the addition of N in the precipitation to an N-poor system.

Although inputs of N and S in acidic precipitation can be a benefit to trees, they can also be a detriment. If present in excess of tree demand, $SO_4$ and $NO_3$ can cause accelerated leaching of base cations from the soil, decreasing soil fertility. Where the soil has a high ability to adsorb these ions, however, leaching would not occur (Johnson and Cole, 1977; Johnson et al., 1982; Johnson and Reuss, 1984). Therefore, leaching might be expected in the soils of New England, where $SO_4$ adsorption capacity is generally low, but not in the soils of the southeast, where $SO_4$ adsorption capacity is high and forests can be deficient in S. In this regard, $NO_3$ is less of a problem because N is often deficient in forests (Abrahamsen, 1984).

### 2. Soil Acidification and Trace Elements

Soil acidification is a natural process in humid regions of the world (Van Breemen et al., 1983; Rechcigl and Sparks, 1985). In areas receiving acidic precipitation, this natural process can be accelerated, altering the composition of the soil solution (Johnson and Cole, 1977; Johnson et al., 1982; Van Breeman et al., 1983; Ulrich, 1983; Rechcigl and Sparks, 1985). Increased soil acidity can result in increased solubility of Al and heavy metals, increasing their availability to plants (Tyler, 1978; Cronan and Schofield, 1979; Abrahamsen, 1983, 1984; McLaughlin, 1985).

Aluminum is normally transported by organic acids from the surface layers of the soil to the B horizon, where it precipitates (Cronan and Schofield, 1979; Nilsson and Bergkvist, 1983). In areas receiving acidic deposition, however, Al concentrations remain high throughout the soil profile (Cronan and Schofield, 1979; Ulrich et al., 1980; Johnson et al., 1981; Nilsson and Bergkvist, 1983; Skeffington, 1983). This has caused concern because Al is potentially toxic to trees and soil microorganisms (McLean, 1976; McCormick and Steiner, 1978; Ulrich et al., 1980; Foy, 1983; Skeffington, 1983). In most acidic soils, high exchangeable Al concentrations, not H ion toxicity, are responsible for poor plant growth (Adams, 1984). Aluminum has been shown to interfere with cell division in roots and with the uptake, transport, and use of Ca, K, P, Mg, and Fe (McLean, 1976; Mugwira et al., 1980; Alam, 1981; Bache and Crooke, 1981; Foy, 1983; Abrahamsen, 1984). Some researchers believe that Al toxicity is responsible for the forest decline and dieback seen in the United States and Europe (Ulrich et al., 1980; Tomlinson, 1983). Several investigators have found that toxicity occurs in trees when the ratio of Ca to Al is less than one (Ulrich et al., 1980; Skeffington, 1983).

Aluminum-induced Ca deficiency is hypothesized to account for declining *Picea rubens* in the northeastern United States. The inferred sequence is that an elevated Al:Ca ratio in the fine root region decreases Ca uptake, decreases wood formation, decreases the amount of functional sap wood and live crown, and makes large trees more vulnerable to other diseases and to insect pests (Shortle and Smith, 1988).

Reductions in soil pH due to acidic precipitation can also increase the solubility, and hence the availability to trees, of other potentially toxic trace elements, including Mn and other heavy metals (Tyler, 1978; Cowling, 1981; Abrahamsen, 1984; McLaughlin, 1985). These elements could also detrimentally affect tree growth.

## C. Changes in Microbial Associations

Mycorrhizae are symbiotic associations between one or more fungi and vascular plant roots. The mycorrhizal symbiosis is often more effective in the acquisition of essential plant nutrients than is the plant root alone, thus offering advantages to the vascular plant on nutrient-poor sites. Any effects of acidic precipitation on this association could affect tree nutrition on these sites. Mycorrhizae are divided into two major types: ectomycorrhizae have fungal structures outside the root that increase the surface area for nutrient absorption; endomycorrhizal roots have fungal structures on the interior of the root.

Relatively little research has been done to determine the effect of acidic rain on mycorrhizae. Most mycorrhizae are acid tolerant and are found on trees growing in nutrient-poor, acidic soils. Potential effects of acidic rain on the mycorrhizal symbiosis include inhibition of fungal infection of the roots, inhibition of metabolism of established mycorrhizae, toxicity from heavy metals, and successional changes in the mycorrhizal flora.

Ectomycorrhizae formation can be inhibited by SAR (Shafer et al., 1985; Stroo and Alexander, 1985; Reich et al., 1985, 1986; Dighton and Skeffington, 1987;

Entry et al., 1987). Nitrogen fertilization by $NO_3$ in acidic rain has been hypothesized as a possible cause (Reich et al., 1985, 1986). Toxicity of Al and other heavy metals at the low soil pH may decrease ectomycorrhizae formation. Decreased infection of *Abies balsamea* was related to increasing Al concentrations in soil (Entry et al., 1987). Mycorrhizal infection in *Pinus banksiana* and in *Picea glauca* decreased upon addition of Ni, Pb, Cd, and Cu (Dixon and Buschena, 1988).

No inhibition of ectomycorrhizae formation was found on tree seedlings planted in soil pretreated with acidic rain, but inhibition occurred where the acidic rain was applied after the seedlings were planted (Stroo and Alexander, 1985). Stroo and Alexander (1985) concluded that decreased ectomycorrhizal infection resulted from changes in carbon allocation in the seedling rather than from direct effects of the acid on the fungus itself.

Shafer and others (1985) suggested that changes in soil chemical and biological characteristics due to exposure to acidic rain may simply result in changes in the fungal symbionts; that is, ectomycorrhizae would still be present, but the fungal species in the association would change. If this is the case, tree nutrition would probably be unaffected.

For a more detailed overview of acidic rain impacts on soil chemistry and biology, see Fernandez (in press) and Reuss and Johnson (1986).

## V. Effects of Acidic Precipitation on Tree Resistance to Secondary Stresses

### A. Other Pollutants

Factorial experiments with various treatment combinations of simulated acidic rain, acidic irrigation, liming, and ozone have been performed in recognition of the fact that trees in the natural environment are not exposed to acidic rain alone, but are simultaneously subjected to multiple and potentially interactive stresses. A factorial experiment with rainfall acidities ranging from 5.6 to 3.0 and ozone levels ranging from 0.01 to 0.14 parts per million addressed these potentially interactive stresses (Reich and Amundson, 1985). Tree species studies included *Acer saccharum, Pinus strobus,* and *Quercus rubra* (Table 1–3). They found no significant effects of acidic rain on gas exchange processes. Ozone treatment reduced photosynthesis. No interaction between ozone and acidic rain was found in this study. However, the responses of *Picea abies* to combined treatments of ozone and acidic mist, summarized in Table 1–4, indicate that ozone can accentuate the negative effects of acidic rain.

Field and laboratory studies of the ecophysiological effects of $SO_2$, $O_3$, other air pollutants, and Mg fertilization on trees are reported in 15 papers in a special issue of *Allgemeine Forst Zeitschrift* (1987; 27/28/29:684–763). Trees included *Picea abies, Pinus silvestris, Pinus nigra, Tsuga canadensis,* and *Abies alba* with some contrasts between trees in healthy and declining stands. Summarizing the findings, Boger and Mohr (1987) say that the "acid rain hypothesis" of direct effects on trees does not offer a convincing explanation of forest decline. They see pollution-induced changes in soil chemistry from increased inputs of $NO_3$ and $NH_4$ and decreased availability of Mg, Ca, K, or Mn as more important factors, but the causes of forest decline are not yet explained.

## B. Insects and Plant Diseases

Interacting effects of acidic deposition, insects, and plant pathogens have been reviewed by Smith et al. (1984) and Bruck and Shafer (1984). Bruck and Shafer (1984) pointed out that facultative parasites are favored by senescent tissue, whereas obligate parasites are favored by healthy tissue. The stress and wounding associated with acidic rain damage tend to accelerate senescence. Acidic rain could therefore be expected to affect negatively tree attack by obligate parasites and positively affect attack by facultative parasites (Bruck and Shafer, 1984).

Trees subjected to stress are frequent targets of insect attacks as well (Hain, 1987). Air pollution, by affecting tree vigor, is likely to alter the interaction between trees and insect pests (Hain, 1987). Smith et al. (1984) concluded that it is not possible to make generalizations about the effects of acidic rain on the stimulation or inhibition of biotic stress agents to trees. They suggested that potential interactions between acidic rain and leaf- or bark-infesting or feeding insects or foliar and stem pathogens that spend a major portion of their life cycles on leaf and stem surfaces would be greater than for soil arthropods and root-infecting microorganisms. Soil microorganisms could be influenced only following acid-induced changes in soil chemistry or altered host resistance.

Ulrich (1983) suggested that acidic rain acts as a stress to trees and predisposes them to insect or pathogen attack. In the case of *Abies fraseri* mortality due to the balsam woolly adelgid (*Adelges piceae*), Hain and Arthur (1985) hypothesized that trees already stressed by acidic rain were less able to resist the insect attack, resulting in greater mortality. This, they concluded, explained the higher mortality rates for fir at the highest elevations, which would have the greatest cloud interception and acidic rain stress. Studies by Shriner (1978) indicated that the interactions between acidic rain and plant pathogens are quite complex. Exposure of *Quercus phellos* L. leaves to pH 3.2 SAR decreased infection by *Cronartium fusiforme*. Experiments on kidney beans showed that treatment of the plants with acidic rain before infection with *Pseudomonas phaseolicola* caused greatest disease development, whereas treatment with SAR following inoculation decreased disease development. Bruck et al. (1981) found reduced rust gall formation on *Pinus taeda* L. treated with low pH SAR for 6 months following inoculation. It is apparent that acidic rain can increase the avenues for parasitic infection due to acid-induced leaf damage. At the same time, acidic rain appears to affect adversely pathogens themselves.

Simulated acidic rain can decrease the survival rate of tree seedlings due to increased pathogenesis. Fungi killed *Pseudotsuga menziesii* seedlings that germinated following treatment at pH 2.0 (McColl and Johnson, 1983). Seedlings of *Acer saccharum* exposed to low pH (2.4 and 3.0) were susceptible to bacterial infection (Raynal et al., 1982b). *Pinus rigida, P. echinata, P. taeda*, and *P. strobus* exposed to pH 3.0 exhibited 100% mortality due to damping-off (a fungal infection).

## C. Temperature

Low-temperature stress has been proposed as a contributing factor in the forest dieback and/or forest decline phenomenon (Friedland et al., 1984; Schutt and Cowling, 1985; Johnson and Siccama, in press). Interactions between cold stress and excess N in acidic precipitation could be responsible for the damage reported in high-elevation forests of the northeastern United States. High foliar N levels are associated with decreased cold hardening (Friedland et al., 1984; Nihlgard, 1985), and the injury symptoms observed are similar to those seen in forests overfertilized with N (Nihlgard, 1985). The damage is seen in trees at high elevations, where N deposition is the greatest (Friedland et al., 1984). Some of these high-elevation forests are developing cold injury at temperatures more moderate than those that have caused damage in the past (McLaughlin et al., 1987), further suggesting that the added stress of acidic rain has increased tree susceptibility to cold stress. Frost shock treatments were combined with treatments of ozone, acidic rain, and Ca + Mg fertilizer in the design of studies with *Picea abies* (Payer et al., 1986; Rehfuess and Bosch, 1986). Preliminary studies with young trees found these factors to be the main cause of *Picea abies* decline on high-elevation acidic soils (Rehfuess and Ziegler, 1986).

## D. Growth Studies in the Field: The Integrated Tree Response

The relation of variability in annual tree growth increments to variation in environmental factors, including acidic rain, has been explored, (LeBlanc et al. 1987a, b; Johnson and McLaughlin, 1986; Federer and Hornbeck, 1987) *Pinus sylvestris* L., *Pinus resinosa* Ait., and *Picea abies* L. (Karst) growing on potentially acidic-deposition-sensitive sites showed synchronous decreased growth after 1960, but this decreased growth coincided with climatic anomalies in drought and in winter temperatures and could not be directly related to variability in acidic deposition (LeBlanc et al., 1987 a, b). Studies reviewed by LeBlanc et al. (1987a, b) did not find direct relations between dendrochronologies and acidic rain.

# VI. Tree Mortality, Forest Decline, and the Forest Dieback Phenomenon

Plant reproductive investment and biomass accumulation integrate both above-ground and belowground plant-environment interactions. These interactions include (1) the net acquisition of $CO_2$ through stomata to the mesophyll of the leaves with a minimum loss of water by the reverse pathway; (2) the net acquisition of solar radiation at rates that avoid overheating of the leaves, thus avoiding increased respiratory carbon loss and possible leaf death; (3) net acquisition of mineral elements while avoiding accumulations of some elements to toxic concentrations; and (4) net acquisition of water from the soil at rates slightly in excess of the transpirational water loss. Acidic rain is obviously only one of the many interacting factors that can influence the acquisition of $CO_2$, solar energy, water, and nutrient resources by plants and the subsequent allocation of those resources to the construction and maintenance of leaves, stems, and roots. Quantifying the contribution of a single factor such as acidic rain to either increased or decreased plant reproduction or biomass accumulation for in situ forest trees is not possible with the present state of ecology and technology.

A brief history of forest decline episodes in North America is given by Manion (1987), who reviews the evolution of conceptual models that have been developed to study tree death. In some simple situations, the cause of tree death can be attributed to the point-source emissions from metal smelters at places like Copper Hill, Tennessee (Seigworth, 1943), and Sudbury, Ontario (Freedman and Hutchinson, 1980). More frequently the quantitative contributions of multiple interacting physical and biological factors in controlling resource acquisition and tree mortality cannot be determined.

A brief overview of forest decline phenomena is provided by Hinrichsen (1987) and is followed by seven more detailed reviews in a special issue of *BioScience* (1987 37:550–610). A conceptual model of the role of air pollution in the decline of *Picea abies* is provided by Schulze (1989).

The complexity of plant-environment interactions, including acidic precipitation phenomena, in the forest decline problem is illustrated in the review by Schutt and Cowling (1985). They explore the history of the problem and its common symptoms and review five hypotheses about its cause. These hypotheses are given below:

1. The acidification-Al toxicity hypothesis holds that the natural acidification of soil is accelerated by acidic deposition. Increased soil acidity results in increased Al solubility, which in turn acts as a predisposing factor damaging roots and leading to water stresses and nutrient stress problems.
2. The ozone hypothesis holds that ozone damage to foliage results in dieback.
3. The Mg-deficiency hypothesis states that the low concentrations of magnesium in soil at high elevations result in extreme Mg deficiency in trees. Acidic rain may create growth disturbances by adding N to the system. Element leaching from foliage is presumably accelerated by ozone or frost damage.
4. The general stress hypothesis holds that air pollution stresses lead to decreased photosynthesis and changed carbohydrate status. Decreased energy to the roots decreases fine root and mycorrhizal production, leading to nutrient and water stress problems.
5. The excess N hypothesis states that N inputs from the atmosphere to forests have increased with industrialization. Increased N increases tree growth with attendant demand for all other nutrients, inhibits mycorrhizae, increases susceptibility to frost, increases susceptibility to root-rot fungi, changes root-shoot ratios, and alters patterns of nitrification, denitrification, and possibly N fixation.

Acidic rain phenomena are part of four of the five hypotheses. The magnitude of the acidic rain contribution to the dieback phenomenon is not suggested by Schutt and Cowling (1985).

Ambient rain in the region of greatest rainfall acidity in North America had a volume-weighted average pH between 4.1 and 4.2 in 1980. This region extends from Illinois and Kentucky and northeastward through New York and Southern Ontario (Stensland et al., 1986, p. 142).

We find no evidence in the refereed scientific literature of a direct negative effect of these ambient levels of acidic rain upon structure and physiological function in trees. Earlier reviews by Morrison (1984) and Evans (1984) draw essentially the same conclusions. Woodman (1987) and Woodman and Cowling (1987) discussed the evidence for pollution-induced injury in North American forests. They found no rigorous proof that regional air pollution, such as acidic rain, has altered tree growth rates, whereas evidence of damage due to local pollution sources has been well documented.

Is quantification of the contribution of acidic rain to variations in tree productivity and tree mortality an important research goal? Although potential negative effects of low-level chronic acidic rain on trees cannot be discounted, there do not appear to be any negative short-term effects of ambient acidic rain on the physiology of trees. Trees do have the property of self-repair. Any long-term effects seem more likely to come from acid-induced changes in soil chemistry that could subsequently affect trees. There are, however, negative physiological effects of ambient concentrations of gaseous pollutants on trees. Investigating the effects of gaseous pollutants appears a more worthwhile research goal. From a landscape or global management perspective, decreasing emissions of oxides of sulfur, oxides of nitrogen, and hydrocarbons from industry and internal combustion engines would decrease the direct effects of these pollutants on trees. Decreasing these emissions would also decrease the accumulation of ozone (National Academy of Sciences, 1977) below phytotoxic concentrations and curtail acidic deposition at the same time.

## VII. Future Research Needs

A. Direct effects, short term

Research results from short-term studies of 1 year or less suggest that some processes in trees are more sensitive to SAR than others. Further research is needed into the effects of SAR on the following:

1. The plant reproductive processes of
   a. Production, germination, and tube growth in pollens of both gymnosperm and angiosperm tree species.
   b. Pollen receptivity or pollen germination and pollen tube growth-promoting properties of stigma surfaces in angiosperm tree species.
   c. Pollen receptivity or pollen germination and pollen tube growth-promoting properties of the pollen droplets in the pollen chambers of gymnosperm tree species.

2. Changes in chemical and physical characteristics of leaves and roots that alter the susceptibility to attack by insects, parasites, microbial pathogens, and microbial symbionts such as N-fixing bacteria and mycorrhizal fungi.

B. Direct effects, long term

Short-term studies of less than 1 year have found no effects of SAR at concentrations close to ambient acidic rain on the physiology of leaves. For evergreen gymnosperms that retain leaves several years, the possibility of cumulative leaf damage from chronic exposure to ambient acidic rain needs investigation.

C. Indirect effects

The direct effects of acidic rain on soil chemistry can have long-term indirect effects on the plant root. Relationships and processes identified in I-B above could be the subjects of long-term studies.

## VIII. Summary

The effects of acidic rain on pollen germination, pollen tube growth, seed production, seed germination, seedling growth, leaf structure and function, and root structure and function have been reviewed. In short-term laboratory studies, pollen germination and pollen growth are susceptible to inhibition by simulated acidic rain (SAR) in the range pH 4.0 to 2.5. Ambient acidic rain in eastern North America has a volume-weighted average pH of 4.1 to 4.2. Damage to leaves and roots occurs usually 1 or 2 pH units below these ambient rainfall pH values. Presently there is no evidence of direct negative effects of acidic rain on tree growth in the field. The potential interactions of acidic rain with other stresses are incorporated into working hypotheses designed by many investigators to account for forest decline/dieback phenomena.

## Acknowledgments

Preparation of this manuscript was supported in part by National Science Foundation grant BSR 85-14328 to the University of Georgia, by the Oak Ridge Associated Universities Faculty Research Participation Program, by Oak Ridge Associated Universities Participation Agreement S-3259, and by contract DE-AC09-76SROO-819 between the U.S. Department of Energy and the University of Georgia.

## References

Abrahamsen, G. 1980. *In* D. Drablos and A. Tollan, eds. *Ecological impact of acid precipitation*, 58–63. SNSF Project, Oslo, Norway.

Abrahamsen, G. 1983. *In* B. Ulrich and J. Pankrath, eds. *Effects of accumulation of air pollutants in forest ecosystems*, 207–218. D. Reidel, Dordrecht, Holland.

Abrahamsen, G. 1984. Phil Trans Royal Soc London B 305:369–382.

Adams, F., ed. 1984. *Soil acidity and liming*. 2nd ed. American Society of Agronomy, Inc., Madison, Wis.

Alam, S.M. 1981. Commun Soil Sci Plant Anal 12:121–138.

Bache, B.W., and W.M. Crooke. 1981. Plant Soil 61:365–375.

Blanpied, G.D. 1979. Hort Science 14:706–708.

Blaschke, H. 1986. Forstwissenschaftliches Centralblatt 105:324–329.

Boger, P., and H. Mohr. 1987. Allgemeine Forst Zeitschrift 27/28/29:691–692.

Bosch, C., E. Pfannkuch, K.E. Rehfuess, K.H. Runkel, P. Schramel, and M. Senser. 1986. Forstwissenschaftliches Centralblatt 105:218–229.

Bruck, R.I., and S.R. Shafer. 1984. *In* R.A. Linthurst, ed. *Direct and indirect effects of acidic deposition on vegetation*, 19–32. Butterworth Publishers, Boston.

Bruck, R.I., S.R. Shafer, and A.S. Heagle. 1981. Phytopathology 71:864.

Cowling, E.B. 1981. *A status report on acid precipitation and its ecological consequences as of April 1981*. North Carolina State University School of Forest Resources, Raleigh, N.C.

Cox, R. 1983. New Phytol 95:269–276.

Cox, R. 1987. *In* T.C. Hutchinson and K. Meema, eds. *Effects of atmospheric pollutants on forests, wetlands, and agricultural ecosystems*, 154–170. Springer-Verlag, Berlin.

Cronan, C.S., and C.L. Schofield. 1979. Science 204:304–306.

Dighton, J., and R.A. Skeffington. 1987. New Phytol 107:191–202.

Dixon, R.K., and C.A. Buschena. 1988. Plant Soil 105:265–271.

Dohmen, G.P. 1986. Forstwissenschaftliches Centralblatt 105:252–254.

Entry, J.A., K. Cromack, Jr., and S.G. Stafford. 1987. Can J For Res 17:865–871.

Evans, L.S. 1984. Ann Rev Phytopathology 22:397–420.

Evans, L.S., and T.M. Curry. 1979. Amer J Bot 66:953–962.

Evans, L.S., N.F. Gmur, and F. Da Costa. 1978. Phytopathology 68:847–856.

Fackler, U., W. Huber, and B. Hock. 1986. Forstwissenschaftliches Centralblatt 105:254–257.

Federer, C.A., and J.W. Hornbeck. 1987. Can J For Res 17:266–269.

Fernandez, I.J. 1989. *In* D.C. Adriano and A.H. Johnson, eds. *Acidic precipitation*, Vol. 2, 61–83. Springer-Verlag, New York.

Forsline, P.L., R. Dee, and R.E. Melious. 1983b. J Amer Soc Hort Sci 108:202–207.

Forsline, P.L., R.C. Musselman, W.J. Kender, and R.J. Dee. 1983a. J Amer Soc Hort Sci 108:70–74.

Foy, C.D. 1983. Iowa State J Res 57:355–391.

Freedman, B., and T.C. Hutchinson. 1980. *In* T.C. Hutchinson and M. Havas., eds. *Effects of acid precipitation on terrestrial ecosystems*, 395–434. Plenum Press, New York.

Friedland, A.J., R.A. Gregory, L. Kaerenlampi, and A.H. Johnson. 1984. Can J For Res 14:963–965.

Hain, F.P. 1987. Tree Physiology 3:93–102.

Hain, F.P., and F.H. Arthur. 1985. Zeitschrift fur angewandte Entomologie 99:145–152.

Haines, B.L., J. Chapman, and C.D. Monk. 1985a. Bull Torrey Bot Club 112:258–264.

Haines, B.L., J.A. Jernstedt, and H.S. Neufeld. 1985b. New Phytol 99:407–416.

Haines, B.L., M. Stefani, and F. Hendrix. 1980. Water Air Soil Pollut 14:403–407.

Haines, B.L., and W.T. Swank. 1987. *In* W.T. Swank and D.A. Crossley, eds. *Forest hydrology and ecology at Coweeta. Ecological Studies*, vol. 66, 359–366. Springer-Verlag, Berlin.

Hinrichsen, D. 1987. BioScience 37:542–546.

Huetterman, A., and B. Ulrich. 1984. Phil Trans Royal Soc London B 305:353–368.

Jacobson, J.S. 1984. Phil Trans R Soc London B 305:327–338.

Johnson, A.H., and S.B. McLaughlin. 1986. *In* National Research Council. *Acid deposition: Long-term trends*. National Academy Press, Washington, D.C.

Johnson, A.H., T.G. Siccama, W.L. Silver, and J.J. Battles. 1989. *In* D.C. Adriano and A.H. Johnson, eds. *Acidic precipitation*, vol. 2, 85–112. Springer-Verlag, New York.

Johnson, D.W., and D.W. Cole. 1977. *Anion mobility in soils: Relevance to nutrient transport from forest ecosystems*. EPA-600/3-77-068. US-EPA, Washington, D.C.

Johnson, D.W., and J.O. Reuss. 1984. Phil Trans Royal Soc London B 305:383–392.

Johnson, D.W., J. Turner, and J.M. Kelly. 1982. Water Resources Res 18:449–461.

Johnson, N.M., C.T. Driscoll, J.S. Eaton, G.E. Likens, and W.H. McDowell. 1981. Geochim Cosmochim Acta 45:1421–1437.

Kreutzer, K., and J. Bittersohl. 1986. Forstwissenschaftliches Centralblatt 105:273–282.

LeBlanc, D.C., D.J. Raynal, and E.H. White. 1987a. J Environ Qual 16:325–333.

LeBlanc, D.C., D.J. Raynal, and E.H. White. 1987b. J Environ Qual 16:334–340.

Lee, J.J., and D.E. Weber. 1979. For Sci 25:393–398.

Lindberg, S.E., G.M. Lovett, D.D. Richter, and D.W. Johnson. 1986. Science 231:141–145.

Linthurst, R.A., ed. 1984. *Direct and indirect effects of acidic deposition on vegetation*. Butterworth Publishers, Boston.

Lovett, G.M., S.E. Lindberg, D.D. Richter, and D.W. Johnson. 1985. Can J For Res 15:1055–1060.

McColl, J.G., and R. Johnson. 1983. Plant Soil 74:125–129.

McCormick, L.H., and K.C. Steiner. 1978. For Sci 24:565–568.

McLaughlin, S.B. 1985. J Air Pollut Control Assoc 35:512–534.

McLaughlin, S.B., D.J. Downing, T.J. Blasing, E.R. Cook, and H.S. Adams. 1987. Oecologia 72:487–501.

McLean, E.O. 1976. Commun Soil Sci Plant Anal 7:619–636.

Magel, E., and H. Ziegler. 1986a. Forstwissenschaftliches Centralblatt 105:234–238.

Magel, E., and H. Ziegler. 1986b. Forstwissenschaftliches Centralblatt 105:243–251.

Manion, P.D. 1987. *In* T.C. Hutchinson and K.M. Meema, eds. *Effects of atmospheric pollutants on forests, wetlands and agricultural ecosystems*, 267–276. Springer-Verlag, Berlin.

Matziris, D.I., and G. Nakos. 1977. Forest Ecol Manage 1:267–272.

Mollitor, A.W., and D.J. Raynal. 1983. J Air Pollut Control Assoc 33:1032–1035.

Morrison, I.K. 1984. For Abstr 45:483–506.

Mugwira, I.M., S.U. Patel, and A.K. Fleming. Plant Soil 57:467–470.

Musselman, R.C., and J.L. Sterrett. 1988. J Environ Qual 17:329–333.

National Academy of Sciences. 1977. *Ozone and other photochemical oxidants: Medical and biologic effects of environmental pollutants*. National Academy Press, Washington, D.C.

Neufeld, H.S., J.A. Jernstedt, and B.L. Haines. 1985. New Phytol 99:389–405.

Nihlgard, B. 1985. Ambio 14:2–8.

Nilsson, S.I., and B. Bergkvist. 1983. Water Air Soil Pollut 20:311–329.

Osswald, W.F., H. Heinsch, and E.F. Elstner. 1986. Forstwissenschaftliches Centralblatt 105:261–264.

Paparozzi, E.T., and H.B. Tukey, Jr. 1983. J Amer Soc Hort Sci 108:890–898.

Parker, G. 1987. *Uptake and release of inorganic and organic ions by foliage: Evaluation of dry deposition, pollutant damage, and forest health with throughfall studies.* Technical Bulletin No. 532. National Council of the Paper Industry for Air and Stream Improvement, New York.

Payer, H.D., C. Bosch, L.W. Blank, T. Eisenmann, and K.H. Runkel. 1986. Forstwissenschaftliches Centralblatt 105:207–218.

Proctor, J.T.A. 1983. Environ Exper Bot 23:167–174.

Pylypec, B., and R.E. Redmann. 1984. Can J Bot 62:2650–2653.

Raynal, D.J., J.R. Roman, and W.M. Eichenlaub. 1982a. Environ Exper Bot 22:377–383.

Raynal, D.J., J.R. Roman, and W.M. Eichenlaub. 1982b. Environ Exper Bot 22:385–392.

Rechcigl, J.E., and D.L. Sparks. 1985. Commun Soil Sci Plant Anal 16:653–680.

Rehfuess, K.E., and C. Bosch. 1986. Forstwissenschaftliches Centralblatt 105:201–206.

Rehfuess, K.E., and H. Ziegler. 1986. Forstwissenschaftliches Centralblatt 105:267–271.

Reich, P.B., and A.G. Amundson. 1985. Science 230:566–570.

Reich, P.B., A.W. Schoettle, H.F. Stroo, and A.G. Amundson. 1986. J Air Pollut Control Assoc 36:724–726.

Reich, P.B., A.W. Schoettle, H.F. Stroo, J. Troiano, and R.G. Amundson. 1985. Can J Bot 63:2049–2055.

Reich, P.B., A.W. Schoettle, H.F. Stroo, J. Troiano, and R.G. Amundson. 1987. Can J Bot 65:977–987.

Reiners, W.A., R.K. Olson, L. Howard, and D.A. Schaefer. 1986. Environ Exper Bot 26:227–231.

Reuss, J.O., and D.W. Johnson. 1986. *Acid deposition and the acidification of soils and waters.* Ecological Studies 59. Springer-Verlag, New York.

Scherbatskoy, T., and R.M. Klein. 1983. J Environ Qual 12:189–195.

Scherbatskoy, T., R.M. Klein, and G.J. Badger. 1987. Environ Exper Bot 27:157–164.

Schier, G.A. 1986. Can J For Res 16:136–142.

Schonwitz, R., and L. Merk. 1986. Forstwissenschaftliches Centralblatt 105:258–261.

Schulze, E.D. 1989. Science 244:776–783.

Schutt, P., and E.B. Cowling. 1985. Plant Disease 69:548–558.

Seigworth, K.E. 1943. Amer For 49:521–523, 558.

Seiler, J.R., and D.J. Paganelli. 1987. For Sci 33:668–675.

Selinger, H., D. Knoppik, and A. Ziegler-Jons. 1986. Forstwissenschaftliches Centralblatt 105:239–242.

Senger, H., W. Osswald, M. Senser, H. Greim, and E.F. Elstner. 1986. Forstwissenschaftliches Centralblatt 105:264–267.

Shafer, S.R., L.F. Grand, R.I. Bruck, and A.S. Heagle. 1985. Can J For Res 15:66–71.

Shortle, W.C., and K.T. Smith. 1988. Science 240:1017–1018.

Shriner, D.S. 1978. Phytopathology 68:213–218.

Skeffington, R.A. 1983. *In* B. Ulrich and J. Pankrath, eds. *Effects of accumulation of air pollutants in forest ecosystems,* 219–231. D. Reidel, Dordrecht, Holland.

Skiba, U., T.J. Peirson-Smith, and M.S. Cresser. 1986. Environ Pollut Ser B 11:255–270.

Smith, W.H., G. Geballe, and J. Fuhrer. 1984. *In* R.A. Linthurst, ed. *Direct and indirect effects of acidic deposition on vegetation,* 33–50. Butterworth Publishers, Boston.

Stensland, G.J., D.M. Whelpdale, and G. Oehlert. 1986. *In* Committee on monitoring and assessment of trends in acid deposition, eds. *Acid deposition, long-term trends,* 128–199. National Academy Press, Washington, D.C.

Stienen, H. 1986. Forstwissenschaftliches Centralblatt 105:321–324.

Stroo, H.F., and M. Alexander. 1985. Water Air Soil Pollut 25:107–114.

Taylor, G.E., Jr., R.J. Norby, S.B. McLaughlin, A.H. Johnson, and R.S. Turner. 1986. Oecologia 70:163–171.

Tomlinson, G.H. 1983. *In* B. Ulrich and J. Pankrath, eds. *Effects of accumulation of air pollutants in forest ecosystems*, 331–342. D. Reidel, Dordrecht, Holland.

Tukey, H.B., Jr. 1970. Ann Rev Plant Physiol 21:305–329.

Tyler, G. 1978. Water Air Soil Pollut 9:137–148.

Ulrich, B. 1983. *In* B. Ulrich and J. Pankrath, eds. *Effects of accumulation of air pollutants in forest ecosystems*, 1–29. D. Reidel, Dordrecht, Holland.

Ulrich, B., R. Mayer, and P.K. Khanna. 1980. Soil Sci 130:193–199.

Ulrich, B., and D. Pirouzpanah. 1986. Forstwissenschaftliches Centralblatt 105:318–321.

Van Breemen, N., J. Mulder, and C.T. Driscoll. 1983. Plant Soil 75:283–308.

Van Ryn, D.M., J.S. Jacobson, and J.P. Lassoie. 1986. Can J For Res 16:397–400.

Weiss, M., and R. Agerer. 1986. Forstwissenschaftliches Centralblatt 105:230–233.

Wertheim, F.S., and L.E. Craker. 1988. J Environ Qual 17:135–138.

Wolters, J.B.H., and J.M. Martens. 1987. Botanical Review 53:372–414.

Wood, T., and F.H. Bormann. 1974. Environ Pollut 7:260–268.

Wood, T., and F.H. Bormann. 1975. Ambio 4:169–171.

Wood, T., and F.H. Bormann. 1977. Water Air Soil Pollut 7:479–488.

Woodman, J.N. 1987. Tree Physiology 3:1–15.

Woodman, J.N., and E.B. Cowling. 1987. Environ Sci Technol 21:120–126.

# Effects of Acidic Precipitation on Crops

Lance S. Evans*

## Abstract

Research on acidic deposition on agricultural crops has shown that chemicals in simulated rain solutions penetrate plant foliage at different rates and that substances in foliage can be leached at various rates. Rates of penetration and leaching of some chemicals are dependent on the acidity of the simulated rainfall solution. To date there are no data to document that nutrients in ambient rainfall can benefit plants or that the acidity levels of ambient rainfall can adversely affect plant growth by producing accelerated leaching of essential plant nutrients. Results of experiments under controlled environmental conditions have been performed to understand mechanisms of plant injury to acidic deposition. However, because the sensitivity of plants to acidic deposition under field conditions is different from the sensitivity of plants under controlled environmental conditions, results of experiments conducted under controlled environments may have a very limited value in an overall assessment of impacts of acidic deposition on crops.

Few experiments have been performed with crops under field conditions. The only crop for which there are adequate data for an assessment of the crop impact of acidic deposition is the soybean. Some highly replicated field experiments have been performed with soybeans; some cultivars such as Amsoy, Asgrow, Corsoy, and Hobbitt are negatively impacted, and other cultivars such as Williams, Davis, and Wells do not appear to be negatively impacted. The degree to which seed yields of the Amsoy, Asgrow, Corsoy, and Hobbitt cultivars are negatively impacted is not well understood. However, research has shown that seed protein contents of seeds of several soybean cultivars are negatively impacted by rainfall acidity.

*Laboratory of Plant Morphogenesis, Manhattan College, The Bronx, NY 10471, and Terrestrial and Aquatic Ecology Division, Department of Applied Science, Brookhaven National Laboratory, Upton, NY 11973 USA.

# I. Introduction

Acidic precipitation, wet or frozen precipitation with a $H^+$ concentration greater than 2.5 $\mu$eq $L^{-1}$ (equivalent to a pH of about 5.6), is a significant air pollution problem in North America and Europe. The northeastern portion of the United States is at the center of the high acidic rainfall area in North America (Evans et al., 1981a). The high $H^+$ concentration of precipitation (rain, snow, fog, sleet, and mist) in the northeastern United States is explained by the presence of strong acids. Sulfuric acid contributes a portion of the acidity (Likens et al., 1972, 1979; Nørdo, 1976; Oden, 1976), and nitrate and chloride are significant anion components of the total acidity in precipitation (Jacobson, et al., 1976; Yue et al., 1976). A significant amount of sulfur dioxide emitted into the atmosphere is converted into sulfuric acid and various ammonium sulfate aerosols (Altshuller, 1976). Particulate sulfur compounds and sulfur oxides may be incorporated into precipitation through conversion to $H_2SO_4$ (Likens et al., 1972). About 90% of the sulfur in the atmosphere of the northeastern United States is contributed by anthropogenic sources (Galloway and Whelpdale, 1980). A budget estimate for nitrogen inputs into the atmosphere of the Adirondack Mountain region indicates that 34% of the anions in rain could be attributed to nitrates (Stensland, 1983). Acidic precipitation is only a portion of the total acidity brought to the earth's surface from the atmosphere. Dry fall plus acidic precipitation is termed *acidic deposition*. The purpose of this chapter is to describe characteristics of ambient precipitation and to review and evaluate the impacts that acidity in precipitation may have on vegetation. The effects of other biotic or abiotic agents will not be considered unless they have a demonstrated interaction with the effects of precipitation acidity.

This document is an attempt to describe the impacts of acidic deposition on agricultural crops. The focus of this report is to document effects and impact mechanisms of acidic deposition on plants. Consideration will be given to those acidic deposition effects that may be present but generally are not or cannot be included in experiments designed to determine plant (crop) productivity. Besides a discussion of direct impacts of acidic deposition, other biotic and abiotic factors that modify plant responses will be considered. Limitations to experimental approaches will also be discussed.

# II. Wet Deposition and Foliage

## A. Chemical Nature of Plant Surfaces

The surfaces of terrestrial vegetation are covered with cuticular and epicuticular waxes of different chemical and physical characteristics. The significance of these differences in characteristics is probably as varied as the climatic and microclimatic environments in which plants must survive.

Leaves and herbaceous stems are covered by the plant cuticle. The cuticle is composed of epicuticular waxes on the surface overlaying a cuticularized layer that is composed of cellulose and pectin encrusted with cutin. Inside this cuticularized layer, a cutinized layer of the cell wall is present. Last, a pectin layer is located between the cutinized cell wall and the plasmalemma of epidermal cells (Martin and Juniper, 1970).

A large variety of epicuticular waxes is present on surfaces of foliage. Many n-primary alcohols and n-fatty acids varying in chain length from 24 to 34 carbon atoms have been characterized. Other waxes such as n-nonacosane, n-hextriacontane, n-hexacosan-lool, and other substituted alcohols have been detected. Nonacosane results from a condensation of two molecules of pentadecanoic acid and subsequent decarboxylation to form nonacosan-15-one, followed by reduction. Many other final product waxes are possible when different fatty acids are subjected to condensation.

Cutin is a large polymer of fatty acids. Leaves contain three fatty acid–oxidizing enzymes for cutin synthesis from fatty acids: lipoxidase, stearic acid oxidase, and oleic acid oxidase. There is evidence that stearic acid, produced from acetate, is converted to oleic acid by stearic acid oxidase. In turn, oleic acid is converted to linoleic acid by oleic acid oxidase. Linoleic acid (and possibly linolenic acid) is oxidized to its hydroperoxides by lipoxidase. In turn, these hydroperoxides can react with a variety of fatty acids to form saturated and unsaturated hydroxy-fatty acid condensation products (Martin and Juniper, 1970).

To produce the entire cutin polymer, peroxide bonds are formed between fatty acids and hydroxy-fatty acids. After these reactions, esterifications between hydroxyl and carboxyl groups can easily occur. The final polyermization step probably involves other enzymes, including catalase that can form various alkoxy and hydroxy radicals by splitting peroxide linkages. This polymerization process is completed when repeated additions of other chains occur in a highly cross-linked pattern (Martin and Juniper, 1970).

Waxes and wax derivatives in plants vary greatly. Many surface waxes contain n-alkanes ($C_{13}$-$C_{34}$), n-primary alcohols ($C_{16}$-$C_{32}$), monomethylalkanes and dimethylalkanes, some alkanes, some fatty acids, and other compounds (Martin and Juniper, 1970). Some unique compounds include tricontaine in *Gossypium hirsutum*, n-tritriacontan-16-18-dione in several *Eucalyptus* species, sterols from sugarcane (*Saccharum officinarum*), and wax and estolides of carnauba wax. Plant waxes frequently are specific for the species and genus in which they occur. Consequently, they have been used extensively in plant taxonomic studies, which reinforces the wide diversity of chemical structures of surface waxes (Martin and Juniper, 1970).

From available evidence, cuticular substances appear to be secreted through the walls of epidermal cells. In some cases, walls on all sides of cells become cutinized to form a cuticular epithelium (Martin and Juniper, 1970). If large cutin deposits occur in the middle lamellae, cutin cystoliths may result.

Cuticles usually develop during early organ ontogeny. The early formed cuticle hardens from oxidation and continued polymerization through the addition of fatty acids, alkanes, and the like. In cuticles of young organs, surface waxes are extruded through and within the cuticular layer. From available evidence, the cuticle appears to develop through leaf expansion only, and, at least for most species, little or no cuticular development occurs after leaf expansion is completed. Young leaves of *Pisum sativum* and *Eucalyptus* species show prominent crystalline and tubular wax formation that last throughout ontogeny (Juniper, 1960).

## B. Interaction of Acidic Precipitation and Leaf Surfaces

### 1. Foliar Wettability

The amount of injury to plant foliage by acidic precipitation may depend upon the area of leaves in contact with rainwater. Moreover, injury may also depend upon the rate of absorption of materials from rainwater per unit area. The amount of water absorbed by foliage depends upon many characteristics. These characteristics vary among plant species (Evans and Curry, 1979; Evans, 1982), and, as a result, may determine relative species-sensitivity to precipitation acidity.

The amount of foliar injury may be a function of foliar wettability. The attraction of a liquid to a solid surface upon impaction is an important measure of the amount of water retained on the surface. However, the area over which a droplet spreads depends upon the relative amount of attraction between the liquid and the surface. The advancing contact angle has been used as a criterion of leaf surface wettability (Martin and Juniper, 1970) and is defined as the angle between the surface of the leaf and the tangent plane of a water droplet at the circle of contact between air, liquid, and leaf. A 0° angle would maximize wettability, and an angle of 180° would provide virtually no wetting. Contact angles of pure water drops may be as small as 31° on *Phragmites communis* and greater than 150° on *Triticum aestivum* and *Lupinus albus* (Hall et al., 1965; Juniper, 1960; Linskens, 1950). In this way, the contact angle is determined mostly by chemical and physical characteristics of epicuticular waxes and to a lesser extent by surface roughness (Martin and Juniper, 1970). Wettability of leaf surfaces increases markedly when the cuticular and epicuticular waxes are removed (Fogg, 1948).

### 2. Interaction of Water Solutions and Cuticles

Solutions must penetrate through cuticular layers or through stomata to reach leaf cells. It has been postulated that cuticles are perforated with micropores (Crafts, 1961). Cuticular pores may be numerous in specialized areas such as the bases of trichomes, hydathodes, glandular hairs (Schnepf, 1965), water-absorbing scales of Bromeliaceae (Haberlandt, 1914), and stigmas (Konar and Linskens, 1966). Although cuticular pores may not be present, solutions penetrate faster at these locations (Martin and Juniper, 1970). These results may explain why injury from acidic precipitation occurs more frequently at bases of trichomes and hydathodes. Penetration of rain or leaf surface solutions through stomata is thought to be infrequent if it occurs at all (Adam, 1948; Gustafson, 1956, 1957; Sargent and Blackman, 1962). Generally, spontaneous infiltration of stomata by water will occur if the contact angle is smaller than the angle of the aperture wall. The degree of stomatal opening from 4 to 10 $\mu$m is of little importance in penetration.

However, cuticular ledges present at the entrance to the outer vestibule and between the inner vestibule and substomatal chamber result in very small wall angles. These small wall angles may not be adequate for water penetration of stomata (Schönherr and Bukovac, 1972). All evidence suggests that the main route of entry of substances should be through the cuticle and not through the stomata.

Nonpolar molecules should penetrate the cuticle at rates faster than those of polar molecules because of the nonpolar nature of both cutin and epicuticular waxes of the cuticle (Norman et al., 1950). More polar compounds, such as inorganic ions and water, may preferentially enter the leaf via pectinaceous channels that traverse the cuticle (Roberts et al., 1948). It is postulated that the ratio of cutin to wax in the cuticle greatly influences the degree of penetration of polar substances (Martin and Juniper, 1970).

## 3. Solution Penetration and Acidity

Solution acidity may also affect cuticular penetration rates (Schönherr, 1976; Schönherr and Schmidt, 1979). In experiments with isolated cuticles of apricot leaves, penetration rates of acidic substances increased with an increase in solution acidity. The penetration rate of basic substances increased with a decrease in solution acidity. These relationships between pH and sorption rates with apricot cuticles (Orgell, 1957) have been verified with isolated cuticles (McFarlane and Berry, 1974) and intact bean leaves exposed to buffered solutions (Evans et al., 1981c).

Cuticles and epicuticular waxes should not be considered to be inert, nondifferential barriers to ion permeability. Relative permeability coefficients of isolated cuticles to various monovalent and divalent cations differ by as much as a factor of six.

The idea of differential or selective uptake of chemical constituents by foliage is reinforced by other experimental data. Penetration of chemicals into foliage is affected by both metabolic and nonmetabolic processes. Reversible diffusion through cuticles is followed by metabolically controlled uptake through cellular membranes (Prasad and Blackman, 1962). There are many examples of light-enhanced foliar uptake (Bennett and Thomas, 1954; Gustafson, 1956; Sargent and Blackman, 1965) being retarded by protein synthesis inhibitors (Jyung and Wittwer, 1964). In this way, uptake of various elements by foliage is selective to an unknown degree and is dependent upon the metabolism of leaf cells.

Some regions of the cuticle are more permeable than others. For example, basal portions of trichomes and guard cells (Dybing and Currier, 1961; Sargent and Blackman, 1962) are preferential sites of absorption. Moreover, absorption of water-soluble materials may be rapid through the cuticle near veins (Leonard, 1958; Linskens, 1950). These studies are germane because about 95% of all foliar lesions following exposure to simulated acidic rain occur near the bases of trichomes, at guard and subsidiary cells of stomata, and along veins (Evans and Curry, 1979; Evans et al., 1977a, 1977b, 1978). More recently, these results have been confirmed with other species (Adams, 1982; Paparozzi, 1981). From these data it seems reasonable to conclude that phytotoxic components of simulated acidic precipitation penetrate the cuticle at faster rates near vascular tissues and subsidiary cells and at the bases of trichomes and hydathodes.

## 4. Changes in Plant Surface Characteristics by Acidic Deposition

Plant cuticles and tissues are subject to physical and chemical weathering. Physical damage may occur by plant tissues rubbing on objects; by other leaves; by rain, hail, and water splash; and by deposition of soot, oils, sand, and dust (Martin and Juniper, 1970). For example, numerous studies have shown that retention of herbicides increased after physical and chemical weathering. Recovery from excessive weathering may occur in expanding leaves in which cutin and wax deposits are still being occluded into the cuticle. It must be emphasized that cuticular weathering affects all surface properties of leaves. These results support the idea that the growth and nutrient balance of plants are the result of their entire environment.

Acidic precipitation may change the surface characteristics of foliage. It has been suggested that acidic precipitation may affect the submicroscopic structure of the epicuticular wax layers of leaves. Shriner (1974) presented scanning electron micrographs that showed that leaves of kidney beans and willow oak exposed to simulated rain of pH 3.2 had eroded superficial waxes, cutin, and cuticular waxes. There was only slight erosion on leaves exposed to rain of pH 6.0. In contrast, Paparozzi (1981) used scanning and transmission electron microscopy and observed no erosion of epicuticular or cuticular waxes on either yellow birch or kidney bean after exposure to simulated rainfalls of pH as low as 2.8. The cuticular waxes were not structurally changed by simulated acidic rain, even though underlying cells were affected. Moreover, the most widely used method of isolating cuticles involves exposure to strong acids (Holloway and Baker, 1968; McFarlane and Berry, 1974). Penetration rates of elements through isolated cuticles were similar to rates through intact cuticles in situ (Evans et al., 1981c). At the present time a relationship between changes in the cuticle, including cutin and epicuticular waxes, and changes in leaf cells remains to be established.

## 5. Changes in Precipitation Chemistry by Leaf Surfaces

Leaf surfaces may also change the chemistry of rain solutions. The acidity of rainwater may be changed by chemicals from leaves as raindrops dry. Adams (1982) demonstrated that leaves of several plant species have differing buffering capacities. The pH of simulated raindrops (50 μL) of pH 5.6, 3.5, and 3.0 usually increased during an observation period of 75 minutes. However, when the initial pH was 2.5, the pH always decreased as time passed. The pH of a Parafilm surface (control) was always lower than that of any of the plant surfaces, suggesting that the leaves produced substances to neutralize the acidity. The presence on leaf surfaces of substances produced by leaves has been recognized for almost a century (Uphof, 1962). Other studies by (Evans et al., 1985b) substantiate the results of Adams (1982). In these experiments pH measurements were obtained from 50 μL droplets of simulated rain solutions. The volume of droplets decreased by about 30% every 30 minutes during exposure periods of 130 to 160 minutes. Droplet volumes were the same on leaflet surfaces as on Teflon plates and had a pH of 3.1 and 2.7, respectively. Acidities of drops of pH 3.1 and 2.7 increased to pH 2.0 and 1.4, respectively, as volume reduction occurred. During this period the number of picoequivalents in droplets of pH 3.1 and 2.7 on leaflets decreased by 62% and 74%, respectively, over 120 minutes. During the same time period, the number of picoequivalents in droplets of pH 3.2 and 2.9 on Teflon decreased by only 47% and 53%, respectively. The pH of droplets of initial pH 4.1 on leaflets did not change significantly during droplet volume reductions while the acidity of droplets on Teflon increased slightly. At pH 5.6, solution pH increased slightly as volume decreased. The change in measured pH and number of picoequivalents per droplet were markedly similar on both leaves and Teflon plates. Although they provided similar data, these experiments represent only a single exposure of a leaf surface to acidic conditions. At present, no data on the effects of multiple exposures are available that relate frequency of precipitation on the same leaves to an acidity level. Possibly, very different responses would be obtained if multiple exposures occurred.

## C. Inputs of Materials into Foliage

Much research has demonstrated that the main penetration route of chemicals in aqueous solutions is through the cuticle (Martin and Juniper, 1960). In general, solution penetration occurs by diffusion, which is dependent upon three factors: the time period of exposure, the surface area of contact, and the chemical nature of the substance absorbed. The first two of these factors are related to the degree of leaf wettability described previously.

Penetration of substances into foliage occurs by a combination of metabolic and nonmetabolic processes. Specifically, substances pass through the cuticle by passive, reversible diffusion, but cellular uptake is under metabolic control at the plasmalemma (Prasad and Blackman, 1962). Evidence demonstrates that the rate-limiting step in uptake of substances is metabolically controlled (Jyung and Wittwer, 1964).

### 1. Differential Penetration Rates

Because a wide variety of chemicals is found in ambient rain in the eastern United States (Galloway et al., 1982; Lindberg, 1982; Lindberg et al., 1981; Reuther et al., 1981), it is necessary to know if materials in rain can be absorbed by foliage and if rainfalls can remove materials from foliage. Recent information suggests that materials can be absorbed at differing rates from rainfalls of varying acidity (Evans et al., 1981c). For example, sulfate penetrated leaves faster at pH 2.7 than at 5.7, and $^{86}Rb^+$ penetrated faster at pH 5.7 than at lower pH levels. Tritiated water entered foliage at similar rates at all pH levels tested. In addition to differences in uptake due to acidity, incorporation rates of various ions may vary markedly. Water penetrated leaves much faster than $^{86}Rb^+$. In addition, water molecules entered foliage about 1,000 times faster than sulfate ions (Evans et al., 1981c). Also, $^{63}Ni$, $^{65}Zn$, and $^{36}Cl$ penetrated leaves faster at lower pH levels (2.7 to 3.0) than at pH 5.7. In general, penetration increased as time of exposure increased. The isotope $^{65}Zn$ was incorporated into foliage more rapidly than the other two isotopes used (Evans et al., 1985). These results suggest that absorption of materials in water on leaf surfaces is a selective process that may be affected by solution acidity, cellular metabolism, and possibly other factors. However, individual ions appear to have their own specific entry that may be affected by many intrinsic and extrinsic factors.

## 2. Foliar Fertilization

Foliar fertilization may occur by inputs of nitrogen and sulfur from ambient rainfalls. Little information is available to evaluate the amount of incorporation of nutrients in rain into foliage. The rate of entry of sulfate is a low percentage of the amount in solution (Evans et al., 1981c). The only other information about the incorporation of nitrogen and sulfur into foliage from solutions may be derived from foliar fertilization experiments. Of course, the application of fertilizers on foliage is usually performed with surfactants that are not present in ambient rain to enhance uptake.

Increases in soybean seed yields may result if large quantities of N (80 kg ha$^{-1}$), P (8 kg ha$^{-1}$), K (24 kg ha$^{-1}$), and S (4 kg ha$^{-1}$) are applied to foliage of soybeans during pod filling only (Garcia and Hanway, 1976). From available evidence the optimum N:P:K:S proportions in the fertilizer solutions for soybeans should be 10:1:3:.5. Inconsistent results occurred with other combinations so that many other nutrient combinations, even with high nitrogen (24 kg ha$^{-1}$) and sulfur (12 kg ha$^{-1}$) applications, resulted in no yield effects (Garcia and Hanway, 1976).

Throughout the 1981 soybean season at Upton, New York, a total of 18.1 kg N ha$^{-1}$ (as nitrate) and 102 kg S ha$^{-1}$ (as sulfate) was applied in simulated rainfalls at pH 2.7, and no obvious growth stimulation occurred (Evans et al., 1983). These rather high inputs of N and S can be compared to the relatively low (1.80 kg N and 2.73 kg S per hectare) amounts applied by ambient rainfalls during the same time period.

No consistent increases in yields of corn (Neumann et al., 1981) and rice (Thom et al., 1981) have been obtained. In addition, two experiments performed with foliar fertilization of Zea mays under the rain-exclusion facility demonstrated no beneficial effect (Harder et al., 1982). During 1 year, a 6.4% reduction in seed yield occurred when plants were exposed to 22.5, 4.5, 4.5, and 2.3 kg ha$^{-1}$ of nitrogen, phosphorus, potassium, and sulfur, respectively, commencing 2 weeks before silking. It would seem reasonable to conclude that plant productivity is not improved through foliar absorption of nitrogen and sulfur from current ambient rainfall and/or simulated rainfall under conditions similar to ambient. Similarly, it is believed that the amounts of nitrogen and sulfur added to soils by rainfall are insignificant compared with amounts added through routine fertilization (Evans et al., 1981a; Evans, Curry, and Lewin, 1981c). This is supported by $^{15}N$ tracer studies showing that the amount of nitrogen incorporated into the foliage of bean plants from simulated rain solutions is insignificant compared with nitrogen inputs from other sources (Evans et al., 1986a). From available evidence, no data show that nutrients can enhance plant growth at levels present in wet deposition.

## 3. Changes in Leaf Cell Permeability of Leaves Exposed to Rainfall Acidity

As $H^+$ ions may penetrate leaves, changes in leaf cell permeability after exposure were tested (Evans et al., 1981c). Isotopes were added to nutrient solutions after a single 20-minute rainfall at various pH levels. Twenty-four hours after the first rainfall, a second 10-minute rain was applied. After the foliage dried, leaf discs were placed in 0.5 mM $CaSO_4$ solutions for 120 minutes. For $^{86}Rb$, $^3H$, and $^{35}SO_4$, there were no significant differences in leaching rates among the acidity treatments.

Cell permeability to $^{63}Ni$ was lowest at a simulated rainfall of pH 3.0. The low permeability observed at pH 3.0 could not be rationalized with high permeability at pH 3.4 and intermediate permeability at both pH 5.7 and 2.7. Between 45% and 68% of the $^{63}Ni$ in the foliage was permeable during the 120-minute time period. Rainfall acidity had no effect on cell permeability to $^{65}Zn$. Moreover, permeability was constant throughout incubation and ranged from 47% to 75% among the pH levels tested. Cell permeability of $^{36}Cl$ from foliage was strongly influenced by simulated rainfall acidity. Permeability at pH 2.7 was about three times higher than that at pH 5.7. Mean cell permeability to $^{36}Cl$ was about 70%. These results taken together suggest that most elements tested did not show any differences in cell permeability. However, the fact that two of the six elements tested showed a significant acidity effect suggests that the acidity of the test solutions in simulated rainfall can alter membrane permeability.

## D. Leaching of Materials from Foliage

### 1. Differential Leaching Rates

Materials can be lost from leaf surfaces by guttation and secretion or by leaching by rain, dew, and mists. Marked effects can result from such losses. For example, Martin and Juniper (1970) reported marked losses of potassium (25–30 kg ha$^{-1}$), sodium (9 kg ha$^{-1}$), and calcium (10.5 kg ha$^{-1}$, annually) from apple (*pyrus malus*) orchards. Most literature confirms that the highest leaching rates occur from older leaves, that is, those that have experienced the most weathering.

Because plants produce substances on their surfaces, it is of interest to determine if the removal of these substances is sensitive to acidity in precipitation. Likewise, substances within leaves may be released after exposure to rainfalls. Moreover, the nutrient levels in harvested portions of crops may affect the quality of foodstuffs. It is conceivable that precipitation acidity could sufficiently influence nutrient leaching from plant surfaces as to alter crop quality. Wood and

Bormann (1975) demonstrated that $K^+$, $Ca^{2+}$, and $Mg^{2+}$ were leached from pinto bean (*Phaseolus vulgaris*) leaves more rapidly at pH levels of rainwater of 3.0 and 3.3 than at pH levels of 4.0 and 5.0. Calcium ions leached faster with rainwater at pH 3.0 than at pH 3.3 from foliage of sugar maple (*Acer saccharum*). Leaching rates of $K^+$ and $Mg^{2+}$ were higher at pH 3.0 than at pH 4.0. In tobacco (*Nicotiana tobacum*) leaves, $Ca^{2+}$ leached faster from foliage exposed to simulated rainfalls of pH 3.0 than from foliage exposed to pH 6.7 (Fairfax and Lepp, 1975). In pinto beans exposed to a 20-minute daily treatment of simulated rain for 5 days, more calcium, nitrate, and sulfate were leached from foliage at pH levels of 2.7, 2.9, and 3.1 than from plants exposed to pH 5.7 (Evans et al., 1981c). In contrast, the amount of potassium leached was greater from leaves exposed to pH 5.7 than from leaves exposed to pH levels between 3.4 and 2.9. The amounts of ammonium, magnesium, and zinc leached were the same at all pH levels tested. Several radioisotopes were detected from leachates from leaflets of *Phaseolus vulgaris* that were exposed to one 20-minute rainfall 24 hours after radioisotope addition. Foliar leaching of Ni was greater at pH 5.7 than at a lower pH, whereas the opposite situation occurred with $^{36}Cl$. The isotope $^{65}Zn$ leached similarly at all pH levels tested. Leaf cell permeabilities of $^{63}Ni$ and $^{36}Cl$ were greater after exposure to low pH rainfalls than at pH 5.7 (Evans et al., 1985). Although the quantities of $^3H_2O$, $^{86}Rb$, and $^{35}SO_4$ were less than 1% of the radioisotopes in leaflets, these results demonstrate that recently absorbed ions may be leached by rainfall. All of the experiments to determine the impacts of rates of nutrient penetration and leaching have been performed under controlled environmental conditions and should be confirmed under conditions of outdoor environments.

## 2. Mass Balance: Foliar Inputs versus Leaching

Cole and Johnson (1977) measured pH and conductivity (a measure of the total cations and anions present) of ambient precipitation and throughfall during two rainfalls in 1973. During the latter portion of the first rainfall, the pH of the throughfall ($\approx 4.0$) was about 1 pH unit below that of ambient precipitation ($\approx 5.0$). Conductivities of the two solutions were similar at about 10 $\mu$mho $cm^{-1}$. During the second rainfall, the pH of ambient precipitation ($\sim 3.5$) was almost 2 units below that of the throughfall ($\sim 5.3$). Over an entire annual cycle, the amount of sulfur in throughfall was greater than in the ambient precipitation. Cole and Johnson (1977) concluded that the differences in sulfur content observed were due to foliar leaching. Moreover, van Breemen and others (1982) demonstrated that ammonium, nitrate, and sulfate volume-weighted concentrations are higher in throughfall and stemflow than in ambient rain from forests at two locations in the Netherlands. These results would suggest that N and S may be leached from trees exposed to rainfalls of pH 4.29 and 4.51. However, washoff of dry deposition may have contributed to higher sulfate concentrations in throughfall. Presently, there are no data that document significant foliar leaching of materials at present levels of precipitation acidity that could produce changes in plant growth or productivity.

# III. Direct Effects on Vegetation

## A. Visible Foliar Injury

### 1. Occurrence

Presently, there is only one documented occurrence of visible foliar injury attributed to ambient rainfalls (Evans et al., 1982a). Visible foliar injury, identical to that caused by simulated acidic rainfalls of pH 4.0 and below, was produced on foliage of field-grown garden beets during a 3-day period in which three short-duration, low-volume rain showers with levels of about pH 3.8 to 3.9 occurred during the 1980 growing season at Brookhaven National Laboratory. Higher-acidity simulated rainfalls resulted in lower beet yields. These lower yields resulted from both a smaller number of saleable roots per plot and a lower fresh mass per saleable root.

Presently, almost all knowledge about visible foliar injury is derived from exposures of plants to simulated acidic rainfalls. Lesions produced by simulated acidic rain occur mostly on leaves and reproductive structures (Evans and Curry, 1979; Evans et al., 1977b, 1978; Gordon, 1972; Jonsson and Sundberg, 1972). Visible lesions on leaves have not been observed at pH values above 3.9. A significant percentage of the leaf area may exhibit lesions. For example, about 0.5%, 2% to 3%, 5% to 10%, and 10% to 15% of the leaf area of pinto beans is injured after one to four exposures to acidic rain at pH levels of 3.0, 2.7, 2.5, and 2.3, respectively (Evans et al., 1977b). In this way, the area showing injury increases with an increase in simulated rainfall acidity.

Visible leaf injury is most pronounced on foliage of some species just prior to full leaf expansion (Evans, 1980, 1982; Evans and Curry, 1979; Evans et al., 1977b, 1978). Densities of trichomes and stomata per unit leaf area are highest in young leaves and lowest in fully expanded leaves. Because the amount of visible leaf injury does not coincide with the density of either stomata or trichomes at full leaf expansion, other factors such as surface wettability must be involved in predisposing leaves to injury. Wood and Bormann (1974) suggested that young (14-day-old) birch seedlings are much more sensitive than older (6-week-old) seedlings. Paparozzi (1981) showed that mature birch leaves were as susceptible to foliar injury as enlarging leaves. Needle elongation is inhibited if simulated acidic rain solutions are applied to immature pine fascicles (Gordon, 1972; Hindawi and Ratsch, 1974).

Leaves of several plant species have reacted to exposure to simulated acidic precipitation by producing galls that elevated the adaxial leaf surface (Evans et al., 1978). Water from subsequent rainfalls did not pool in these elevated locations, so less injury (total area) resulted. Galls were produced from abnormal cell proliferation (hyperplasia) and abnormal cell enlargement (hypertrophy). These results have been confirmed on *Artemisia tilesii, Phaseolous vulgaris,* and *Spinacia oleracea* (Adams, 1982). In this manner, the hyperplastic and hypertrophic responses of foliage may have alleviated extensive foliar injury and thus may be responsible for species sensitivities to visible foliar injury.

## 2. Relative Species Sensitivity

Experiments were performed to rank species sensitivities to simulated acidic rain. Based upon visible effects on foliage, it has been suggested that plants rank in sensitivity from high to low in the following order: herbaceous dicots, woody dicots, monocots, and conifers (Evans, 1980; Evans and Curry, 1979). Within each species, the amount of visible leaf injury appears to relate linearly to the hydrogen ion concentration of the simulated rain solution (Evans, 1982).

## 3. Relation to Plant Productivity

Although there appears to be a linear relationship between hydrogen ion concentration and amount of visible foliar injury, there is no relationship between the presence of visible foliar injury and plant yields or productivity. This latter relationship is true because many field experiments with soybeans have demonstrated seed yield reductions between pH 3.4 and 4.1 without visible foliar injury. Because changes in plant productivity can occur in the absence of visible foliar injury, the utility of foliar injury as a meaningful experimental parameter to measure under field conditions may be questioned.

## B. Crop Growth and Yield (Controlled Environments)

Reductions in dry weight of trifoliate leaves and dry weights of pods and seeds of bush beans (*Phaseolus vulgaris*) (19% and 11%, respectively) occurred after exposure to acidic mists (pH 3.0) with no visible leaf injury (Hindawi et al., 1977). In contrast, Wood and Bormann (1974) showed that simulated acidic mists of pH 3.0 did not reduce growth rates of yellow birch (*Betula lutea*) even though all leaves exhibited some leaf pitting and curling. However, plant productivity (crop yields) and/or individual survival and so forth may be determined by development and survival of reproductive organs as well as by cumulative injuries to foliage. Because no clear view of yield effects has been demonstrated among various plant species exposed to acidic precipitation, a clear relationship between foliar injury and seed yield cannot be predicted with present knowledge.

Lee and others (1981) conducted experiments with 28 crops grown in pots in field chambers. Simulated acidic rain was applied at pH levels of 5.6, 4.0, 3.5, and 3.0. Marketable yield was inhibited for five crops (radish (*Raphanus sativa*), beet (*Beta Vulgaris*), carrot (*Daucus carota*), mustard greens (*Brassica alba*), and broccoli (*Brassica oleracea*), and stimulated for six crops (tomato (*Lycopersicon Esculentum*), green pepper (*Piper nigrum*), strawberry (*Frageria* sp.), alfalfa (*Medicago sativa*), orchard grass (*Dactylis glomerata*), and timothy (*Phlenm pratense*)). No consistent effects were observed for 16 other crops. The experimental results suggest that the probability of yield being affected by acidic rain depends upon the plant species as well as the part of the plant utilized. Foliar injury was not always related to effects on crop yields. The marked decreases in root yields of greenhouse-grown radish (Lee et al., 1981) were verified by other experiments (Evans et al., 1982b). In contrast, yields of alfalfa, wheat, and lettuce grown under controlled environmental conditions were not markedly affected by applied simulated rainfalls between pH values 5.6 and 3.1 (Evans et al., 1982b).

In recent experiments, simulated acidic rain of pH 3.1 and below decreased dry mass of seeds, leaves, and stems of pinto beans exposed to 45 20-minute rainfalls throughout the growing season under greenhouse conditions (Evans and Lewin, 1981). The decrease in seed yield noted was similar to the decrease in biomass of both leaves and stems. The decrease in yield of pinto beans by simulated acidic rain was attributed to both a decrease in the number of pods per plant and a decrease in the number of seeds per pod. In soybeans, application of 78 simulated acidic rainfalls decreased the dry mass of both stems and leaves. Seed yield also decreased after treatment with rain at pH 2.5. However, an increase in seed yield occurred when plants were exposed to rainfalls of pH 3.1. A larger dry mass per seed was responsible for the larger dry mass of seeds per plant exposed to simulated rainfalls of pH 3.1 (Evans and Lewin, 1981).

Significant effects of rainfall acidity were not observed on apple seedlings unless the pH was below 2.75 as compared with controls (pH 5.6). Moreover, both leaf and seedling growth rates were not different between pH 7.0 and 2.75 when 14 plants per treatment were harvested after nine weekly (amount of water applied not given) rainfalls (Forsline et al., 1983a). The pH levels of ambient rainfalls did not affect fruit quality or production in grapes or apples. However, reductions in pollen germination, fruit set, and fruit quality were observed below pH 3.1 for some cultivars (Forsline et al., 1982, 1983b).

## C. Crop Growth and Yield (Field Conditions)

Because radish root yields were reduced after exposure to simulated acidic rainfalls under controlled conditions (Lee et al., 1981), two experiments have been performed with field-grown radishes (Evans et al., 1982a; Troiano et al., 1982). Troiano et al. (1982) exposed four plots per treatment and demonstrated significantly higher root yields (13% of dry mass) at high-acidity rainfalls as compared with simulated rain of pH 5.6. In contrast, no statistically significant yield effects were demonstrated in an experiment that utilized 45 plots per treatment (Evans et al., 1982a). The mass of individual roots harvested in the former experiment (Troiano et al., 1982) was three to four times greater than yields of the latter experiment (Evans et al., 1982a) and from roots purchased from commercial markets. The results of these two field studies indicate that the response of plants to rainfall acidity under field conditions may differ markedly from those obtained under controlled environmental conditions (see above for the effects with radishes grown under controlled environmental conditions). The reason for this difference is not known.

Significant reductions in garden beet yields were obtained with plants receiving applications of simulated rainfalls in addition to ambient rainfalls. Plants exposed to simulated rainfalls of pH 2.7, 3.1, and 4.0 gave yields 73%, 77%, and 65% of the root yields of plants exposed to rainfalls of pH 5.7 (Evans et al., 1982a). No significant yield reductions were obtained for kidney beans and alfalfa.

Several field experiments have been performed with soybeans. In one experiment (Experiment 1) at Yonkers, New York, plants were grown in open-top chambers (OTC), one treatment per chamber (Jacobson et al., 1980; Troiano et al., 1983). The experimental treatments were simulated rainfalls at pHs 4.0, 3.4, and 2.8. A clear plastic tarpaulin was loosely secured over the top of each chamber. Within the OTCs, light transmission was decreased up to 25% and temperatures were elevated up to 3°C above ambient conditions. Eighteen simulated rainfalls of 12.7 mm each for a duration of 60 minutes were applied over a period of 50 days during the latter portion of the growing season. These simulated rainfalls provided a means of 4.6 mm of water daily.

A second experiment (Experiment 2) was performed in Kendall County, Illinois, in 1977 and 1978 (Irving and Miller, 1981). Experimental plots were placed at two different locations in the same field during the 2-year study. Only one experimental plot (four 5-m rows in 1977 and four 6-m rows in 1978) per treatment was used each year. In both years, simulated rain and ambient rain treatment plots were also exposed to both ambient and artificially elevated concentrations of sulfur dioxide. The treatments consisted of simulated rains with pH 5.25, 4.09 (ambient only), and 3.06 (1978; 3.17, 1977). Simulated rainfalls were distributed from a single nozzle for each treatment plot, whereas sulfur dioxide was administered from a gas distribution system above the plant canopy level.

Eight and 11 simulated rainfalls provided 3.38 and 4.46 cm of water in 1977 and 1978, respectively. These treatments provided 36% and 49%, respectively, more water to the experimental plots than the control plot, which received only 9.39 cm of water during the treatment period of 1978. Thus, not all plots received the same amount of water. Plants were exposed to 11 simulated rainfalls 1.4 times weekly for 54 days during the second half of the growing season of 1978. The simulated rainfalls of pH 5.25 and 3.17 provided calculated mean weighted pH values of 4.25 and 3.47 as compared with an ambient rainfall pH of 4.09.

A third experiment (Experiment 3) was performed at Upton, New York (Evans et al., 1983). Two simulated rainfalls of 12.7 mm each provided 25.4 mm of water weekly for 15 weeks. Each simulated rainfall occurred during 1.75 hours between 1800 and 2400 hours. The crop received no measurable water from ambient rain because it was covered during every ambient rainfall. All treatments received the same amount of water. Four replicates of four treatments of simulated rainfalls of pH 5.6, 4.1, 3.3, and 2.7 were arranged into Latin squares. The field-plot design included eight complete Latin squares yielding a total of 32 plots per treatment with four rows per plot. In this experiment, plants were shielded from all ambient rainfalls by two 30 × 10-mm moveable rainfall-exclusion shelters. All four rows of each plot were exposed to simulated rainfalls.

A fourth experiment (Experiment 4) was also performed at Upton, New York, during the summer of 1981 (Evans et al., 1983). Four replicates of four experimental treatments of simulated rainfalls of pH 5.6, 4.1, 3.3, and 2.7 were arranged into Latin squares. The field-plot design included eight Latin squares, yielding a total of 32 plots per treatment. These plants were not shielded from ambient rainfalls. The weighted mean acidities of all precipitation on plants exposed to ambient rain plus simulated rainfall treatments of pH 5.6, 4.1, 3.3, and 2.7 were equivalent to pH 4.07, 4.04, 3.92, and 3.64, respectively. Plants received three weekly exposures (between 0800 and 1200 hours) to simulated precipitation in sufficient quantities to wet the foliage only. These simulated rainfalls added only 16.7 mm more water than ambient rainfalls (211.6 mm) throughout the growing season. Each simulated rainfall provided water similar to the mode of all ambient rain showers.

In a fifth experiment (Experiment 5) performed in Raleigh, North Carolina, in 1979 and 1980, soybeans of the Davis cultivar were exposed to simulated rainfalls in addition to ambient rainfalls (Heagle et al., 1983). In 1979, plants were kept potted, placed in a field, and exposed to simulated rainfalls of pH 5.3, 4.0, 3.3, and 2.8, which yielded weighted mean acidities of all precipitation equivalent to pH 4.4, 4.2, 3.6, and 3.2, respectively. In 1980, seeds were planted into the field and exposed to simulated rainfalls of pH 5.4, 4.1, 3.2, and 2.4, which yielded weighted mean acidities of all precipitation equivalent to pH 4.4, 4.1, 3.5, and 2.8, respectively. In 1979 and 1980, 30 and 25 simulated rainfalls were applied during 15 and 13 weeks in which the soybean growing season was 24 and 23 weeks, respectively. In 1979 and 1980, the simulated rainfalls provided an additional 22.2 and 21.3 cm of water compared with 40.3 and 33.0 cm of water provided by ambient precipitation, respectively. Six plots per treatment were present. In 1980, the plant population density in the field was about 178,000 plants per hectare.

In a sixth experiment (Experiment 6) performed at Urbana, Illinois, during 1984, soybeans were exposed to simulated rainfalls of pH 5.6, 4.6, 4.2, 3.8, 3.4, and 3.0 under automatically moveable rainfall-exclusion shelters that excluded all ambient rainfalls (Banwart et al., 1984). Simulated rainfalls of 1.04 cm were applied twice weekly. This rate was the same as the average amount of rainfall in the region for that growing season. The field plot design consisted of 12 plots per treatment.

Different conclusions have been drawn from results of these individual experiments with field-grown soybeans. In Experiment 1, soybean seed yields were not significantly related to rainfall acidity in either filtered or unfiltered air. In Experiments 2 and 3, simulated acidic precipitation also produced no statistically significant effect on seed yield. However, decreases in soybean yields due to acidic precipitation were observed in Experiments 3, 4, and 6. An analysis of variance of selected subsets of one, two, four, and eight Latin squares of Experiment 3 provides some insight on field variability (Evans and Thompson, 1984). The eight Latin squares analyzed as individual experiments (four plots per treatment) yielded erratic conclusions. Statistically significant linear comparisons were detected in four of the eight Latin squares. When two randomly selected

Latin squares (eight plots per treatment) were used in the analysis of variance, the linear component was just significant at the 1% level. When four Latin squares (16 plots per treatment) were used in the analysis of variance, the linear component was just significant at the 1% level. When four Latin squares (16 plots per treatment) were used, the linear component was highly significant ($<0.001$). When eight Latin squares (32 plots per treatment) were combined, the evidence that the treatment differences may be attributed to a linear gradient is overwhelming ($<0.001$). The low probability value ($<0.0033$) for differences among Latin squares indicates that field locations are a legitimate source of stratification. These results demonstrate that inadequate replication may result in a failure to detect biologically significant differences.

The most important difference in these experiments is their statistical design. Two of the field layouts (Experiment 1, Troiano et al., 1983; Experiment 2, Irving and Miller, 1981) used only one plot per treatment. These provided no replication except in the sense of plants, rows, or sectors within the single plot. Such studies are subject to the criticism that treatment effects cannot be separated from other microenvironmental variables peculiar to a specific plot location. Experiments 3 and 4 (Evans et al., 1983) were highly replicated. They were designed to detect differences of 10% or less among treatment means. Type 1 error, $\alpha$, was predetermined, and replication was sufficient to keep type 2 error low. This design was made possible because a preliminary experiment was available to estimate the expected components of variance. Despite differences in the procedures and protocols among the four experiments, it is primarily the quality of the experimental design that determines their validity and relative utility for crop-loss assessment.

No statistically significant pod yield differences were shown in Experiment 5. Several conclusions are possible in the evaluation of this experiment: (1) The Davis cultivar is not sensitive to acidic rain; (2) the six plots per treatment were not sufficient to detect significant treatment differences; and/or (3) the large amount of moisture to the plants (62.5 cm in 1979 and 54.3 cm in 1980) provided exceptionally good growth, which overshadowed the rainfall acidity effects. These possibilities and other possible conclusions cannot be evaluated further.

In Experiment 6, lower seed yields were obtained from Amsoy soybeans exposed to simulated acidic rainfall compared with controls (pH 5.6). For example, seed yields exposed to pH 3.0 were about 9% lower than plants exposed to pH 5.6. Approximately one-half of the yield decrease occurred between the control and the pH 4.2 treatment, which is near the normal acidity in rain for Illinois.

It appears that the Williams cultivar is not sensitive to rainfall acidity in three separate experiments (Evans et al., 1983, 1984a; Banwart et al., 1984). Since Amsoy soybeans were consistently affected by rainfall acidity while Williams was not (Evans et al., 1981b, 1983, 1984a, 1986, in press; Banwart et al., 1984), experiments were performed to determine if other soybean cultivars were sensitive to rainfall acidity.

Asgrow 3127 soybeans exposed to simulated rainfalls of pH 4.4, 4.1, and 3.3

exhibited yields of 14.5, 12.2, and 9.0%, respectively, below yields of plants exposed to simulated rainfalls of pH 5.6 under rainfall exclusion conditions in any experiment during 1984 (Evans et al., 1986b). In this same experiment, Corsoy plants exposed to rainfalls of pH 4.4., 4.1, and 3.3 had yields of 13.7%, 12.7%, and 7.8%, respectively, below yields of plants exposed to simulated rainfalls of pH 5.6. For these two cultivars, the linear component treatment-response function of seed yield versus rainfall pH was not statistically significant. As with Amsoy, yields of Asgrow and Corsoy were most closely related to the number of pods per plant.

For the Hobbit cultivar, yields of plants exposed to simulated rainfalls of pH 4.4, 4.1, and 3.3 were 9.2%, 6.2%, and 16.6%, respectively, below yields of plants exposed to simulated rainfalls of pH 5.6 under rainfall-exclusion conditions in an experiment during 1984 (Evans et al., 1986a). For Hobbit, the linear treatment-response function of seed yields versus rainfall pH was $y = 3.65 + 0.40x$, and had a correlation coefficient of 0.27. The linear component of pH treatment differences showed a significant decrease in yield ($p = 0.0276$). Differences in yield appeared again to be due to differences in pod number per plant. During an experiment in 1985, the responses of Amsoy, Asgrow, Corsoy, and Hobbit were very similar to the responses obtained during the experiment of 1984. These data taken together demonstrate that soybean cultivars exhibit a variety of responses to simulated rainfall acidity.

## D. Altered Protein Content of Soybean Seeds

Protein analyses of seeds harvested from the various soybean field experiments show that the protein content of such seeds can be reduced when plants are exposed to high levels of rainfall acidity. In the field experiments of 1979 and 1981, statistically significant decreases in protein content were observed. However, in the 1982 experiment no significant differences in seed protein content resulted (Evans et al., 1984c).

In an experiment conducted in 1983, seed yields of plants exposed to long-duration rainfalls (70 minutes, twice weekly) showed no significant differences in seed protein contents among the acidity levels tested. In contrast, significant differences in protein contents were found for seed samples from plants exposed to daily rainfalls of short duration (21 minutes). Seeds of plants exposed to simulated rain of pH 5.6 had average protein contents of 37.9% compared with a range between 30.6% and 32.1% from plants exposed to lower pH values. Overall, the protein contents of seeds of acidic rainfall treatments, expressed on a per plant basis, ranged from 23% to 34%. The results suggest that the response of soybean seed yields to acidity is affected by the duration and frequency of simulated rainfalls.

In the 1984 growing season, four commercial cultivars were tested under conditions of longer-duration rainfalls (70 minutes, twice weekly). Significant effects were observed only with the Amsoy cultivar and not with Asgrow, Corsoy, or Hobbit. For the latter three cultivars, the protein contents of samples were all relatively high (the means were between 35% and 39%). The relatively low protein content of Amsoy soybeans exposed to simulated rainfalls of pH 2.7 was apparently responsible for the statistical significance among treatments (Evans et al., 1986b).

Several generalizations are evident from the data of the seven experiments. First, the results show that the yields of both seed and seed protein are generally lower in plants exposed to high acidity than in those exposed to low acidity. Second, decreases in protein content as a result of increased rainfall acidity are greatest when expressed on a per plant basis. Third, seed yields and protein contents vary from year to year. Finally, the effects of rainfall acidity on seed yields are independent of seed protein content. This latter statement appears true because, for example, in the rainfall experiment of 1981 the decrease in protein content per plant was due mostly to changes in protein content per seed mass, whereas for the rainfall exclusion experiment of 1982 the decrease was due to differences in seed mass per plant.

From available evidence it appears that the seed protein content of Amsoy soybeans is sensitive to simulated rainfall acidity. In contrast, results of 1 year of experimentation with Asgrow 3127, Corsoy 79, and Hobbit suggest that the protein content of the seeds of these cultivars is not sensitive to simulated rainfall acidity. These different responses to acidity among cultivars are similar to seed yield responses to rainfall acidity (Evans et al., 1986b). Although differences in protein content and seed yields among cultivars of soybeans occur, the underlying reasons for differences in responses remain to be addressed.

## E. Comparison of Results from Controlled and Field Environments

In order to establish cause-and-effect relationships that characterize agricultural environments, manipulative experiments should be performed under conditions that most closely approximate the agricultural field environment. This situation is particularly germane in acidic deposition–vegetation response research because there is sufficient evidence that the effects of simulated rain applications on vegetation under controlled-environment conditions may differ markedly from effects obtained under field conditions. Two examples illustrate this situation. Root (hypocotyl) yields of radishes were reduced after exposure to simulated acidic rainfalls under controlled environmental conditions (Lee et al., 1981; Evans et al., 1982b). In contrast to these results, Troiano et al. (1982) demonstrated higher yields at high acidity rainfalls compared with controls with radishes grown under field conditions and no significant yield effects were demonstrated in another experiment that utilized 45 plots per treatment (Evans et al., 1982a). In addition, seed yields of Amsoy soybeans showed no consistent effects due to acidic deposition under controlled environmental conditions (Evans and Lewin, 1981). However, experiments with field-grown Amsoy soybeans have shown a negative effect of acidic deposition on yield (Evans et al., 1983, 1984a, 1986b, in press; Banwart et al., 1984). This information places limitations on the ability of results of experiments performed under controlled environmental conditions to be used to predict effects of acidic deposition under ambient conditions.

Differences in vegetation responses to acidic deposition between controlled environments and more normal field conditions may arise because of differences in macroclimate or microclimate. Important plant physiological functions that may be affected in this way are photosynthesis, respiration, translocation, and water and nutrient uptake.

## F. Seed Germination and Seedling Establishment

Lee and Weber (1979) exposed seedlings of 11 woody species to simulated rainfalls (23 mm weekly) of pH 3.0, 3.5, 4.0, and 5.6 for up to 9 months. Seedling emergence of eastern white pine (*Pinus strobus* L.), yellow birch (*Betula alleghaniensis* Britton), eastern red cedar (*Juniperus virginiana* L.), and Douglas fir (*Pseudotsuga menziesii* [Mirbel] Franco var. *menziesii*) was significantly higher during exposure to high-acidity rainfalls as compared with controls. In contrast, emergence was lower for staghorn sumac (*Rhus typhina* L.) under acidic rain exposures as compared with controls. No significant treatment effects were detected for seven other species tested. In similar experiments by Raynal et al. (1980), seedling emergence of yellow birch and red maple (*Acer rubrum*) was inhibited at low pH levels as compared with controls, whereas emergency of white pine was highest after exposure to simulated acidic rain. No significant treatment effects were present for sugar maple (*A. saccharum*). Moreover, Baldwin (1934) found better germination of red spruce (*Picea rubra*) under more acidic conditions. From available data, it seems that seedling germination and emergence can occur over a wide range of substrate pH levels.

At present, it is not possible to evaluate the ecological significance of acidic rain treatment effects as they relate to normal agricultural systems. This statement is true because experiments have not been conducted under field conditions in which rainfall acidity is the only variable that could affect productivity.

## G. Plant Reproduction

### 1. Higher Plants

From available evidence from soybeans, documented decreases in seed yields from acidic deposition exposure resulted from a decrease in pod number per plant, suggesting that plant reproduction may be affected (Evans et al., 1983, 1984a). Because the number of seeds per pod and the mass of individual seeds per plant did not vary among treatments, the decrease in soybean seed yields was attributed to a decrease in the number of mature pods. A decrease in the number of pods per plant may result from either a decrease in flower pollination (and fertilization), a decrease in pod retention, or an inadequate development of young pods. A decrease in pod number may also result from pod abortion caused by lack of nutrients.

### 2. Lower Plants

Fertilization and spermatozoid motility in gametophytes of bracken fern are very sensitive to low pH and additions of sulfate, nitrate, and chloride (Evans, 1979).

Spermatozoid motility at pH levels 6.0, 5.5, and 5.1 was about 70%, 50%, and 30%, respectively, of values obtained at pH 6.1 at 2 to 4 minutes after exposure. Motility at sulfate concentrations of 43.3, 86.6, and 173.2 $\mu$M at pH 6.1 was about 60%, 45%, and 35%, respectively, of values obtained with no sulfate additions. Fertilization decreased as pH decreased below pH 6.1 (buffers of pH 6.1 in the absence of sulfate, nitrate, and chloride were considered the experimental control). Fertilization after 3.5-hour exposure to pH 5.5, 5.1, and 4.5 was about 90%, 75%, and 60%, respectively, of values obtained at pH 6.1. Fertilization values at sulfate concentrations of 43.3, 86.6, and 173.2 $\mu$M at pH 6.1 were about 85%, 75%, and 60%, respectively, of values obtained with no sulfate additions (Evans, 1979). Similar results were present in a forest at Brookhaven National Laboratory (Evans and Conway, 1980). Experimental results with bracken fern were similar to results of megaspore germination and sporophyte formation in the heterosporous fern *Marsilea vestita* (Mahlberg and Yarus, 1977). Germination of megaspores was much more sensitive to low pH than was sporophyte development.

## H. Interactions with Plant Pathogens

Many microorganisms inhabit the surfaces of higher plants. A particular microorganism may have a positive (beneficial), negative (pathogenic), or neutral influence upon normal plant growth, development, and reproduction (see reviews by Dickinson and Preece, 1976; Preece and Dickinson, 1971). Leaf-surface bacteria within a canopy of Douglas fir may fix significant amounts of nitrogen (Jones, 1976). Many pathogens can reduce crop yields substantially (Ridgway et al., 1978; Shaw, 1979). Effects of air pollutants on relationships between host plants and their leaf-surface microbes have been reviewed (Smith, 1976). An air pollutant may differentially influence host and microorganism as well as their interactions (De Sloover and Le Blanc, 1968; Saunders, 1973; Smith, 1976).

Germane experiments with simulated rainfall highlight these interactions. Simulated acidic rain (pH 3.2) produced an 86% inhibition of telia of *Cronartium fusiforme* on willow oak (*Quercus phellos*), compared with pH 6.0 under greenhouse conditions (Shriner, 1977). Halo blight caused by *Pseudomonas phaseolicola* on leaves of greenhouse-grown *Phaseolus vulgaris* was stimulated or inhibited by simulated acidic rain, depending upon the stage of disease development when rainfall occurred. Simulated acidic rain inhibited initial disease infection but stimulated disease development infection.

In field experiments (Shriner, 1977), kidney bean plants that received simulated rainfalls of pH 3.2 had 34% as many nematode eggs per root system as plants that had received simulated rain of pH 6.0. All plants received ambient rainfalls in addition to simulated rainfalls. Plants that had received the low-pH rainfalls had only 48% as much root surface galling as plants that received pH 6.0. Biomass of roots, shoots, and pods as well as soil pH were not significantly affected by the simulated rain treatments. Bean leaf area injury caused by bean rust (*Uromyces phaseoli*) was not consistently affected by simulated rainfall acidity under field conditions. There are no experimental field data that demonstrate that rainfall acidity affects host-pathogen relations that result in significant changes in plant productivity or survival.

# IV. Factors Affecting Vegetation Responses

## A. Genetic Factors

### 1. Species Differences

Plant species differ in response to simulated acidic rain. In section III, A, 2 above, it was noted that plant species vary markedly with respect to sensitivity to visible foliar injury. As changes in plant productivity may occur without visible foliar injury, however, there is no method available to predict relative plant species sensitivity in the ambient environment.

### 2. Cultivar Differences

At present, little information is available to determine relative cultivars sensitive for all crops, mostly because so few crops and cultivars of crops have been studied. Most complete data are available for soybeans. Numerous field experiments have shown that the Amsoy cultivar is sensitive, whereas the Williams cultivar is insensitive to rainfall. Soybean cultivars Asgrow 3127, Corsoy, and Hobbit were adversely affected by acidic deposition—a response more similar to Amsoy compared with Williams. However, data for only one growing season are available for Asgrow, Corsoy, and Hobbit (see section III, C). Unfortunately, these are the only reliable data concerning acidic deposition responses of various cultivars of a crop grown under field conditions.

## B. Environmental Factors

### 1. Moisture Availability

Moisture availability and utilization impact considerably on plant growth and reproduction. Results shown (Evans et al., 1985a) indicate that productivity of plants exposed to longer-duration, twice-weekly simulated rainfalls is greater than with shorter-duration, daily rainfalls, even though the plants exposed to daily rainfalls received more water. These differences probably result from the fact that the long-duration rainfalls more effectively supply moisture to the plant roots. It is noteworthy that even though the yield differences between the two treatment frequencies were greater than the effect of rainfall acidity, the slope of the dose-response function for the two frequencies for each cultivar were markedly similar. These results show that simulated rainfall frequency and duration can significantly affect soybean yields independently of rainfall-acidity effects at the frequencies and durations tested. At present, these are the only data that vary moisture availability and rainfall acidity.

## 2. Climate

At present, there are no data to indicate interactive effects of acidic deposition with any climatic factor. In general, however, visible foliar injury is more pronounced under controlled environmental conditions, compared with field conditions. As a result, there is poor understanding of the effects of acidic deposition and other environmental influences.

An estimate of differences in climate or acidic deposition effects may be obtained from a comparison of year-to-year variations in yields of Amsoy soybeans exposed to ambient deposition, yields of plants exposed to simulated rainfalls of pH 5.6 under rainfall-exclusion conditions, and the overall responses of Amsoy soybeans exposed to simulated rainfall acidity and may provide perspective of how climate may influence plant responses to acidic deposition. Data from four consecutive years (1981 through 1984) will be used.

Yields of Amsoy plants exposed to ambient rainfalls for 1981, 1982, 1983, and 1984 were 4,460, 4,000, 4,820, and 3,650 kg ha$^{-1}$, respectively. Differences in yields of plants grown in successive years under ambient conditions may be attributed, in part, to annual weather variations, planting dates, and/or as differences in plant populations. During three of the four growing seasons mentioned above, ambient rainfalls varied markedly, totaling 212, 457, 361, and 411 mm for 1981, 1982, 1983, and 1984, respectively. Moreover, planting dates also varied from year to year depending upon climate during the planting season. Bean yields for Amsoy plants shielded from ambient rainfalls and exposed to twice-weekly simulated rainfalls of pH 5.6 were consistent among the growing seasons of 1981 (4,600 kg ha$^{-1}$, 1982 (4,590 kg ha$^{-1}$), 1983 (4,260 kg ha$^{-1}$), and 1984 (4,030 kg ha$^{-1}$). It is apparent that yields of plants grown under rainfall-exclusion conditions are much more consistent on a year-to-year basis than plants grown under ambient conditions. These results would suggest that total rainfall amount markedly affects soybean yields. The slope of the dose-response function relating soybean yields to rainfall acidity is rather uniform among many experiments with different exposure conditions (Evans et al., in press). These results suggest that Amsoy soybean response to acidic deposition is consistent, even though exposure conditions were not similar and experiments occurred during five growing seasons.

## 3. Influence of Gas Exchange

Several years ago, it was proposed that acidic precipitation might affect the rate of gas exchange by stomata (Tamm and Cowling, 1977), and that epidermal cells may be injured preferentially upon initial exposure to acidic rain (Evans et al., 1977b, 1978). Because guard cells are preferentially affected, it was postulated that acidic precipitation might affect gas exchange rates. In *Phaseolus vulgaris*, diffusion resistance was much lower in foliage exposed to simulated rain of pH 2.7 to 3.4, compared with foliage exposed to rain of pH 5.7 (Evans et al., 1981c). These results suggest that foliage exposed to acidic rain may be more subject to wilting or water stress. Moreover, such foliage may be more sensitive to exposures of gaseous air pollutants. This decrease in resistance may also increase carbon dioxide uptake for photosynthesis. Knowledge of the effects of rainfall acidity in combination with gaseous pollutants such as ozone and sulfur dioxide is needed in order to understand the total impact acidic rain might have on vegetation.

## 4. Gaseous Pollutant Interactions

At the present time, there are no experimental data to demonstrate a statistically significant interactive effect of acidic deposition with any gaseous air pollutant. Although some experiments have been conducted, no interaction has been demonstrated. This lack of documentation may result because there was no actual interaction, the experimental designs and procedures were inadequate, or other unforeseen factors were involved.

# V. Limitations to Experimental Approaches

## A. Scientific Approach

Science is the process by which scientists seek to understand the natural world. To accomplish this process, scientists attempt to focus upon one or a few phenomena of nature. A single phenomenon may be studied in relative isolation, and the scientist gains some knowledge about how other naturally occurring (or artificially applied) phenomena affect this single phenomenon. During the experimentation process, it should be remembered that increasing the measurement of one phenomenon increases the uncertainty with which other phenomena may be known, the Heisenberg uncertainty principle. Simple cause-and-effect relationships may be established between the focal phenomenon and other phenomena that are systematically varied. This new knowledge gained in relative isolation must then be considered in the context of the natural world. Knowledge about the natural world is gained if, and only if, the information obtained in relative isolation is consistent with information from the natural world.

## B. Cause-and-Effect Linkages

The word *cause* is defined as "something that produces an effect, result, or consequence" or "the person, event, or condition for an action or result" (Morris, 1982). At the same time, the word *effect* is defined as "something brought about by a cause or agent; result" (Morris, 1982). To determine if an air pollutant such as acidic deposition causes a particular effect such as a change in plant growth, development, and/or reproduction, a linkage of cause and effect must be established. For an adequate cause-and-effect linkage, the effect should occur when the causative event occurs, on a repetitive basis. For example, in experiments aimed at determining the effects of acidic deposition on vegetation, specific plant responses should occur repeatedly after exposure to acidic deposition.

Information from many scientific disciplines must be considered in an evaluation of research data to establish a knowledge linkage between acidic deposition and terrestrial vegetation. Knowledge is necessary of air pollution sources, meteorology and atmospheric chemistry, pollutant deposition phenomena, plant-atmosphere interactions, plant physiology, and agronomy, to mention a few subject areas. The purpose of this paper is to evaluate the state of knowledge that has been generated in order to establish a knowledge linkage between acidic deposition and terrestrial vegetation. Emphasis will be given to the scientific uncertainties associated with experimental approaches used, data available, statistical analyses used, and evaluation procedures.

## C. Role of Associative Surveys

The purpose of survey approaches is to document the mutual occurrence of two or more naturally occurring events. Such approaches are helpful because they tend to exclude certain factors from further consideration and, as a result, lead to hypothesis construction. However, such associative surveys, which document only the co-occurrence of events, do not establish cause-and-effect relationships.

## D. Need for Manipulative Experiments

In order to establish if cause-and-effect relationships are present, manipulative experiments are required in which experimental treatments are implemented. Experimental treatments should be established in which various intensities of the causative agent (various levels of deposition acidity) are administered and one or more responses (effects) are monitored. To date, most researchers have varied concentrations of the strong acid of solutions while keeping the ratio of sulfate and nitrate among solutions relatively constant.

## E. Characterization of Ambient Deposition

To administer experimental treatments, the characteristics of the independent variable under natural conditions must be considered. To administer acidic deposition treatments, knowledge of the spatial, temporal, and other characteristics of ambient acidic deposition (rain, snow, cloud moisture impaction, rime ice) must be understood (Evans, 1984), and suitable technical equipment must be available to administer such treatments. The interpretation of experimental data to establish cause-and-effect relationships must at least consider both the natural variabilities of acidic deposition and the degree to which technology can implement suitable treatment regimens.

The characteristics of atmospheric deposition have been studied extensively. Much is known about the chemical composition and temporal and spatial patterns of rain and snow (Galloway and Likens, 1976; Likens et al., 1976; Semb, 1976; Evans et al., 1981a). However, little is known about the characteristics of cloud water deposition at high elevations. At the very least, it must be recognized that there is a large amount of variability in all characteristics of precipitation deposition (Evans et al., 1981a; Evans et al., 1984b). This large range in variability suggests that an assignment of a mean value for a parameter may have little meaning. If this is true, then it should be realized that application of simulated rainfalls with fixed acidity levels may not be representative of natural conditions.

Most researchers have used simulated rain chemistries that are similar to the chemistry of ambient rainfalls. Simulated rainfall acidity has been used as the experimental variable under most conditions. Some experiments have used a simulated rain composed only, or mostly, of major cations and anions (Troiano et al., 1983; Irving and Miller, 1981; Evans et al., in press), whereas other experiments (Evans et al., 1981b, 1982a) have included concentrations of heavy metals that are within the range of concentrations found in nature (Galloway et al., 1982). Similar reductions in soybean seed yields were observed in the presence of high (Evans et al., 1981b, 1982a) and low (Evans et al., in press) concentrations of heavy metals, suggesting that soybean yields are dependent upon rainfall acidity but are independent of metal concentrations in the range found in ambient precipitation.

At present, simulated rainfalls are most frequently characterized by acidity or pH. In most experiments, sulfuric and nitric acids are usually used to provide the majority of the strong acid component in simulated rainfalls. Usually the ratio of sulfate to nitrate varies between 3:1 and 1:3 on an equivalent basis. However, no significant differences in treatment effects have been presented in the peer-reviewed literature. As a result, total solution acidity or pH has usually been used to express the independent variable in data expressions.

## F. Exclusion of Ambient Deposition

In order to adequately administer experimental treatments, it may be necessary to isolate the system, at least partially, from the natural environment. For example, automatically moveable rainfall-exclusion shelters have been employed in acidic deposition experiments conducted out-of-doors (Lewin and Evans, 1984). These shelters have been used to cover vegetation when natural precipitation occurs and when artificial treatments are imposed. Exclusion of ambient precipitation is necessary in order to characterize accurately the effects of the imposed treatments. In other words, if ambient precipitation is not excluded, the researcher cannot separate the effects of imposed treatments from the effects of ambient precipitation.

## G. Minimal Perturbation of Ambient Environments

To establish cause-and-effect relationships to characterize natural environments, manipulative experiments should be performed under conditions that most closely approximate natural environments. This situation is particularly germane in acidic deposition–vegetation response research because there is sufficient evidence that the effects of simulated rain applications on vegetation under controlled environmental conditions may differ markedly from effects obtained under field conditions. Two examples illustrate this situation. Root (hypocotyl) yields of radishes were reduced after exposure to simulated acidic rainfalls under controlled conditions (Lee et al., 1981; Evans et al., 1982b). In contrast to these results, Troiano et al. (1982) demonstrated higher yields at high-acidity rainfalls compared with controls with radishes grown under field conditions, and no significant yield effects were demonstrated in another experiment that utilized 45 plots per treatment (Evans et al., 1982a). In addition, seed yields of Amsoy soybeans showed no consistent effects due to acidic deposition under controlled environmental conditions (Evans and Lewin, 1981). However, many experiments with field-grown Amsoy soybeans have shown a negative effect of acidic deposition on yield (Evans et al., 1983, 1984a, in press). This information places limitations on the ability of results of experiments performed under controlled environmental conditions to predict effects of acidic deposition under ambient conditions.

Differences in vegetation responses to acidic deposition between controlled environments and more natural conditions may arise because of differences in macroclimate and/or microclimate. Important plant physiological functions such as photosynthesis, respiration, translocations, and water and nutrient element uptake processes are affected by climate. For example, the experimental conditions should not significantly alter dew formation conditions because in eastern North America, dew formation occurs frequently when plants are actively growing; for example, dew formed at Brookhaven National Laboratory 65 times for a total of 755 hours during the 1979 soybean growing season (Evans et al., 1981b).

Moveable rainfall-exclusion shelters that have been used in many studies have been shown to minimize changes in a plant's microclimate (Dugas and Upchurch, 1984; Legg et al., 1978). Such moveable shelters have been used to assess the effects of acidic deposition on field-grown soybeans for 4 years (Evans et al., 1983, 1984a, in press). Such structures have a minimum impact on overall plant microclimate because they cover the crop only when ambient rainfalls occur or when simulated rainfalls occur.

Most commonly, crop coverage by the shelter will decrease wind velocity, increase temperature, and increase relative humidity. During ambient rainfalls, solar radiance is usually low and relative humidity is high so the impact of shelters would be minimal. Air temperature and humidity should not differ significantly under the covers versus in open air at night. Most rainfalls during the growing season in the northeastern United States occur during the dark period (Evans et al., 1984b). The major difference in the microclimate would be wind velocity, which should have, by itself, minimal impact on a plant's microclimate, considering the fact that ambient precipitation normally occurs during a small fraction of the season.

In contrast to this condition with rainfall-exclusion shelters, open-top chambers (OTC) have also been used to study acidic deposition effects (Jacobson et al., 1980; Troiano et al., 1983). Open-top chambers have been shown to alter yields of crops grown within them compared with crops grown in ambient conditions (Heagle et al., 1973). The microclimate within such chambers differs from the ambient conditions outside the chambers (Olszyk et al., 1980), which may influence yields (Clarke et al., 1983). Troiano and McCune (1984) suggested that the "experimental conditions [in such chambers] are probably closer to greenhouse than to field experimentation."

## H. Need for Adequate Treatment Replication

In addition to minimally perturbing the environment of experimental vegetation, adequate replication of experimental plots is absolutely necessary in such experiments (Nelson and Rawlings, 1983; Hurlbert, 1984; Evans et al., 1984d). Adequate replication is necessary because plant microclimates, local soil fertilization levels, soil moisture contents, and localized soil texture heterogeneities may vary significantly within a small area of experimentation. Significant variation within a small area is supported by the fact that statistically significant differences among Latin squares have usually occurred in field experiments with soybeans at Brookhaven National Laboratory over many years. In addition, sources of variation are also highly significant for both pooled among rows within squares and the pooled among columns within squares, indicating that the two stratifications within each Latin square were doubly effective in reducing the residual error term (Evans and Thompson, 1984). This reduction in the residual error term indicates that there is significant variation within individual Latin squares (e.g., within localized areas of fields). The lack of adequate treatment replication is a limiting factor in many experiments under both controlled and field conditions. In experiments with inadequate treatment replication, an inappropriate error term may be used for tests of statistical significance and/or the tests may not differentiate between effects and field-position or random effects (Evans et al., 1984d).

# VI. Summary and Conclusions

Presently there is evidence that the acidity in precipitation, at ambient levels, is having deleterious effects upon terrestrial vegetation. This statement is mostly derived from data of crops grown under field conditions. Experiments have been conducted under controlled-environment conditions but, as plant responses to acidic rain grown under such conditions differ from responses under field conditions, data from controlled environments may only point to mechanisms. Few field experiments have been conducted that are statistically capable of detecting significant differences among treatments. Although decreases in yields of various crops have been documented in controlled-environment and field experiments at pH levels of 4.0 or below, only well-designed field experiments will document changes in plant productivity or survival that may be expected from actual acidic rainfall exposures. It is essential that experiments using standard agronomic practices or total ecosystem considerations with large numbers of replicates be conducted so results can be extrapolated to actual crop or ecosystem situations, respectively. Large numbers of treatment replicates are needed to demonstrate differences (statistically significant) among treatments with means that differ by 10%.

To date, there is only one documented occurrence of visible foliar injury attributed to ambient rainfalls (Evans et al., 1982a). Visible foliar injury, identical to that caused by simulated acidic rainfalls of pH 4.0 and below, was produced on foliage of field-grown garden beets during a 3-day period in which three short-duration, low-volume rain showers with pH levels of about 3.8 to 3.9 occurred. Simulated rainfalls of higher acidity resulted in lower beet yields, which were attributed to both a smaller number of saleable roots per plot and a lower fresh mass per saleable root.

In order to standardize results of future experiments, a uniform expression of data should be utilized. To date, the most meaningful relationships between rainfall acidity and plant responses have been obtained when responses are plotted as a function of the hydrogen ion concentration or pH of the simulated rainfalls. This expression, or some other more meaningful expression, should be used so that data can be compared with ambient precipitation acidities.

Acidic precipitation may affect the productivity of crop and forest plants by direct or indirect means (necrosis or soil nutrient effects). If significant alterations in productivity of either crops or forest occur, significant economic impacts may result. Currently, evidence linking existing and/or anticipated levels of rainfall acidity to crop yield reductions is meager. From available data, it is clear that the amount of injury, if injury occurs, is less than year-to-year changes in other abiotic and biotic factors. This does not mean, however, that possible injuries from acidic precipitation should be ignored. It is important to determine even small changes in productivity.

At the present time, the best available data to understand the effects of acidic deposition can be derived from experiments using moveable rainfall-exclusion shelters. These shelters are important because they minimize changes in a plant microclimate compared with a normal agricultural field setting, and a large number of plots per treatment, essential to detect treatment effects if present, may be used. The usefulness of data from other experimental approaches is questionable because ample data demonstrate that plants react to acidic deposition differently in controlled environments compared with field situations. Other experiments may not be useful for a quantitative assessment because their experimental designs are not capable of detecting statistically significant treatment effects unless differences among means are greater than 25%.

# References

Adam, N.K. 1948. Discuss Faraday Soc 3:5–11.

Adams, C.M. 1982. The response of *Artemisia tilesii* to simulated acid precipitation. M.S. Thesis, University of Toronto, Ontario.

Altshuller, A.P. 1976. J Air Pollut Control Assoc 26:18–324.

Baldwin, H.I. 1934. Plant Physiol 9:491–532.

Banwart, W.L., J.J. Hassett, and B.L. Vasalis. 1984. Proceedings of the Illinois Fertilizer and Chemical Dealers' Conference, 19–21.

Bennett, S.H., and W.D. Thomas. 1954. Ann Appl Biol 41:484–500.

Clarke, B.B., M.R. Henninger, and E. Brennan. 1983. Phytopathology 73:104–113.

Cole, D.W., and D.W. Johnson. 1977. Water Resources Res 13:313–317.

Crafts, A.S. 1961. *The chemistry and mode of action of herbicides*. Interscience Publishers, New York.

De Sloover, J., and F. LeBlanc. 1968. *In* R. Misra and B. Gopal, eds. *Tropical plant ecology* (*Proceedings of the symposium on recent advances tropical ecology*), Varanasi Press, Varanasi, India. 409 pp.

Dickinson, C.H., and T.F. Preece, eds. 1976. *Microbiology of aerial plant surfaces*. Academic Press, New York.

Dugas, W.A., Jr., and D.R. Upchurch. 1984. Agron J 76:867–871.

Dybing, C.D., and H.B. Currier. 1961. Plant Physiol 36:169–174.

Evans, L.S. 1979. J Air Pollut Control Assoc 29:145–148.

Evans, L.S. 1980. *In* T.Y. Toribara, M.W. Miller, and P.E. Morrow, eds. *Polluted rain* (Twelfth annual Rochester International Conference on Environmental Toxicity), 239–257. Plenum Publishing, New York.

Evans, L.S. 1982. Environ Exp Bot 22:155–169.

Evans, L.S. 1984. Botanical Review 50:449–490.

Evans, L.S., D.C. Canada, and K.A. Santucci. 1986c. Environ Exp Bot 26:143–146.

Evans, L.S., and C.A. Conway. 1980. Amer J Bot 67:866–875.

Evans, L.S., and T.M. Curry. 1979. Amer J Bot 66:953–962.

Evans, L.S., T.M. Curry, and K.F. Lewin. 1981c. New Phytol 88:403–420.

Evans, L.S., L. Dimetriadis, and D.A. Hinkley. 1984c. New Phytol 97:71–76.

Evans, L.S., N.F. Gmur, and F. Da Costa. 1977b. Amer J Bot 64:903–913.

Evans, L.S., N.F. Gmur, and F. Da Costa. 1978. Phytopathology 68:847–856.

Evans, L.S., N.F. Gmur, and J.J. Kelsch. 1977a. Environ Exp Bot 17:145–149.

Evans, L.S., N.F. Gmur, and D. Mancini. 1982b. Environ Exp Bot 22:445–453.

Evans, L.S., G.R. Hendrey, G.J. Stensland, D.W. Johnson, and A.J. Francis. 1981a. Water Air Soil Pollut 16:469–509.

Evans, L.S., G.R. Hendrey, and K.H. Thompson. 1984d. J Air Pollut Control Assoc 34:1107–1114.

Evans, L.S., and K.F. Lewin. 1981. Environ Exp Bot 21:102–113.

Evans, L.S., K.F. Lewin, C.A. Conway, and M.J. Patti. 1981b. New Phytol 89:459–470.

Evans, L.S., K.F. Lewin, E.A. Cunningham, and M.J. Patti. 1982a. New Phytol 91:429–441.

Evans, L.S., K.F. Lewin, E.M. Owen, and K.A. Santucci. 1986a. New Phytol 102:409–417.

Evans, L.S., K.F. Lewin, and M.J. Patti. 1984a. New Phytol 96:207–213.

Evans, L.S., K.F. Lewin, M.J. Patti, and E.A. Cunningham. 1983. New Phytol 93: 377–388.

Evans, L.S., K.F. Lewin, K.A. Santucci, and M.J. Patti. 1985a. New Phytol 100: 199–208.

Evans, L.S., G.S. Raynor, and D.M.A. Jones. 1984b. Water Air Soil Pollut 23:187–195.

Evans, L.S., K.A. Santucci, and M.J. Patti. 1985b. Environ Exp Bot 25:31–40.

Evans, L.S., M.J. Sarrantonio, and E.M. Owen. 1986b. New Phytol 103:689–693.

Evans, L.S., and K.H. Thompson. 1984. Agron J 76:81–84.

Fairfax, J.A.W., and N.W. Lepp. 1975. Nature 255:324–325.

Fogg, G.E. 1948. Discuss Faraday Soc 3:162–169.

Forsline, P.L., R.J. Dee, and R.E. Melious. 1983a. J Amer Hort Sci 108:202–207.

Forsline, P.L., R.C. Musselman, W.J. Kender, and R.J. Dee. 1982. Amer J Enol and Vitic 34:17–24.

Forsline, P.L., R.C. Musselman, W.J. Kender, and R.J. Dee. 1983b. J Amer Soc Hort Sci 108:70–74.

Galloway, J.N., and G.E. Likens. 1976. Water Air Soil Pollut 6:241–258.

Galloway, J.N., J.D. Thornton, S.A. Norton, H.L. Volchok, and R.A. McLean. 1982. Atmos Environ 16:1677–1700.

Galloway, J.N., and D.M. Whelpdale. 1980. Atmos Environ 14:409–417.

Garcia, R.L., and J.J. Hanway. 1976. Argon J 68:653–657.

Gordon, C.C. 1972. *Interim report to the Environmental Protection Agency.* U.S. Environmental Protection Agency, Washington, D.C.

Gustafson, F.G. 1956. Amer J Bot 43:157–160.

Gustafson, F.G. 1957. Plant Physiol 32:141–142.

Haberlandt, G. 1914. *Physiological plant anatomy.* Macmillan Publishing Co., London.

Hall, D.M., A.I. Matus, J.A. Lamberton, and H.N. Barber. 1965. Aust J Biol Sci 18:323–332.

Harder, H.J., R.E. Carlson, and R.H. Shaw. 1982. Agron J 74:759–761.

Heagle, A.S., D.E. Body, and W.W. Heck. 1973. J Environ Qual 2:365.

Heagle, A.S., R.B. Philbeck, P.F. Brewer, and R.E. Ferrell. 1983. J Environ Anal 12:538–543.

Hindawi, J.J., and H.C. Ratsch. 1974. Presentation at the 67th annual meeting of the Air Pollution Control Assoc. Abstract 74:252.

Hindawi, I.J., J.A. Rea, and W.L. Griffis. 1977. Presentation at the 70th annual meeting of the Air Pollution Control Assoc. Abstract 77:304.

Holloway, P.J., and E.A. Baker. 1968. Plant Physiol 43:1878–1879.

Hurlbert, S. 1984. Ecol Monogr 54:187–198.

Irving, P.M., and J.E. Miller. 1981. J Environ Qual 10:473–478.

Jacobson, J.S., L.I. Heller, and P. Van Leuken. 1976. Water Air Soil Pollut 6:339–349.

Jacobson, J.S., J. Troiano, L.J. Colavito, L.L. Heller, and D.C. McCune. 1980. *In* T.Y. Toribara, M.W. Miller, and P.E. Borrow, eds. *Polluted rain* (Twelfth annual Rochester International Conference on Environmental Toxicity), 291–299. Plenum Press, New York.

Jones, K. 1976. *In* C.H. Dickinson and T.F. Preece, eds. *Microbiology of aerial plant surfaces,* 451–463. Academic Press, New York.

Jonsson, B., and R. Sundberg. 1972. *In* B. Bolin, ed. *Supporting studies to air pollution across national boundaries. The impact on the environment of sulfur in air and precipitation. Sweden's case study for the United Nations conference on the human environment.* Royal Ministry of Foreign Affairs, Royal Ministry of Agriculture, Stockholm. 46 pp.

Juniper, B.E. 1960. Ph.D. Dissertation, University of Oxford, Great Britain.

Jyung, W.H., and S.H. Wittwer. 1964. Amer J Bot 51:437–444.

Konar, R.N., and H.F. Linskens. 1966. Planta 71:356–371.

Lee, J.J., G.E. Neely, S.C. Perrigan, and L.S. Grothaus. 1981. Environ Exp Bot 21:171–185.

Lee, J.J., and D.E. Weber. 1979. For Sci 25:393–398.

Legg, B.J., W. Day, N.J. Brown, and G.J. Smith. 1978. J Agric Soc 91:321–336.

Leonard, O.A. 1958. Hilgardia 28:115–160.

Lewin, K.F., and L.S. Evans. 1984. Design and performance of an experimental system to determine the effects of rainfall acidity on vegetation. ASAE 29:1654–1658.

Likens, G.E., F.H. Bormann, J.S. Eaton, R.S. Pierce, and N.M. Johnson. 1976. Water Air Soil Pollut 6:435–445.

Likens, G.E., F.H. Bormann, and J.M. Johnson. 1972. Environment 14:33–44.

Likens, G.E., R.F. Wright, N.N. Galloway, and T.G. Butler. 1979. Sci Amer 241:43–51.

Lindberg, S.E. 1982. Atmos Environ 16:1701–1709.

Lindberg, S.E., R.R. Turner, D.S. Shriner, and D.D. Huff. 1981. In S.E. Lindberg and R.R. Turner, eds. *Third international conference on heavy metals in the environment*, 306–309. Oak Ridge National Laboratory, Oak Ridge, Tenn.

Linskens, H.F. 1950. Planta 38:591–600.

McFarlane, J.C., and W.L. Berry. 1974. Plant Physiol 53:723–727.

Mahlberg, P.G., and S. Yarus. 1977. J Exp Bot 28:1137–1146.

Martin, J.T., and B.E. Juniper. 1970. *The cuticles of plants*. St. Martin's Press, New York.

Morris, W.I. 1982. *The American heritage dictionary*. 2d ed. Houghton Mifflin Co., Boston.

Nelson, L.A., and J.O. Rawlings. 1983. J Agron Ed 12:100–109.

Neumann, P.M., Y. Ehrenreich, and Z. Golab. 1981. Agron J 73:979–982.

Nørdo, J. 1976. Water Air Soil Pollut 6:199–227.

Norman, A.G., C.E. Minarik, and R.L. Weintraub. 1950. Ann Rev Plant Physiol 1:141–168.

Oden, S. 1976. Water Air Soil Pollut 6:137–166.

Olszyk, D.M., T.M. Tibbitts, and W.M. Hertsberg. 1980. J Environ Qual 9:610.

Orgell, W.H. 1957. Proc Iowa Acad Sci 64:189–197.

Paparozzi, E.T. 1981. Ph.D. Thesis, Cornell University, Ithaca, N.Y.

Prasad, R., and G.E. Blackman. 1962. Plant Physiol 37(Suppl.):xiii.

Preece, T.F., and C.H. Dickinson, eds. 1971. *Ecology of leaf surface microorganisms*. Academic Press, New York.

Raynal, D.J., A.L. Leaf, P.D. Manion, and C.J.K. Wang. 1980. *Project Report to the New York State Energy Research and Development Authority*. SUNY, Syracuse.

Reuther, R., R.F. Wright, and U. Förstner. 1981. In S.E. Lindberg and R.R. Turner, eds. *Third international conference on heavy metals in the environment*, 318–321. Oak Ridge National Laboratory, Oak Ridge, Tenn.

Ridgway, R.L., J.C. Tinney, J.T. Macgregor, and N.J. Starler. 1978. Environ Health Perspec 27:103–112.

Roberts, E.A., M.D. Southwick, and D.H. Palmiter. 1948. Plant Physiol 23:557–559.

Sargent, J.A., and G.E. Blackman. 1962. J Exp Bot 13:348–368.

Sargent, J.A., and G.E. Blackman. 1965. J Exp Bot 16:24–47.

Saunders, P.J. 1973. Pestic Sci 4:589–595.

Schnepf, E. 1965. Ztschr Pflanzenphysiol 53:245–254.

Schönherr, J. 1976. Planta 128:113–126.

Schönherr, J., and M.J. Bukovac. 1972. Plant Physiol 49:813–819.

Schönherr, J., and J.W. Schmidt. 1979. Planta 144:391–400.

Semb, A. 1976. Water Air Soil Pollut 6:231–240.

Shaw, W.C. 1979. *Integrated pest management systems technology: past, present, and future*. Pest management workshop, January 29–30. University of Missouri, Columbia.

Shriner, D.S. 1974. Effects of simulated rain acidified with sulfuric acid on host-parasite interactions. Ph.D. Thesis, North Carolina State University, Raleigh.

Shriner, D.S. 1977. Water Air Soil Pollut 8:9–14.

Smith, W.H. 1976. *In* C.H. Dickinson and T.F. Preece, eds. *Microbiology of aerial plant surfaces*, 75–105. Academic Press, New York.

Stensland, G.J. 1983. *In* A.P. Altschuller, ed. *Atmospheric sciences (the acidic deposition phenomenon and its effects)*, vol. 1, chap. 4. EPA-600/8-83-016A.

Tamm, C.O., and E.B. Cowling. 1977. Water Air Soil Pollut 7:503–511.

Thom, W.O., T.C. Miller, and D.H. Borman. 1981. Agron J 73:411–414.

Troiano, J., L. Colavito, L. Heller, D.C. McCune, and J.S. Jacobson. 1983. Environ Exp Bot 23:113–119.

Troiano, J., L. Heller, and J.S. Jacobson. 1982. Environ Pollut 29:1–11.

Troiano, J., and D.C. McCune. 1984. JAPCA 34(11):1115–1116.

Uphof, J.D.Th. 1962. *Plant hairs*. Gebrüder Borntraeger, Berlin-Nokolassee.

Van Breemen, N., P.A. Burrough, E.J. Velhorst, H.F. van Dobben, T. de Wit, T.B. Ridder, and H.F. Reijnders. 1982. Nature 299:548–550.

Wood, T., and F.H. Bormann. 1974. Environ Pollut 7:259–268.

Wood, T., and F.H. Bormann. 1975. Ambio 4:169–171.

Yue, G.K., V.A. Mohnen, and C.S. Kiang. 1976. Water Air Soil Pollut 6:277–294.

# Effects of Acidic Precipitation on Soil Productivity

Ivan J. Fernandez*

## Abstract

One of the earliest concerns for the potential effects of acidic deposition on forest ecosystems was for accelerated soil acidification and nutrient depletion. Although research on this topic has resulted in significant advances in our understanding of soil response to S, N, and trace metal deposition, little can be concluded at this time as to the consequence of soil changes for forest health and productivity on a regional basis. It is essential to define our understanding in terms of (1) effects on soil productive capacity versus (2) soil-mediated effects on forest health and growth. In the case of soil productivity, we can say with some confidence that the modern chemical composition of the atmosphere has resulted in changes to soils. Nitrogen cycling in forest soils has been altered, as well as the cycling of other nutrients; soils in acidic deposition–affected regions often contain significant amounts of $SO_4^{2-}$ in soil solution; trace metals have accumulated in forest soils; and selected ecosystems in high-deposition areas of Europe have exhibited nutrient-deficiency symptoms thought to be partially a result of acidic deposition leaching. Although many uncertainties remain, there is little question that air pollution has altered the chemical and biological character of forest soils. How these changes have, or will have, an effect on forest health and productivity remains undefined. Some of the difficulty in understanding linkages between soil changes and resulting consequences for forests is due to the lack of long-term data on forest ecosystems and the lack of suitable control sites for experimental research. In addition, complex interactions among varying natural and anthropogenic stresses, which vary in time and space, must be included to assess realistic consequences of air pollution on forests.

*Department of Plant and Soil Sciences, University of Maine, One Deering Hall, Orono, ME 04469, USA.

# I. Introduction

Over the past two decades we have witnessed the evolution of an environmental issue often referred to in the popular press as *acidic rain,* but clearly encompassing a more comprehensive concern for the influence of the modern chemical composition of the atmosphere on the health of our environment. The main focus of this chapter is the effects of acidic deposition on the productivity of soils in forested ecosystems, and it intentionally excludes related issues regarding natural stress factors and interactive stresses on forests, which are covered by other chapters in this series. Agricultural soils are not considered sensitive to acidic deposition effects, as they are buffered by their intensive management schemes.

As might be expected, many questions remain unanswered on the effects of acidic deposition on forest soils. This results from the chronic nature of acidic deposition exposure, the long life cycle of typical forest communities, the lack of suitable "control" field sites for experimentation, and most importantly the highly variable and complex nature of forest ecosystem function. Most of the critical hypotheses regarding the potential effects of acidic deposition on forest soils were delineated early in the development of research on this subject, as evidenced by the studies included in the first large-scale program on acidic precipitation effects established in Norway (Overrein et al., 1981). Similarly, we can refer to the first international conference on this topic (Dochinger and Seliga, 1976) and see that issues regarding soil effects identified at that time remain the central theme of much of the research today on acidic deposition.

During the later 1970s the research community focused its attention on the possible role of acidic deposition in surface water acidification, with a secondary focus on terrestrial effects including agricultural and forest soil-plant systems. In agriculture, no regional direct effect of acidic precipitation to plant surfaces at ambient levels could be demonstrated, and the intensive management of agricultural soils minimized the likelihood of detecting subtle changes in soil productivity attributable to acidic deposition. Although surface water acidification continues to be the focus of a great deal of research, we have made noteworthy progress in identifying mechanisms of aquatic response and identifying the critical role the soil component of a watershed plays in those mechanisms.

Research and public concern over the possible consequences of acidic deposition for forest health grew dramatically in the early 1980s with the identification of an unexpected forest deterioration in central Europe and in high-elevation coniferous forests in eastern North America. The characteristics of the modern forest health problem were chronicled by several key publications, such as the papers of Johnson and Siccama (1983) and Schutt and Cowling (1985). Numerous documents have surveyed the literature on soil effects, and no attempt is made here to repeat such efforts. Several of the more recent publications dealing specifically with acidic deposition effects on soil productivity include Fernandez (1985a, 1985b, 1986), Johnson (1984), McLaughlin (1985), Morrison (1984), Rechcigl and Sparks (1985), Tabatabai (1985), and Ulrich (1987). The following discussion is intended to provide an overview of our understanding of acidic deposition effects on forest soil productivity from my viewpoint, including suggestions for concepts that should be further developed with future research.

## II. Potential Soil Effects

Numerous hypotheses have been proposed over the course of the last two decades as to the possible effects of acidic deposition on soils. All evolve from the concept that acidic deposition may alter the chemical and biological behavior of soils and ultimately have indirect effects on the resources of concern (i.e., aquatic or forest ecosystem health). Although in different time periods attention has been given to one hypothesis over another, none has clearly been dismissed or proven to be a causative factor in forest health declines. Modern changes in the chemistry of certain freshwater ecosystems does appear to be related to atmospheric deposition of sulfur compounds, and major progress has been achieved in defining the mechanisms for these effects (NRC, 1986; Malanchuk and Turner, 1987). All potential hypotheses for the effects of atmospheric deposition on forest soil productivity are interrelated, but several major mechanisms of effect can be conceptually distinguished as follows:

### A. Soil Acidification

The potential for acidic deposition to acidify soils was clearly the earliest concern identified for detrimental effects that may indirectly influence forest health via changes in the soil chemical environment. This hypothesis more specifically proposes that the addition of strong mineral acids (e.g., $H_2SO_4$ and $HNO_3$) via wet and dry deposition may reduce soil pH and base saturation, leading to declines in soil fertility and possibly detrimental effects of elevated hydrogen ion activity in soil solutions. Bockheim (1984) reviewed the literature on experimental acidic leaching studies for soils that show acidic solutions result in soil cation leaching and acidification. The difficulty in drawing conclusions based on those types of studies is due to the short time periods and often unrealistic levels of acidic treatments employed. Widespread soil acidification or base saturation declines due to acidic deposition under field conditions have not been conclusively shown, although recent studies such as the work of Tamm and Hallbacken (1988) strongly suggest soil acidification is occurring due to atmospheric deposition. Although the scientific community has not discarded a concern for soil acidification under field conditions, it has become clear that these types of soil effects are not likely to develop rapidly, given that most forest soils are already highly acidic with low base saturation and have relatively large buffer capacities when compared to the amount of hydrogen ions derived from deposition. Adequately answering the acidification question will also require the quantification of mineral weathering rates in soils, which remains a difficult research problem to address.

Therefore, only a small subset of soil types is likely to be subject to rapid effects of acidic deposition, these being soils we could define according to Turner et al.

(1986) as poorly buffered (C.E.C. $< 15$ cmol($+$)/kg), moderate base saturation soils (e.g., 20% to 60%) with a pH $> 4.5$. Other factors would also be considered, but these criteria provide useful guidelines to suspect that most forest soils are not highly sensitive to a rapid acidification effect. Ulrich (1987) provides a framework for identifying the major buffering mechanisms for different pH ranges in mineral soils exposed to acidic deposition (Table 3–1). However, as discussed above, most forest soils are already acidic and fall in the aluminum or cation exchanger buffering range due to natural pedogenic processes. The question with regards to acidic deposition is whether modern air pollution has accelerated the natural processes of soil acidification. When base saturation is low, the cation available for exchange is mostly aluminum, resulting in either of these buffer ranges releasing aluminum in response to acidic deposition. Therefore, although this classification scheme can be used to assess soil sensitivity to acidic deposition on a global scale, it has limited use for evaluating forest soils in cool, humid, temperate, and boreal climates where soils are already acidic. Binkley and Richter (1987) discussed the complexity of determining long-term changes in soil acidity and pointed out the types of detailed information required on hydrogen ion budgets in order to draw meaningful conclusions on acidification processes.

A parallel concern intimately associated with soil acidification is the potential mobilization and toxicity of aluminum in soil solutions due to acidic deposition, often referred to as the *Aluminum hypothesis*. Acidic deposition provides a source of hydrogen ions that can undergo cation exchange with adsorbed cations, as well as anions (e.g., $SO_4^{2-}$ and $NO_3^-$), which must be accompanied by cations as they move through soils in solution. Aluminum may be released by dissolution and hydrolysis or cation exchange reactions when soil base saturation is low. Wolt (1987) recently reviewed the main factors controlling aluminum mobilization in soils (e.g., pH, mineralogy, organic matter) and pointed out the importance of knowing the speciation of aluminum when considering soil-plant interactions. Ulrich et al. (1980) first presented evidence of the possibility of aluminum toxicity in forest ecosystems and proposed acidic deposition was likely the major contributing cause. Since then, researchers have sought to test this hypothesis, and major research programs have been developed specifically for that purpose (Cronan et al., 1987). It is well known that soil acidity promotes aluminum toxicity and nutritional disorders within agricultural soil-plant systems (Pearson

**Table 3-1.** Classification of soil buffering ranges to acidic deposition according to Ulrich (1987).

| Buffering range | pH range |
|---|---|
| Carbonate buffer range | 6.5 to 8.3 |
| Silicate buffer range | All pH (typical at pH $>5$) |
| Exchanger buffer range | 4.2 to 5.0 |
| Aluminum buffer range | $>4.2$ |
| Aluminum/iron buffer range | $>3.8$ |
| Iron buffer range | $>3.2$ |

and Adams, 1967), but demonstrating similar effects in forest soils has not been possible to date. Forests have evolved under acidic soil conditions with high levels of mobile aluminum relative to circumneutral agricultural soils. Greenhouse experiments have shown coniferous species, particularly *Picea,* to be relatively tolerant to soil solution aluminum (McCormick and Steiner, 1978; Schier, 1985). The literature does not indicate that levels of forest soil solution aluminum in eastern North America are typically as high as those reported by Ulrich et al. (1980), and it is extremely difficult to extrapolate from greenhouse dose-response experiments to field settings, where soil solution chemistry and ecosystem processes are much more complex. In addition, we have little information on how lysimeter solution chemical composition compares with solution chemistry in the rhizosphere actually experienced by roots. Complex interactions, such as imbalances between aluminum and calcium in the fine root environment reported by Shortle and Smith (1988), are even more difficult phenomena to define yet may be more likely consequences of chronic exposure to acidic deposition. Therefore, the long-term consequences of chronic acidic deposition for soil acidification, base cation leaching, and aluminum-induced declines in forest health and composition remain undefined.

## B. Nitrogen Cycle

Forested ecosystems throughout the world are typically nitrogen limited. Acidic deposition is caused by compounds of S and N, both of which are essential plant nutrients. Much of the concern for acidic deposition effects on soil acidification described above has focused on the effects of $SO_4^{2-}$ leaching, because most of the $NO_3^-$ delivered to forest soils from atmospheric deposition is quickly removed from soil solution through uptake by soil organisms and higher plants. Both the S and N cycles are altered by atmospheric inputs of these nutrients. However, N deposition seems likely to have direct effects on forest functioning in most N-limited ecosystems, whereas S is more likely to have indirect effects on forests, as they are rarely S limited.

From a nutritional aspect, nitrogen supplied from acidic deposition has the potential to increase forest growth. Some evidence exists suggesting this has occurred in places (Nilsson, 1986), but widespread increases in forest growth rates attributable to nitrogen deposition are not evident. On further consideration, changes in forest ecosystem processes are possible due to elevated N deposition that may adversely affect forest health or may simply change the character of a forest community. Increased N availability in soil solutions could alter the physiological status of trees, create element imbalances, or change the dynamics of interspecific competition. Whether a change in species composition without a reduction in forest growth is a negative effect on forests is entirely a subjective determination, depending on the viewpoint of the individual.

Friedland and others (1984a) advanced the hypothesis that high-elevation red spruce decline might be attributable to the predisposition of those forests to winter damage as a result of N-induced changes in tree physiology. This article marked the early stages of a continuing debate in the scientific community on the implications of N deposition for changes in forest health, including concerns for both physiological and nutritional mechanisms of effect. Nihlgard (1985) proposed that N "saturation" could be a key hypothesis for explaining the forest deterioration phenomenon in Europe. Although the annual loading of N from atmospheric deposition in low-elevation eastern U.S. forests is not likely to "saturate" soils with N in the near future, high-elevation forest ecosystems may be much more subject to this effect, given the elevated loadings they receive due to orographic and canopy interception phenomena. For example, wet deposition of $NH_4^+$ and $NO_3^-$ at the Hubbard Brook NADP station in 1982 was 1.8 and 15.8 kg $ha^{-1} yr^{-1}$, respectively (NADP, 1985). However, Lovett et al. (1982) estimated the deposition of $NH_4^+$ and $NO_3^-$ to be 20.5 and 124.9 kg $ha^{-1} yr^{-1}$, respectively, for combined cloud droplet capture and bulk deposition in a nearby subalpine balsam fir forest at 1220-m elevation. In their study they calculated cloud deposition was responsible for approximately 80% of the total combined cloud and bulk deposition.

## C. Trace Metal Enrichment

Along with S and N, many other substances are released as a result of fossil fuel combustion, including trace metals such as Pb, Zn, Cd, Ni, and Mn, to name a few. Some of these metals are essential plant nutrients (e.g., Zn and Mn), and all can be potentially toxic to organisms at high concentrations in the environment. With respect to forest soil productivity, the major concern is that these metals may accumulate in soils following deposition from the atmosphere and inhibit soil microbial activities (and thus nutrient cycling), reduce forest health by direct toxicities, or inhibit nutrient availability. Some of these mechanisms are similar to the concerns discussed for Al, although trace metals are usually considered separately because the proportion of total soil metal content derived from the atmosphere is large as compared to the predominant source of soil (i.e., the soil itself), and they are likely to be toxic in much smaller concentrations when compared to Al. We can consider the trace metal issue in four steps as deposition → accumulation → mobilization → toxicity. A significant body of literature now exists to support the conclusion that selected metals are transported over long distances, are deposited on forested ecosystems, and are accumulating in forest soils (Friedland et al., 1984b; Hanson et al., 1982; Johnson et al., 1982a; Moyse and Fernandez, 1987; Siccama et al., 1980). The primary cause for this accumulation is the tenacity with which soil organic matter binds these metals, particularly in forested ecosystems where a surficial accumulation of organic matter develops, resulting in what is referred to as the *forest floor*. This highly organic surficial material is the first soil layer to intercept trace metals deposited from the atmosphere and serves as a very effective sink for these pollutants. Much less is known about the speciation and rates of release for trace metals from soils (i.e., mobilization) or the potential biological toxicity of these metals under ambient conditions either singly or interactively. Some studies indicate that trace metal accumulations in forest soils may inhibit microbial processes (Ruhling and Tyler, 1973; Tyler, 1975) or higher plant processes such as root growth (Burton et al., 1984; Godbold and Hutterman, 1985). However, there is little evidence to suggest that trace metal burdens in forest soils under ambient conditions are currently a primary cause for forest deterioration. Although long-term trace metal enrichment in forest soils may alter soil microbial processes, current levels of enrichment even in relatively polluted sites do not appear to be of concern (Friedland et al., 1986).

### D. Organic Matter Decomposition and Microbial Processes

Forest soil microflora are extremely important to the health and function of forests, being responsible for critical processes such as organic matter decomposition, mineral transformations, N fixation, and the formation of symbiotic relationships with higher plants (e.g., mycorrhizae). Francis (in this volume) has reviewed the available evidence for microbiological effects of acidic precipitation and concludes that acidic rain can affect microbial populations and diversity, organic decomposition and transformation, and certain steps in the nitrogen cycle, especially nitrification. Clearly our understanding of the complexities of soil biological processes is less than complete, and adding the evaluation of subtle pollutant effects makes the assessment task even more difficult.

Sulfur added to forest soils quickly becomes biologically active, affecting both soil fauna and microflora (David and Mitchell, 1987; Mitchell et al., 1983; Swank et al., 1984). Morrison (1984) reviewed the literature on this topic and was unable to draw any conclusions regarding soil microbial effects of acidic deposition. Since his review, experimental research continues to indicate the possibility of effects on processes such as nitrification, decomposition, or more generally on soil respiration and the diversity of the microbial community (Berg, 1986; Mancinelli, 1986; Skiba and Cresser, 1986; Thompson et al., 1987). However, no evidence exists to suggest these effects are causative in currently observed unexplained forest deterioration phenomena. The possible consequences of trace metal accumulation in forest soils for soil microbial processes was mentioned earlier and should also be identified as part of the regional air pollutant effects issue.

It seems likely that alterations to soil biota due to acidic deposition would be subtle and would take long periods of time to affect a change in forest health or composition. The effect, it should be noted, is perhaps beyond our ability to identify conclusively.

## III. Linkages and Effects

As the environmental issue regarding the effects of atmospheric deposition on forest soils evolves, it will be critical to identify effects on forest soils that can be demonstrated and to differentiate clearly these findings from work showing causative linkages between demonstrable soil effects and forest health. Although it may be possible, for example, to prove that S deposition changes soil microbial processes, this does not prove that these changes will result in detrimental effects on the forest community. There is sometimes a great temptation, once any change in the ecosystem is shown, to conclude that such changes must be inherently detrimental to the health of the ecosystem. However, we must pursue research that identifies changes in forest soils due to air pollutants as a first step in testing hypotheses as to potential mechanisms of effects on both forest and aquatic resources. In this context, several observations and conclusions can be drawn at this time.

## A. The Knowns

### 1. Modern Chemistry of the Atmosphere

Ample evidence exists that forests are exposed on regional and even global scales to a modern atmospheric chemistry that has been altered by anthropogenic activities. Regional patterns of precipitation composition are strongly related to the distribution of emissions of S and N (NADP, 1987), and long-term temporal trends in the chemical composition of the atmosphere reflect the evolution of industrialized societies (NRC, 1986). Sulfur and nitrogen are only two of the many chemical species found in the atmosphere, including numerous organic substances (Lioy and Daisey, 1986; Mazurek and Simoneit, 1986), of which there exists relatively little information as to the deposition and effects on forests. It is useful to note this fact in a discussion of soil productivity effects, as it implies that even without changes in soil productivity forests are likely to be exposed to increased environmental stresses that may be reflected in the demands forests make on their supporting soils. Under increased pollutant stress, trees once tolerant of sites that are marginal due to poor soil conditions may no longer be able to survive. This could be of concern where trees have been planted "off-site" through artificial regeneration practices.

### 2. Altered Nutrient Cycles

Many of the atmospherically derived materials deposited on forest ecosystems are essential plant nutrients. As such, it can be considered a fact that the cycling of many of the elements in forests essential to tree growth has been altered due to atmospheric deposition. Swank (1984) and Johnson and others (1982b, 1985b) demonstrate the quantitative significance of atmospheric sources for some of the macronutrients by using forest nutrient budgets. Liebig's Law of the Minimum, although somewhat simplistic and best applied to steady-state systems, clearly suggests that physiological processes such as suggested by Friedland et al. (1984a) or unexpected consequences of altered nutrient balances can result in negative effects on forest health. The issue of nutrient balances may become key to evaluating N because complex ecosystems such as forests should reveal the greatest change when the most limiting factors for growth are altered. Nitrogen is well known to be the most commonly limiting nutrient in forests and might therefore be considered the atmospherically derived nutrient most likely to alter forest function significantly. From the commercial forest management standpoint, it is well established that forest growth and composition can be altered by N fertilization (Bengston, 1979; Allen, 1987). Table 3–2 illustrates a conservative estimate of the N contribution from precipitation for selected forest ecosystems as compared with N uptake. These data largely omit the dry deposition contribution of N, which Grennfelt and Hultberg (1986) point out can result in significantly underestimating the atmospheric contribution of N to the ecosystem. They suggest that for low N deposition sites greater than 95% of the deposited N is retained by plants via uptake and soils via fixation, but areas of high N deposition, such as those in central Europe, may receive up to 69 kg/ha and become N saturated in a matter of decades.

Given that N, along with S, is a major component in atmospheric deposition and that N is limiting to forest growth, it is reasonable that N deposition is the most likely component of acidic deposition to cause early, regionally important, and detectable changes in forest growth or composition. Increased growth rates would be economically beneficial in commercial forests. However, alterations in deposition affect other nutrients because improved N status will create increased demands on the supply of all other essential nutrients in N-limited forests. These statements are also simplistic in that they ignore the potential effects of N deposition on more subtle processes such as interspecific competition (i.e., stand composition) or soil microbial processes. Therefore, we find that acidic deposition has a demonstrable effect on the critical and often growth-limiting N cycle of forests, but the consequences of this effect for forest growth, composition, or health have yet to be determined.

## 3. Soil Solution Sulfate

Like N, S is also an essential plant nutrient, but it is rarely limiting to forest growth. Therefore, much of the atmospherically derived S is retained by soil processes or leached as $SO_4^{2-}$ in soil solution. The mobile $SO_4^{2-}$ anion is the focal point of concerns for acidic precipitation–induced declines in soil fertility, as well as for the transport of Al to freshwater ecosystems. Linkages between soil solution $SO_4^{2-}$ concentrations and forest health or aquatic acidification are not yet fully understood, although many of these linkages have been fairly well defined for aquatic ecosystems. Morrison's (1984) discussion of S cycling demonstrates that in relatively unpolluted areas bicarbonate is the dominant anion, whereas in regions such as the northeastern United States that are subject to acidic deposition $SO_4^{2-}$ becomes an important or even major anion in soil solutions, along with organic acids. Although the consequences of this modern $SO_4^{2-}$ burden on soil productivity and forest health remain undefined, it can be considered fact that air pollution has fundamentally changed the chemical composition of soil–soil solution systems on a regional scale.

## 4. Trace Metal Accumulation

A significant body of literature exists, some of which was referred to earlier, to demonstrate that trace metal deposition occurs in forests and that the accumulation of these metals is a long-term and continuing process. Trace metal residence times

**Table 3-2.** Precipitation-derived nitrogen as a percentage of nitrogen uptake for selected forests.*

| Location | Forest Type[†] | Uptake[‡] (kg/ha/yr) | Precipitation (kg/ha/yr) | Precipitation as % of uptake |
|---|---|---|---|---|
| Karelia, USSR | Spruce (12) | 34.4 | 1.1 | 3 |
| Alaska, USA | Birch (4) | 25.0 | 2.1 | 8 |
| New Hampshire, USA | Maple/Birch/Beech (27) | 74.3 | 6.5 | 9 |
| Solling, FRG | Beech (32) | 75.6 | 21.8 | 29 |
| Solling, FRG | Spruce (12) | 55.6 | 21.8 | 39 |

* Adapted from Cole and Rapp (1980).

† Number in brackets refers to stand number in Cole and Rapp (1980).

‡ Uptake = bole increment + branch increment + litterfall + leaf wash + stem flow.

in forest ecosystems and toxic thresholds for forest health are poorly understood. Available data for North America suggest that direct toxicity to trees from the commonly studied metals is not evident under ambient conditions, and possible trace metal effects on mineralization processes or inhibitory effects of soil trace metals on nutrient availability have not been shown. Although these issues remain the subjects of hypothesis testing, little question remains as to the modern accumulation of trace metals in forest soils subjected to atmospheric deposition.

## 5. Nutrient Deficiencies

The most long-standing concern for acidic precipitation effects on forests has been the possibility of accelerated base cation leaching resulting in reduced soil fertility and forest growth. No evidence exists to show this has occurred on a regional basis in North America, although speculation that this indirect and subtle effect is occurring remains strong. In Europe, the forest health issue referred to as *Waldsterben* or the "new type" forest decline represents a great diversity in symptomatology (Schutt and Cowling, 1985). Many of the cationic nutrients (Ca, Mg, K, Zn, Mn) have been found at low concentrations in the foliage of damaged trees, which Huettl and Wisniewski (1987) point to as support for the role of acidic deposition in the leaching loss of nutrients from foliage and soils. This proposition appears strongly supported by fertilization studies using Norway spruce (*Picea abies* (L.) Karst.) at high-elevation sites in West Germany where chlorosis has been observed (Zoettl and Huettl, 1986; Huettl and Wisniewski, 1987). In those studies, fertilization with Mg was effective in eliminating chlorotic symptoms and in improving foliar nutrient status. The authors indicate that the development of nutritional disorders were highly correlated with soil types and the existence of other stresses such as ozone may very well play a decisive role in these forest health phenomena. No evidence is yet available that conclusively identifies the causal mechanisms for these nutritional disorders. One or more of the following mechanisms could result in acidic deposition–induced nutritional disorders:

1. Acidic deposition may accelerate the leaching of nutrients from foliage, which can lead to nutrient deficiencies where supply of that nutrient from the soil is naturally limited. The possible interactions of acidic deposition with other stress factors such as ozone or drought could enhance the vulnerability of trees to nutrient losses.
2. Acidic deposition may accelerate the leaching of nutrients from the soil, which reduces their availability. This becomes an even greater concern when tree uptake demands are increased due to foliar leaching induced by pollutant stress.
3. Acidic deposition may cause unfavorable nutrient balances in the soil and soil solution by inducing the mobilization of antagonistic cations (e.g., Al) that inhibit nutrient availability to tree roots and may reduce fine root biomass.
4. Acidic deposition provides significant amounts of N to typically N-limited forests, which can be expected to increase tree growth and the subsequent demand for all other essential nutrients. This N-related nutrient imbalance might result in deficiencies of cationic nutrients in soils where these conditions have previously not been found.

Fertilization experiments do not prove that acidic deposition has caused nutritional disorders or that nutrient deficiencies are primary causal factors in forest health declines. However, they do show that for certain forest ecosystems experiencing selected symptoms of the "new type" forest decline (e.g., chlorosis) there is a direct relationship between the availability of specific nutrients (e.g., Mg) and the expression of specific symptoms. In a scientific issue of such complexity, this type of empirical evidence is significant and suggests that for specific sites, nutrient disorders are a component of forest health deterioration.


## B. The Unknowns

### 1. Causative Mechanisms

Although there is sufficient evidence to conclude that acidic deposition and related air pollutants have resulted in changes in forest soils on a regional basis, no conclusive evidence yet exists to demonstrate that these changes have, or will have, a negative influence on forest health. We can demonstrate "effects" on forest soils, but not a potential linkage with forest health. Sufficient information has resulted from research on terrestrial-aquatic linkages to describe the role of soils in acidic deposition–induced changes in aquatic ecosystems. It now appears that, for certain ecosystems, freshwaters may become acidified with essentially no effect on "soil productivity." A lake may become acidified due to the mobilization of soil Al by $SO_4^{2-}$ leaching, with essentially no change in the base saturation of the associated soils. This is likely to occur in watersheds dominated by highly acidic forest soils with low base saturation and high levels of exchangeable Al.

The concept of soil productivity is typically viewed in the context of the ability of a soil to support plant growth. Concerns for acidic deposition effects on soil productivity largely deal with the possibility of declining soil fertility or toxic effects of mobilized soil metals. As defined by Grady (1984), *soil fertility* is "the status of a soil with respect to the amount and availability to plants of elements necessary for plant growth." It is useful to point out that in commercial forestry even improved soil fertility that might result from increased nutrient deposition (e.g., N) could have negative economic consequences. Where land is managed to grow coniferous species for pulp and paper production, herbicides are commonly used to suppress deciduous competition. Changes in forest composition—or increased herbicide use due to improved N supply to the site that encourages greater deciduous competition—could result in increased costs to the land manager. Studies such as those reported by the Swedish scientists Falkengren-Grerup (1986) and Tyler (1987) attributing floristic changes in forest composition to acidic and N deposition have begun to appear in the literature. This suggests the importance of defining all significant changes in soil productivity that can result from acidic deposition, regardless of whether those changes are viewed as positive or negative from a traditional soil productivity standpoint.

The lack of long-term data and control sites for field experimentation limits our ability to demonstrate mechanisms of chronic acidic deposition stress on soil productivity and forest health. Site-specific research such as that reported by

Johnson et al. (1985a) in the United States and Ulrich et al. (1980) and Zoettl and Huettl (1986) in the Federal Republic of Germany demonstrates that progress is being made on understanding acidic deposition effects on soil-related forest health concerns, but these data are not directly applicable to large geographic regions. We can say with confidence that regional-scale changes in soil properties have occurred due to atmospheric deposition, but how these changes relate to current forest health and how they may affect future forest health remain unknown.

## 2. Long-Term Trends

A major obstacle in our ability to determine relatively subtle changes in soil properties as a result of atmospheric deposition is the lack of long-term studies that characterize forest soil properties and quantify their variability over periods of decades or centuries. Declining soil fertility results in lower base saturation and soil pH. Natural processes of soil genesis and afforestation themselves acidify soils, so acidic deposition–induced changes in the rates of these processes are understandably difficult to determine. Several studies have reported recent declines in forest soil pH over periods of 40 to 50 years and partially attribute this acidification to acidic deposition (Hallbacken and Tamm, 1986; Falkengren-Grerup, 1987). Andersen (1987) reported soil pH declines for Adirondack forest soils in New York that could be explained by changes in Ca cycling attributable to natural processes. Krug and Frink (1983) hypothesized on the causative role of changing land use patterns in soil and freshwater chemical trends, and Brand et al. (1986) demonstrated how coniferous afforestation acidifies soils in Ontario. However, Linzon and Temple (1980) reported no changes in Canadian soils over an 18-year period. Johnston et al. (1986) illustrated that acidic deposition can be a significant component of acidification processes in soils and that its importance has increased over the last century. What emerges from these studies is the reality of our relatively poor understanding of temporal variability in forest soil properties and processes. Good examples of the complexity of even the pH change question are shown by Andersen et al. (1987), who reported pH changes only where original soil pH in 1930 was greater than 4.0. Alban (1982) found that nutrient accumulation by spruce and aspen resulted in a redistribution of elements in soils, with pH both increasing and decreasing, depending on the horizon examined. Therefore, we can conclude that forest soils are dynamic and complex, that they typically acidify due to natural processes, and that the potential role of acidic deposition in regional soil acidification processes is greater over the last few decades than ever before. We are not able to draw more definitive conclusions on regional soil pH effects at this time, and long-term research will be required to describe confidently air pollution effects on complex phenomena such as stand dynamics and N cycling.

## 3. Interactions

This chapter has alluded to many stress factors that could alter forest health and character. Natural factors such as drought, cold temperatures, poor soil drainage, or infertile soils, as well as pollutant stresses including acidic deposition, trace metals, ozone, and organic compounds, are examples of the multiple stress factors that must be considered. Even for issues such as forest harvesting, where a substantial database on nutrient budgets already exists (e.g., Kimmins et al. 1985), it has not been possible to eliminate conclusively or identify cause for concern regarding acidic deposition's role in nutrient depletion at most of these intensively studied sites. As we better understand individual processes of soil response to acidic deposition, there will remain the need for more sophisticated analyses of ecosystem function that incorporate multiple stress scenarios.

## IV. Evolving Concepts on Soil Effects

Few, if any, of the hypotheses related to atmospheric deposition effects on forest soil productivity have been disproved as a result of the research done to date on this issue. Rather, an evolution in thinking has taken place, encompassing a more complex and more realistic approach to determining soil response. Simple cause-and-effect hypotheses regarding dilute $H_2SO_4$ solution effects on soil fertility have grown to encompass multiple-stress hypotheses with complex effects on the whole ecosystem. Several concepts regarding forest soil effects and the research to evaluate them deserve recognition due to their importance in understanding potential changes in soil productivity.

### A. Spatial and Temporal Variability

Basic to our ability to evaluate soil effects is the need to understand adequately, in a quantitative manner, the variability of ecosystem properties. Much of the research providing insight on potential soil productivity changes due to atmospheric deposition has been of limited duration and encompasses various levels of resolution, including the watershed, stand, tree, soil, horizon, rhizosphere, and even cellular levels. Great care must be exercised in the synthesis of information from such a diversity of research approaches. Even simple assumptions such as the relationship between base saturation and pH must be cautiously applied, depending on the soil system in question. For example, Figure 3–1 shows mean chemical characteristics from a sampling of well-drained and moderately well-drained podzolic forest soils in Maine. These data illustrate that each morphologically distinct horizon has a unique chemistry and that the highest base saturation is evident in the lowest pH horizon due to the influence of organic acids. Much of the fine root biomass for the shallow-rooted coniferous species in this region is found in the forest floor, yet little research is available to define the effects on tree growth and interactions among horizons in these types of soils. Therefore, generalizations about a "soil" response are difficult to make in soils encompassing such a morphological and chemical diversity within a single pedon. If we conclude that acidic deposition has an effect on soil fertility, we then need to determine if the whole pedon responds similarly or differently by compartment and how those changes influence tree growth.

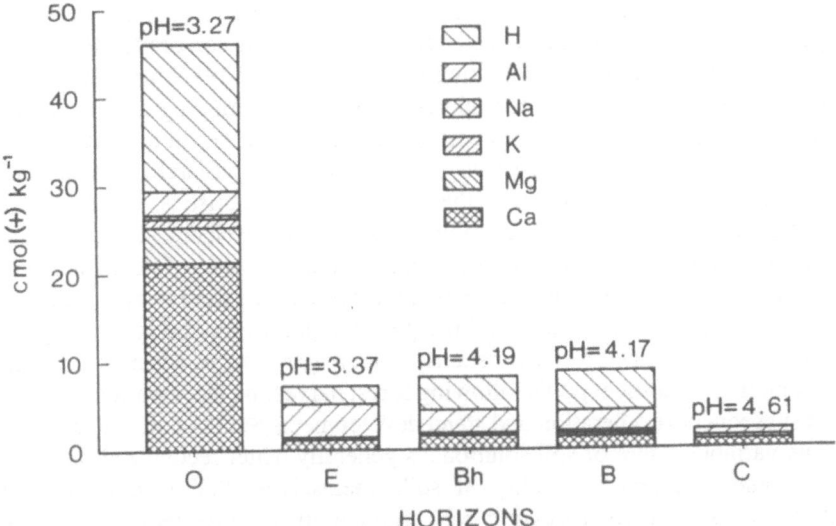

**Figure 3–1.** Exchangeable cations (1 N NH₄Cl extractable) and pH (0.01 M CaCl₂) from a sampling ($n = 23$) of undisturbed well-drained and moderately well-drained pedons under forest cover in Maine.

Other considerations must be included in a discussion of soil responses to atmospheric deposition and our ability to define them. Beyond the morphological distinctions mentioned above, soils can respond to changes in deposition inputs on very narrowly defined spatial scales. Banin and others (1987) showed that the very top few millimeters of soil in a roadside forest in Israel had 1.5, 6, and 15 times the natural background levels of Zn, Cd, and Pb, respectively. Fernandez (1987) demonstrated that the top 2 cm of the B horizon in reconstructed forest soil microcosms responded to simulated acidic rain treatments significantly differently from B horizon soil material below for properties such as soluble sulfate and exchangeable cations. These studies illustrate that conventional soil sampling approaches in "effects" research may often be inadequate to identify changes in critical soil properties.

Spatial variability across plots or landscape units has long been recognized as a significant concern in forest soils research (Mader, 1963). Wilding and Drees (1983) provide an excellent discussion of this topic, within which they identify soil pH as a parameter of low variability but exchangeable cations as moderately variable and water-soluble sulfate or organic matter content as some of the most variable. It is essential to account for anticipated variability in soil sampling designs, given the importance of these parameters in atmospheric deposition research. In addition, Mausbach et al. (1980) showed that soil chemical variability followed the sequence spodosols > ultisols = Inceptisols = Entisols > Alfisols = Mollisols > Vertisols. From this series it is evident that the most highly variable soils tend to be the dominant forest soil orders for the eastern United States. The influence of distance from individual trees on surrounding soil properties has been demonstrated by studies such as those of Riha et al. (1986) and Skeffington (1983), further emphasizing the importance of careful sampling designs in order adequately to distinguish treatment effects and characterize soil properties.

Soil properties are often viewed as static in ecological research, which can lead to erroneous assumptions about the chemical characteristics of ecosystems. Significant changes occur in soil properties, such as exchangeable cations and extractable $NH_4^+$ or $NO_3^-$ that over time can result in two to fourfold differences in soil nutrient pools, depending on the month in which soils were sampled. Peterson and Hammer (1986) conclude that this type of temporal variation in soil properties is typical in temperate forest ecosystems. Also, major changes occur in forest floor and mineral soil nutrient pools following forest harvesting that continue to change with subsequent regeneration and stand development (Smith et al., 1986).

The variable nature of soil solutions is generally better recognized than is the spatial and temporal variability of soils themselves. Soil solution data are extremely valuable for assessing soil productivity because they represent the dynamic character of soil systems that most directly influence tree growth. Yet available methods for studying soil solutions are largely qualitative due to spatial and temporal variability and the practical limitations of adequate sampling strategies. Soil solution chemistries vary on temporal scales from minutes to years and on spatial scales from less than millimeters within soil pores across the range of soil pore sizes up to the watershed level of resolution. In addition, the type of soil solution sampler (e.g., lysimeter) employed affects the portion of the soil solution sampled. Studies continue to appear in the literature comparing commonly used soil solution samplers (Rasmussen et al., 1986; David and Gertner, 1986) and generally point to the utility of these collection devices as long as their limitations are adequately understood. David and Gertner (1987) showed that limitations on the use of lysimeters could include the need for increased replication of oil pits and lysimeters or increased sampling frequency, depending on the soil solution parameters being studied.

Figure 3–2 shows soil solution sulfate concentrations from the 1987 growing season from an ongoing study at the University of Maine. The data represent soil solution chemistries for four tension cup lysimeters in the upper B horizon in each of two 5- × 5-meter plots. Soils are typic haplorthods under mixed coniferous cover, and both plots were treated biweekly with the same dilute $H_2SO_r$ solutions, with soil solution samples collected on alternating weeks. This figure indicates that treatment effects are evident but also demonstrates the large diversity in response among lysimeters that can be expected. No data are available from the dry periods of the year because lysimeter tensions typically are not adequate to extract soil water under moisture stress conditions, although soils obviously contained moisture and trees were still growing. Indeed, solutions in soil micropores during this period of a growing season may ultimately be most important for determining potential effects on forest growth. Nevertheless, this type of manipulative study continues to provide useful information on soil

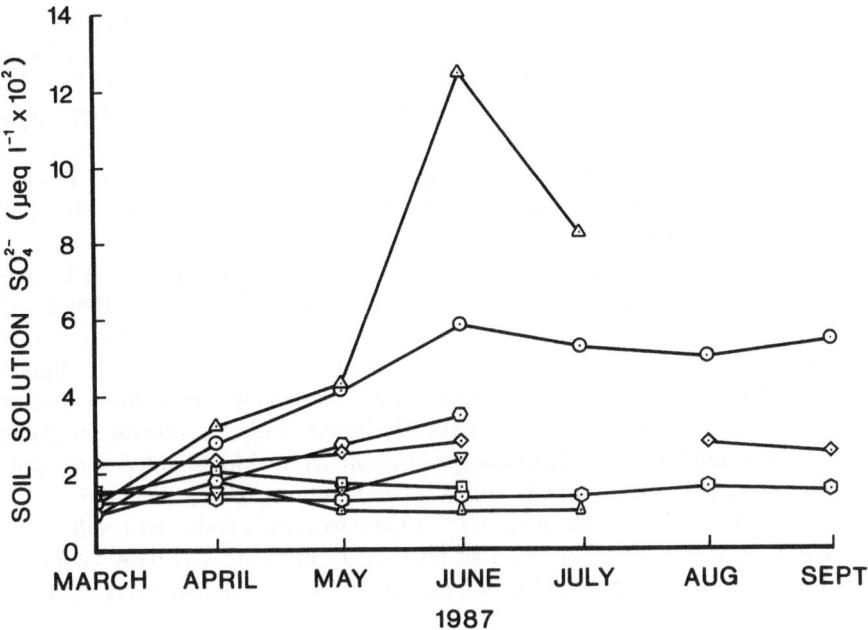

**Figure 3–2.** Soil solution $SO_4^{2-}$ concentrations from eight upper B horizon tension lysimeters on two $25^2$m plots treated biweekly with dilute $H_2SO_4$ for the 1987 growing season. Missing data points reflect no collections during dry periods of the growing season. From Cole and Rapp (1980), with the permission of Cambridge University Press.

processes as long as interpretations are conceived within the limitations of the measurements.

## B. The Role of Anions

The evolution in our approach to studying the effects of atmospheric deposition on soil productivity is well represented by the focus on more complex mechanisms involved in soil and soil solution interactions seen in the recent literature. Atmospheric deposition not only results in the addition of $H^+$ to soil cation exchange reactions but also provides additional anions to soil solutions, thus changing the ionic strength of the system. In acidic soils typical of many temperate forest ecosystems, total $H^+$ inputs via acidic deposition are relatively small compared to natural $H^+$ generation. However, increased soil solution ionic strength due to acidic deposition results in the phenomenon known to soil scientists as the *salt effect*, which increases soil solution $H^+$ activity due to cation exchange reactions (Wiklander, 1975; Seip, 1980). Where soils have limited ability to retain an anion, accelerated leaching of cations will occur due to a requirement for electroneutrality in solution. As forests are typically N limited, undisturbed forest ecosystems retain $NO_3^-$ through biological uptake. In contrast, atmospherically derived $SO_4^{2-}$ is typically in excess of plant requirements and may readily leach through soils with a counter cation. Soils differ markedly in their ability to retain $SO_4^{2-}$ (Johnson and Todd, 1983), which has often been a key characteristic in delineating soils thought to be "sensitive" to acidic deposition (Turner et al., 1986). Soils in the northeastern United States are considered sensitive to accelerated leaching due to acidic deposition largely as a result of their relatively small capacity to retain $SO_4^{2-}$.

Whether base cations or acidic cations (e.g., Al) accompany the mobile anion is dependent on the base saturation and buffer capacity of the soils. Reuss and Johnson (1985) developed a model of soil–soil solution dynamics that incorporated both the influence of anions and the role of dissolved $CO_2$ in soil solution chemical responses to acidic inputs. Their model was significant in that it appears to have been the first that was theoretically based, included several inorganic constituents simultaneously in exchange phenomena, and highlighted the importance of $CO_2$ partial pressures in soil solution reactions. Using their model it is possible to describe how elevated $SO_4^{2-}$ inputs to certain soils can result in the acidification of soil solutions without necessarily reducing soil base saturation (i.e., acidic deposition can acidify soil solutions without reducing soil productivity).

The importance of the anion in soil–soil solution interactions is further emphasized by evidence that marine-derived neutral salt (i.e., NaCl) accumulation in soils can lead to acidification events for associated freshwaters due to mobile $Cl^-$ leaching. Episodic acidification of streams in both Norway (Norton et al., 1987) and Maine (Kahl et al., 1985) due to $Cl^-$ leaching demonstrates that HCl leaches from these coastal soils following NaCl accumulation, with subsequent exchange of $Na^+$ for $H^+$. Johnson et al. (1986) showed similar effects on forest soils in column studies using several different $Cl^-$ salts. These comments should not be interpreted as minimizing the overall importance of $NO_3^-$ deposition, as the changes in N cycling discussed earlier are significant. Yet chronic accelerated leaching due to mobile $NO_3^-$ has not been found in most undisturbed forested ecosystems of North America at this time. Exceptions to this statement should include forest ecosystems where N-fixing species occur (Van Miegroet and Cole, 1984) and certain areas in the northeastern United States and central Europe where N deposition (i.e., $NH_4^+ + NO^{3-}$) may now be resulting in "N saturation" of forests (Nilsson, 1986). In addition, natural events such as spring snowmelt or ecosystem changes due to fire or forest harvesting can result in periods of high $NO_3^-$ release and leaching.

## C. Element Balances

Our current understanding of nutrient availability in agricultural soils has been developed over the course of many years, and much of what was learned is now being drawn upon to determine atmospheric deposition effects on forest soils. The historical accounting of ion exchange research by Thomas (1977) points out the early recognition of the importance of multiple ions in soil–soil solution reactions.

Likewise, Schofield (1955) introduced the concept of nutrient potentials as a theoretically based approach to determine P availability, and Beckett (1972) discussed the evolution of cation activity ratios as measures of cation availability in soils. Beckett's (1972) paper cites many workers who developed "Q/I" relationships that relate the quantity of a nutrient in the soil ($Q$) to the chemical activity or intensity ($I$) of that nutrient in soil solution. The atmospheric deposition research community has adopted similar terminology in recent years, using the term *capacity* rather than *quantity*, and *intensity* in much the same context as these agricultural researchers.

These comparisons with earlier work using agricultural soils suggest that we can expect to find similar relationships operative in forest soils subject to atmospheric deposition. Simple approaches to determining metal toxicities, such as with Al, or nutrient deficiencies, such as with Mg, are appropriate first steps in the assessment process. However, we now have enough of an understanding of forest soil response to air pollutants to identify and evaluate more comprehensive mechanisms of effect on forest health. Empirical relationships such as Ulrich's (1987) Ca/Al ratios are an important contribution to developing a more mechanistic understanding of soil productivity effects. Where such empirical indices of forest soil productivity can be identified in the field, theoretical models such as the one developed by Reuss and Johnson (1985) can be used to predict future changes in soil productivity and to extrapolate over carefully defined landscape units. Ratios between N and other macronutrients in soils and plants are being considered (Nilsson, 1986), and ratios with N are likely to play a useful role in evaluating the sensitivity and status of forest soils in relation to N deposition. Studies such as those of Truman et al. (1986) demonstrate that P nutrition in forest soils can be altered by changes in Ca and Al activities in soil solutions. In short, it appears that hypotheses regarding the effects of atmospheric deposition on forest soil productivity must be developed to include multiple elements and, where possible, be theoretically based.

## V. Research Needs

The previous discussion has alluded to numerous opportunities for research on the influence of atmospheric deposition on forest soil productivity. No attempt is made here to develop a listing of these possible hypotheses or experimental approaches. It is useful, however, to note two general research needs that deserve emphasis in the future.

1. The need for long-term studies is clear, and careful consideration must be given to future funding criteria in order to preserve key research programs initiated over the last decade, while still directing new funding to the most pressing information needs. A number of ecosystem-level studies have been initiated that will begin to provide critical information on the temporal and spatial variability of ecosystem behavior in the years to come. These study sites also

provide a useful foundation for experimental short-term research that can benefit from the data already in place. Eliminating support for such research must be considered within a larger context than just the narrowly defined objectives of the original program for which it was originated. Long-term studies are essential and the foundations for this type of information developed to date must be preserved where possible. Emerging issues such as global climate change only underscore this need.

2. Monitoring and characterization research is limited by the lack of long-term data records and appropriate control sites for comparative assessments. Manipulative studies are the only means to determine the behavior of identical soil systems under differing deposition scenarios and will be required to test and refine models as well as to define mechanisms of response. A clear hazard in manipulative research is the risk of creating artificial effects due to high treatment rates, short time periods, or unnatural means of treatment application imposed by practical limitations. Nevertheless, carefully designed manipulative soils studies can provide otherwise unavailable information that is essential for predicting the potential effects of regulatory decisions. When manipulation experiments are conducted, plans to assess recovery processes should be incorporated into the overall timetable of the study.

## Acknowledgments

The author thanks Mary Thibodeau for her assistance in the preparation of this chapter. Maine Agricultural Experiment Station Publication No. 1282.

## References

Alban, D.H. 1982. Soil Sci Soc Am J 46:853–861.
Allen, H.L. 1987. J For 85:37–46.
Andersen, S.B., A.H. Johnson, and T.G. Siccama. 1987. Agron Abst, 250.
Banin, A., J. Narrot, and A. Perl. 1987. The Sci Total Env 61:145–152.
Beckett, P. 1972. Adv Agron 24:379–409.
Bengston, G.W. 1979. J For 77:222–229.
Berg, B. 1986. Scand J For Res 1:317–322.
Binkley, D., and D. Richter. 1987. Adv Ecol Res 16:1–51.
Bockheim, J.G. 1984. *In Proceedings of U.S.-Canadian Conference on Forest Responses to Acidic Deposition*, 19–35. Land and Water Resource Center, University of Maine, Orono.
Brady, N.C. 1984. *The nature and properties of soils*. Macmillan, New York. 750 pp.
Brand, D.G., P. Kehoe, and M. Connors. 1986. Can J For Res 16:1389–1391.
Burton, K.W., E. Morgan, and A. Roig. 1984. Plant Soil 78:271–282.
Cole, D.W., and M. Rapp. 1980. *In Dynamic properties of forest ecosystems*, 341–409. Cambridge University Press, London.
Cronan, C.S., J.M. Kelly, C.L. Schofield, and R.A. Goldstein. 1987. In R. Perry, R.M. Harrison, J.N.B. Bell, and J.N. Lester, eds. *Acid rain: Scientific and technical advances*, 649–656. Selper Ltd., London.

David, M.B., and G.Z. Gertner. 1987. Can J For Res 17:190–193.

David, M.B., and M.J. Mitchell. 1987. JAPCA 37:39–44.

Dochinger, L.S., and T.A. Seliga. 1976. *Proceedings of the First International Symposium on Acid Precipitation and the Forest Ecosystem.* USDA For. Serv. Gen. Tech. Rep. NE-23. 1074 pp.

Falkengren-Grerup, V. 1986. Oecologia 70:339–347.

Falkengren-Grerup, V. 1987. Environ Pollut 43:79–90.

Fernandez, I.J. 1985a. *In Proc. air pollutants effects on forest ecosystems,* 238–250. Acid Rain Foundation, St. Paul, Minn.

Fernandez, I.J. 1985b. *In Acid deposition,* 223–239. Plenum Publishing, New York.

Fernandez, I.J. 1986. *In Stress physiology and forest productivity,* 217–239. Martinus Nijhoff Pub., Dordrech, The Netherlands.

Fernandez, I.J. 1987. *Vertical trends in the chemistry of forest soil microcosms following experimental acidification.* Maine Ag. Exp. Sta. Tech. Bull. No. 126, Orono, Maine, 19 pp.

Francis, A.J. 1989. *In* D.C. Adriano and A.H. Johnson, eds. *Acidic precipitation,* vol. 2. Springer-Verlag, New York.

Friedland, A.J., R.A. Gregory, L. Karenlampi, and A.H. Johnson. 1984a. Can J For Res 14:963–965.

Friedland, A.J., A.H. Johnson, and T.G. Siccama. 1986. Can J Bot 64:1349–1354.

Friedland, A.J., A.H. Johnson, T.G. Siccama, and D.L. Mader. 1984b. Soil Sci Soc Am J 48:422–425.

Godbold, D.L., and A. Huttermann. 1985. Environ Pollut (Ser A) 38:375–381.

Grennfelt, P., and H. Hultberg. 1986. Water Air Soil Pollut 30:945–963.

Hallbacken, L., and C.O. Tamm. 1986. Scand J For Res 1:219–232.

Hanson, D.W., S.A. Norton, and J.S. Williams. 1982. Water Air Soil Pollut 18:227–239.

Huettl, R.F., and J. Wisniewski. 1987. Water Air Soil Pollut 33:265–276.

Johnson, A.H., and T.G. Siccama. 1983. Environ Sci Technol 17:294–306.

Johnson, A.H., T.G. Siccama, and A.J. Friedland. 1982a. J Environ Qual 11:577–580.

Johnson, D.W. 1984. Biogeochem 1:29–43.

Johnson, D.W., D.W. Cole, H. Van Miegroet, and F.W. Horng. 1986. Soil Sci Soc Am J 50:776–783.

Johnson, D.W., J.M. Kelly, W.T. Swank, D.W. Cole, J.W. Hornbeck, R.S. Pierce, and D. Van Lear. 1985a. *A comparative evaluation of the effects of acid precipitation, natural acid production, and harvesting on cation removal from forests.* ORNL Env. Sci. Div. Pub. No. 2505, Oak Ridge, Tenn. 107 pp.

Johnson, D.W., D.D. Richter, G.M. Lovett, and S.E. Lindberg. 1985b. Can J For Res 15:773–782.

Johnson, D.W., and D.E. Todd. 1983. Soil Sci Soc Am J 47:792–800.

Johnson, D.W., J. Turner, and J.M. Kelly. 1982b. Water Resourc Res 18:449–461.

Johnston, A.E., K.W.T. Gouding, and P.R. Poulton. 1986. Soil Use Manag 2:3–5.

Kahl, J.S., J.L. Andersen, and S.A. Norton. 1985. *Water resource baseline data and assessment of impacts from acidic precipitation, Acadia National Park, Maine.* National Park Service Technical Rep. No. 16. 123 pp.

Kimmins, J.P., D. Binkley, L. Chatarpaul, and J. de Catanzaro. 1985. *Whole-tree harvest-nutrient relationships: A bibliography.* Petawa Nat. For. Inst. Info. Rep. PI-X-60E/F, Canada. 377 pp.

Krug, E.C., and C.R. Frink. 1983. Science 221:520–525.

Linzon, S.N., and P.J. Temple. 1980. *In* D. Drablos and A. Tollan, eds. *Ecological impact of acid precipitation* (Proc. Int. Conf.) SNSF Proj., 176–177. Oslow, Norway.

Lioy, P.J., and J.M. Daisey. 1986. Environ Sci Technol 20:8–14.

Lovett, G.M., W.A. Reiners, and R.K. Olsen. 1982. Science 218:1303–1304.

McCormick, L.H., and K.C. Steiner. 1978. For Sci 24:565–568.

McLaughlin, S.B. 1985. JAPCA 35:512–534.

Mader, D.L. 1963. Soil Sci Soc Am Proc 27:707–709.

Malanchuk, J.L., and R.S. Turner. 1987. *In NAPAP interim assessment (vol. IV): effects of acidic deposition,* Chap. 8. The National Acid Precipitation Assessment Program, Washington, D.C.

Mancinelli, R.L. 1986. Arctic Alp Res 18:269–275.

Mausbach, M.J., B.R. Brasher, R.D. Yeck, and W.D. Nettleton. 1980. Soil Sci Soc Am J 44:358–363.

Mazurek, M.A., and B.R.T. Simoneit. 1986. *In Critical reviews in environmental control,* 1–140. CRC Press, Boca Raton, Fla.

Mitchell, M.J., M.B. David, and C.R. Morgan. 1983. *In* P. Lebrum, H.M. André, A. De Medt, C. Gregoire-Wibo, and G. Wauthy, eds. *New Trends in Soil Biology: Proceedings of the VIII International Colloquium of Soil Zoology,* 75–85.

Morrison, I.K. 1984. Forestry Abst 45:483–506.

Moyse, D.W., and I.J. Fernandez. 1987. Water Air Soil Pollut 34:385–397.

National Atmospheric Deposition Program. 1985. *NADP annual data summary: Precipitation chemistry in the United States, 1982.* National Atmospheric Deposition Program, Colorado State University, Fort Collins.

National Atmospheric Deposition Program. 1987. *NADP/NTN annual data summary: Precipitation chemistry in the United States, 1986.* Natural Resources Ecology Laboratory, Colorado State University, Fort Collins. 363 pp.

Nihlgard, B. 1985. Ambio 14:2–8.

Nilsson, J., ed. 1986. *Critical loads for nitrogen and sulfur.* The Nordic Council of Ministers, Rep. 11, Copenhagen, Denmark. 232 pp.

Norton, S.A., A. Henrikson, R.F. Wright, and D.F. Brakke. 1987. *In* B. Moldan and T. Paces, eds. *International Workshop on Geochemistry and Monitoring in Representative Basins.* Extended Abst., 148–150. Geological Survey, Prague, Czechoslovakia.

NRC. 1986. *Acid deposition: Long-term trends.* National Research Council, National Academy Press. Washington, D.C. 506 pp.

Overrein, L.N., H.M. Seip, and A. Tollan. 1981. *Acid precipitation: Effects on forests and fish. Final report of SNSF project 1972–1980.* Oslo, Norway. 175 pp.

Pearson, R.W., and F. Adams, eds. 1967. *Soil acidity and liming.* Agronomy 12. American Society of Agronomy, Madison, Wisc. 274 pp.

Peterson, D.L., and R.D. Hammer. 1986. For Sci 32:318–324.

Rasmussen, L., P. Jorgensen, and S. Kruse. 1986. Bull Environ Contam Toxicol 36:563–570.

Rechcigl, J.E., and D.L. Sparks. 1985. Commun Soil Sci Plant Anal 16:653–680.

Reich, P.B., A.W. Schoettle, H.F. Stroo, and R.G. Amundson. 1986. JAPCA 36: 724–726.

Reuss, J.O., and D.W. Johnson. 1985. J Environ Qual 14:26–31.

Riha, S.J., B.R. James, G.P. Senesac, and E. Pallant. 1986. Soil Sci Soc Am J 50:1347–1352.

Ruhling, A., and G. Tyler. 1973. Oikis 24:402–416.

Schier, G.A. 1985. Can J For Res 15:29–33.

Schofield, R.K. 1955. Soils Fert 18:195–198.

Schutt, P., and E.B. Cowling. 1985. Plant Disease 69:548–558.

Seip, H.M. 1980. *In* D. Drablos and A. Tollan, eds. *Ecological impact of acid precipitation*, 358–366. Johs Grefslie Trykkeri A/S, Mysen, Norway.

Shortle, W.C., and K.T. Smith. 1988. Science 240:1017–1018.

Siccama, T.G., W.H. Smith, and D.L. Mader. 1980. Environ Sci Technol 14:54–56.

Skeffington, R.A. 1983. *In* B. Ulrich and J. Paukrath, eds. *Effect of accumulation of air pollutants in forest ecosystems*, 219–231. D. Reidel, Hingham, Mass.

Skiba, U., and M.S. Cresser. 1986. Environ Pollut (Series A) 42:65–78.

Smith, C.T., W.C. Martin, and L.M. Tritton, eds. 1986. *Proceedings of the 1986 symposium on the productivity of northern forests following biomass harvesting*. NE-GTR-115. U.S.D.A. Forest Service, Broomall, Pa. 104 pp.

Swank, W.T. 1984. Water Res Bull 20:313–321.

Swank, W.T., J.W. Fitzgerald, and J.T. Ash. 1984. Science 223:183–184.

Tabatabai, M.A. 1985. Crit Rev Environ Control 15:65–110.

Tamm, C.O., and L. Hallbacken. 1988. Ambio 17:56–61.

Thomas, G.W. 1977. Soil Sci Soc Am J 41:230–237.

Thompson, I.P., I.L. Blackwood, and T.D. Davies. 1987. Environ Pollut 43:143–154.

Truman, R.A., F.R. Humphreys, and P.J. Ryan. 1986. Plant Soil 96:109–123.

Turner, R.S., R.J. Olsen, and C.C. Brandt. 1986. *Areas having soil characteristics that may indicate sensitivity to acidic deposition under alternative forest damage hypotheses*. ORNL Env. Sci. Div. Pub 2720. Oak Ridge, Tenn. 63 pp.

Tyler, G. 1975. *In* T.C. Hutchinson and M. Havas, eds. *Effects of acid precipitation on terrestrial ecosystems*, 252–282. Plenum Press, New York.

Tyler, G. 1987. Flora 179:165–170.

Ulrich, B. 1987. *In Ecological studies*, vol. 61, 11–49. Springer-Verlag, New York.

Ulrich, B., R. Mayer, and P.K. Khanna. 1980. Soil Sci 130:193–199.

Van Miegroet, H., and D.W. Cole. 1984. J Environ Qual 13:586–590.

Wiklander, L. 1975. Geoderma 14:93–105.

Wilding, L.P., and L.R. Drees. 1983. *In* L.P. Wilding, N.E. Smeck, and G.F. Hall, eds. *Pedogenesis and soil taxonomy: I. Concepts and interactions*, 83–116. Elsevier Science Publishing, Amsterdam, The Netherlands.

Wolt, J. 1987. *Effect of acidic deposition on the chemical form and bioavailability of soil aluminum and manganese*. National Council of the Paper Industry for Air and Stream Improvement Tech. Bull. 518. New York. 46 pp.

Zoettl, H.W., and R.F. Huettl. 1986. Water Air Soil Pollut 31:449–462.

# Effects of Acidic Precipitation on Stream Ecosystems

J.W. Elwood
and P.J. Mulholland*

## Abstract

Review of the literature demonstrates that almost all stream communities, including microbial (bacterial, meiofaunal), macroinvertebrate, and fish communities, are adversely affected by acidification. Fungal diversity may be enhanced at low pH; the effect of acidification on fungal activity is unclear. Although many species of epilithic algae common at pH >6 are reduced in abundance with acidification, the biomass and productivity of epilithic algae and byrophytes, however, appear to increase with declining pH in streams that are chronically acidic to a pH <5 for extended periods. The higher biomass and areal productivity of benthic algae and bryophytes at low pH coincide with low grazer density, suggesting that low grazer pressure accounts for the higher biomass of autotrophic communities at low pH. Microbial decomposition of organic matter is also adversely affected by low pH, most likely resulting from reductions in the biomass and activity of bacteria. Benthic macroinvertebrate species richness and diversity and functional group richness show a consistent decline with decreasing pH, particularly at pHs <6.0. Experimental additions of Al to streams show an additive or synergistic toxic effect on macroinvertebrates at low pH. Macroinvertebrate taxa in streams that are most sensitive to low pH are grazing mayflies, molluscs, and some stone fly and caddis fly species. Streams with a pH <5.5 generally have no fish, and reproduction of some fish species is impaired in streams with a pH <6.0. Depressions in pH during spring snowmelt are acutely toxic to some fish taxa; the overall effect of such events on fish populations and communities in streams is unknown. Limited evidence indicates that dissolved P and dissolved organic carbon (DOC) concentrations are reduced in stream sites with a pH >5 that are located downstream of more acidic headwaters. This pattern of P and DOC concentrations is consistent with the pH-dependent adsorption and coprecipitation of these materials with aluminum hydroxide particles. Additional comparative studies of whole ecosystems, coupled with controlled experimental

*Environmental Sciences Division, Oak Ridge National Laboratory, Oak Ridge, TN 37831-6036, USA.

studies, are needed to understand the direct and indirect effects of stream acidification at the population, community, and ecosystem levels.

## I. Introduction

Streams, particularly those of low order in upland areas, are among the surface waters potentially most susceptible to acidification from acidic deposition. This is because the chemical and physical characteristics of stream water are tightly constrained by both precipitation quality and weathering and chemical exchange reactions in the terrestrial environment of the catchment. In areas where the capacity or opportunity for chemical reactions to occur is low as a result of geochemical or hydrologic factors or both, the susceptibility of streams to acidification from acidic deposition will be high. Those streams considered to be most susceptible characteristically have clear water and drain low-order upland watersheds with high precipitation, thin soils, and relatively unreactive soils and bedrock.

A substantial amount of research on the biological and ecological effects of acidification on stream organisms, communities, and ecosystems in Europe and North America has been published since the effects of acidification on stream biota were last reviewed by Burton and others (1982). Further, the earlier review did not address whether the reported "acidification effects" on stream biota and ecology resulted from acidic deposition or from natural internal source(s) of acidity. In this review, we assess the evidence for causal links between acidic deposition and biological effects on stream organisms, communities, and processes. The effects are examined in terms of taxonomic-, population-, community-, and process-level changes that can be attributed, either directly or indirectly, to alterations in the acid-base chemistry of stream water. We also suggest areas of research needed to better understand and predict the biological and ecological impacts of acidic deposition on stream ecosystems.

The review is divided into seven major subject areas dealing with effects on both structural and functional aspects of stream organisms and communities: (1) bacterial and fungal biomass and productivity; (2) the species composition, biomass, and productivity of autotrophic organisms and communities (e.g., periphyton, moss, and macrophytes); (3) heterotrophic decomposition of organic matter; (4) nutrient supplies and nutrient cycling in streams; (5) species composition and abundance of microconsumers or meiofaunal taxa; (6) species and functional group composition and abundance of benthic macroinvertebrate communities; and (7) species composition, abundance, and productivity of fish populations and communities.

With a few exceptions, we have excluded from this review studies dealing with the biological or ecological effects of acid mine drainage because the potential effects on stream biota of other physical and chemical changes in streams receiving acid mine drainage (for example, siltation, iron precipitation) are generally not typical of lotic ecosystems receiving acidic runoff resulting from atmospheric deposition. Results of acid mine drainage studies are included only when the results appear to be germane because (1) such confounding physical and chemical differences appear to be absent, or (2) the physical or chemical differences do not seem to be a factor in the observed responses of stream biota.

Throughout this chapter, pH or $H^+$ activity is used as the primary indicator of the acid-base status of water. Acidic streams are chemically defined as those with a pH <5.6, the equivalence point for carbonate alkalinity. This chemical definition is preferable in making comparisons of the acid-base chemistry of streams because the carbonate ion systems generally account for most of the acid-neutralizing capacity in clear water streams. However, the distinction between acidic (pH ≤5.6) and nonacidic (pH >5.6) streams is less relevant in a biological or ecological context because biological effects of lowered pH can occur before a stream becomes acidic by the bicarbonate definition; that is, a pH of 5.6 may not be the threshold for pH-caused effects on stream organisms and communities.

## A. Types of Evidence of Acidification Effects on Streams

Evidence for biological and ecological effects of acidification on streams falls into three major categories: (1) comparative studies or surveys in which results of population-, community-, and/or process-level measurements are compared over time, within streams with changing acid-base chemistry, or within time among streams or stream reaches with a gradient in their acid-base chemistry, that is, different concentrations of acidic cations ($H^+$, $Al^{n+}$), acidic anions; (2) field experiments or manipulations involving, for example, the acidification of natural streams or the transfer of taxa or portions of communities between streams with a different pH; and (3) laboratory bioassays in which stream taxa or communities are exposed to different concentrations of one or more acidic cations.

By themselves, comparative studies and significant correlations derived therefrom provide only circumstantial evidence of the causal effects of acidification because the streams sampled are not identical in all respects except in the acid-base status of the water. Similarly, field experiments or manipulations involving pH depressions may demonstrate an effect of acidification, but these methods alone do not prove that biological differences among streams with contrasting acid-base chemistry result from acidification. This problem of proving a causal link between water quality and biotic/ecological differences is not unique to the topic of acidification.

# II. Effects of Acidification on Populations and Communities

## A. Heterotrophic Microbes

### 1. Bacteria

Findings of two recent studies have indicated that bacterial communities associated with stone surfaces (epilithic community) and decomposing leaves in streams are substantially affected by acidification resulting from acidic precipitation.

Palumbo and others (1987a, b) reported that adenosine triphosphate (ATP) levels and area-specific rates of production (as measured by tritiated thymidine uptake) of epilithic bacteria in streams of the southern Appalachian Mountains were significantly correlated with pH over a pH range from 4.6 to 6.8 ($p$ <0.05, $r$ = 0.49, $N$ = 23), being lower at the more acidified sites and higher at the sites with higher pH (Figure 4–1). In a study of bacterial abundance and production in streams in the Adirondack Mountains of New York with large longitudinal pH gradients resulting from acidic precipitation (pH range 4.5 to 7.0), Osgood (1987) found that although epilithic ATP and bacterial numbers were correlated with stream pH in only one of the three streams studied, epilithic bacterial production per unit area was substantially lower at the more acidic stream sites. Experimental transplants of rocks from higher (pH 6.14) to lower (pH 5.67) pH sites in one Adirondack stream resulted in reduced bacterial production, and the reciprocal transplant resulted in increased bacterial production (Osgood, 1987). These results are in agreement with the findings of Palumbo and co-workers.

Studies comparing bacterial biomass and production associated with decomposing leaves in streams of differing pH have indicated results very similar to those for

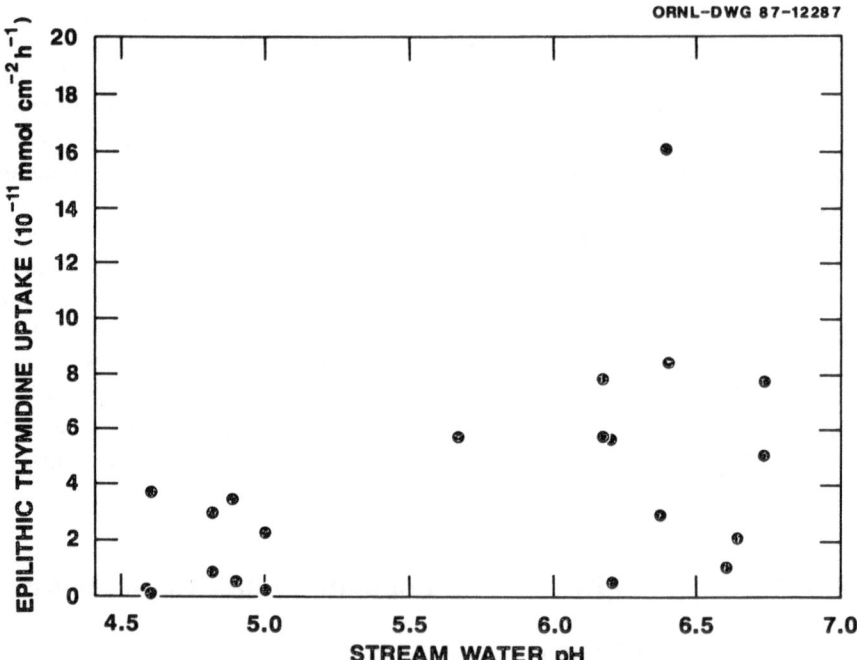

**Figure 4-1.** Epilithic bacterial production as measured by thymidine uptake in southern Appalachian streams ranging in pH from 4.6 to 6.8 (data from Palumbo et al., 1987a, 1987b, and A.V. Palumbo, personal communication, Oak Ridge National Laboratory, Oak Ridge, Tennessee).

the epilithic communities. In the southern Appalachians, microbial biomass (as measured by ATP), respiration rates, and area-specific rates of bacterial production (thymidine uptake) were significantly lower on decomposing leaves in streams with pH <6 than in a stream with pH 6.0 to 6.5 (Palumbo et al., 1988; Mulholland et al., 1987). Somewhat similar results were reported for streams in the Adirondacks (Osgood, 1987). Although microbial ATP levels and respiration rates were not significantly correlated with stream pH, bacterial production rates on decomposing leaves were lowest at the Adirondack site with lowest pH (4.75) and highest at the site with highest pH (6.14). In a study of microbial degradation of leaves in streams in the English Lake district, Chamier (1987) reported that bacterial populations on leaves were greater in an upland stream with pH 6.6 than in an upland stream with pH 4.9.

Bacterial communities in fine-grained stream sediments do not appear to be affected by acidic precipitation. In both the southern Appalachian and Adirondack stream studies, bacterial numbers and/or production in fine-grained sediments were significantly correlated with the organic content of the sediment and not with the pH of the stream water (Palumbo et al., 1987a; Osgood, 1987). These results are consistent with studies of lake sediments, which have indicated little effect of the pH of overlying water on microbial abundance and activity within sediments (Gahnstrom and Fleischer, 1985; Kelly et al., 1984). Alkalinity generation via microbially mediated nitrate and sulfate reduction has been shown to be important in pH buffering of lake sediments (Rudd et al., 1986) and may also maintain pH levels above 6 in fine-grain stream sediments.

Numbers of suspended bacteria in streams, entrained as a result of sloughing from various stream bottom habitats, also do not appear to be influenced by acidic precipitation. In the southern Appalachian and Adirondack streams studied, no correlation between suspended bacteria and pH were found (Palumbo et al., 1987a; Osgood, 1987). Conversely, in a study comparing acidified (pH 4.6 to 5.4) and less acidic (pH 5.6 to 7.1) British streams, Minshall and Minshall (1978) reported fewer suspended bacteria in the acidified stream. It was unclear, however, whether water quality changes associated with acidification or differences in organic matter input from riparian vegetation between the two streams were responsible for the bacteria differences. Also, bacterial numbers, which were computed from plate counts on a yeast-agar-leaf extract medium may have substantially underestimated living cells.

Rimes and Goulder (1986) compared suspended bacterial number and activity in calcareous headwater streams (pH 7.5 to 7.8) and acidic headwater streams (pH 4.6 to 5.3) thought to be acidified as a result of acidic precipitation and poor buffering by the granite and graywacke in the catchments. They found that, whereas the mean concentration of bacteria was marginally greater in the calcareous streams, the metabolic activity of these bacteria (glucose uptake) was greater by an order of magnitude or more, suggesting that acidification affects the metabolic activity more than it does the number of suspended bacteria.

Several mechanisms may be responsible for the decline in epilithic and leaf-associated bacterial activity in acidified streams. These include physiological stress from high concentrations of $H^+$ or dissolved metals, interferences from precipitation of hydrous metal oxides, and indirect effects from reduced cropping of the bacterial community by invertebrates. In acidic environments, bacteria must maintain a cytoplasmic environment against a strong gradient in $H^+$ and must maintain cation uptake with $H^+$ ions occupying a relatively large fraction of the cell surface exchange sites (Krulwich and Guffanti, 1983). High concentrations of dissolved metals, particularly Al, may also inhibit bacterial metabolism in acidified streams. In an experiment involving the introduction of acid to stream-side troughs in eastern Tennessee (reducing stream water pH from 6.5 to 4.2), the addition of 0.25 μg/L of Al resulted in greater reductions in bacterial production rate (thymidine incorporation) than did acid addition alone (A. V. Palumbo, personal communication). The addition of Al to the acidified trough did not exceed the solubility of $Al(OH)_3$; consequently, the effect was most likely the result of physiological interference from the dissolved Al.

In streams receiving highly acidic runoff with high Al concentrations, within-stream increases in pH due to dilution or acid-consuming reactions may result in precipitation of hydrous aluminum oxide coatings on streambed materials that could interfere with microbial attachment or cover attached microbial cells. Mulholland and others (1987) found much greater Al levels on decomposing leaves in acidic streams in the southern Appalachians exhibiting low levels of microbial ATP and low rates of respiration and production. In a study involving transplants of leaf material from a stream with pH of 6.4 to one with pH 4.9, Palumbo and others (1987b, 1988) found significantly reduced bacterial production associated with the leaves after 1 week. However, in the reciprocal transplant (low-pH stream to higher-pH stream), an increase in bacterial production was delayed until 4 weeks later, indicating a persistent negative effect, perhaps the result of the higher leaf surface Al that also persisted when the leaves were transplanted to the higher pH stream.

Indirect effects of acidification in acidified streams, such as the reduction in abundance of microinvertebrate grazers (see section II, F), may have an adverse effect on the bacterial community. Grazing of microbial communities by both microinvertebrates and macroinvertebrates may stimulate microbial activity in aquatic ecosystems (Fenchel and Harrison, 1975; Hanlon and Anderson, 1979; Morrison and White, 1980). In an experimental enclosure study conducted in an eastern Tennessee stream, epilithic bacterial production (thymidine uptake) was greater in the presence of grazing mayflies than in their absence (A.V. Palumbo and P.J. Mulholland, unpublished data), suggesting that the lower bacterial production in the acidic stream may result from lower rates of grazing on the epilithic microbes.

Different functional components of the bacterial community may be affected to differing degrees by acidification. For example, nitrification was shown to cease at pH <5 (Bick and Drews, 1973). However, few studies have addressed the effect of stream acidification on different microbially mediated processes, with the exception of detritus decomposition as discussed in section III,B.

## 2. Fungi

Stream fungi, predominantly the aquatic hyphomycetes, are important components of the microbial community decomposing leaves in streams (Kaushik and Hynes, 1971; Barlocher and Kendrick, 1974; Suberkropp and Klug, 1976). Evidence that stream fungi are either positively or negatively affected by increased acidity seems inconclusive. Comparing the fungal spores found in the water of two soft water streams (pH averages of 6.5 and 7.3) in the German Black Forest and two hard water streams (pH averages of 8.4) in the Swiss Jura, Barlocher and Rosset (1981) reported a significantly greater number of fungal species in the soft water streams. In subsequent studies of streams in western Europe (Wood-Eggenschwiler and Barlocher, 1983) and eastern Canada (Barlocher, 1987), fungal species richness was found to be highest in streams with pH 5.4 to about 7 and to decline rapidly with pH at pH values >7 (no streams with pH <5.4 were studied). Barlocher (1980) and Barlocher and Rosset (1981) speculate that the higher fungal species richness found in soft water streams with low pH could be the result of fewer leaf-eating invertebrates, which in effect compete with fungi for leaf substrate, or could be the result of more favorable water chemistry in soft water streams (including lower pH).

In contrast, studies of the fungal communities of streams in Great Britain by Iqbal and Webster (1973, 1977) indicated that hard water streams with pH values ranging from 6 to 9 carried twice the spore concentration and number of species of soft water, moorland streams (pH 4.1 to 4.5). In another British study, Shearer and Webster (1985) reported that the upstream portion of the River Teign, which originates in moorland overlying granite and is slightly acidic (pH 5.4 to 6.0), had fewer fungal species and fewer spores per liter of water and per unit area of leaf detritus compared to a downstream lowland section with higher pH (7.0 to 7.2). However, the differences in fungal abundance between hard water and moorland streams may be the result of differences in riparian vegetation, specifically the absence of deciduous trees and consequently lower amounts of leaf litter in the latter streams.

Chamier (1987) measured fungal abundance directly on leaf detritus in a study of microbial degradation of leaves in streams in the English Lake district ranging in pH from 4.9 to 6.8. Using an induced sporulation technique (relating the conidia produced to the extent of fungal colonization of the leaves), Chamier found that the density of aquatic hyphomycetes on leaves was dependent on type of riparian vegetation, landscape position (lowland versus upland), and pH. Larger fungal populations were found in lowland streams as compared to upland streams of the same pH and vegetation type, and larger populations in upland streams of higher pH (6.6 to 6.8) compared to upland streams of low pH (4.9 to 5.5) with similar riparian vegetation.

Hall and others (1980) found reductions in aquatic hyphomycete populations in an experimentally acidified stream (to a pH of 4) in the Hubbard Brook Experimental Forest, New Hampshire. Hall and co-workers reported that the aquatic hyphomycete community colonizing leaves in the acidified reach was less diverse, and there were smaller numbers of spores than leaves collected in the control reach. They also reported an apparent shift in the fungal community and noted the presence of a basidiomycete growing on moss and tree roots in the acidified stream reach that was absent in the control reach.

Laboratory studies of the effect of various water chemistry parameters, including pH, on growth and enzyme activity of fungi isolated from streams are also in conflict with regard to optimum pH. It is believed that many of the enzymes produced by aquatic hyphomycetes that degrade the structural polysaccharides of leaf cell walls typically have acidic pH optima (Chamier, 1985). Experiments conducted on ten species of aquatic hyphomycetes isolated from a hard water Swiss stream and a soft water German stream indicated that all grew best on artificial media at pH values between 4 and 5, although increased levels of calcium also increased fungal growth (Rosset and Barlocher, 1985). In contrast, a study of enzyme activity and growth of seven species of aquatic hyphomycetes isolated from a circumneutral river in Great Britain indicated that only two could produce pectinases and grow on solid pectic medium at pH 5, whereas all could at pH 7 (Chamier and Dixon, 1982). Evidence from an experiment conducted with one of these species indicated that its metabolism on pectic substrates at pH 7 is dependent on the $Ca^{2+}$ concentration in the water (Chamier and Dixon, 1982). These studies do agree on the positive role of $Ca^{2+}$ on fungal growth and enzyme activity. The apparent conflict with regard to pH optimum may indicate that fungal growth is influenced by a set of biotic and abiotic factors that are altered in different ways by acidification.

From the studies conducted to date, it is difficult to generalize about the effect of acid precipitation on fungal communities in streams. Certainly there would be changes in the species composition of the fungal community with acidification. However, additional studies are needed to determine the effects of acidification on fungal biomass and enzyme activity in soft water streams and to define the mechanisms responsible for these effects. In particular, studies are needed to compare the direct response of fungi to water chemistry changes accompanying acidification and indirect responses to changes in the leaf-eating invertebrate community.

## B. Periphyton

### 1. Chlorophyll and Biomass

Heavy growths of filamentous algae in acidified Norwegian streams were among the earliest reported observations of aquatic effects of acid precipitation (Hendrey et al., 1976; Leivestad et al., 1976; Overrein et al., 1980). Since then, several experimental acidification studies have documented increased stream periphyton biomass (measured either as chlorophyll $a$ or dry mass per unit area) when stream pH was reduced to < 5 (Hendrey, 1976; Hall et al., 1980; Allard and Moreau, 1985; Parent et al., 1986; Planas and Moreau, 1986).

In a study of four soft water streams in eastern Tennessee with pH ranging from 4.5 to 6.5, Mulholland and others (1986) reported that epilithic chlorophyll *a* levels were significantly higher in two streams with pH <5 than in two streams with pH >5.5. A somewhat more extensive survey of nine eastern Tennessee streams conducted during October, when stream flow was more stable, further indicated the inverse relationship between stream pH and periphyton chlorophyll *a* (Figure 4–2). Although it is likely that much of the acidity in the low-pH eastern Tennessee streams is the result of internal sources (oxidation of bedrock sulfides, high rates of nitrification) rather than acidic deposition and poor soil buffering, streamwater chemistry is quite similar to that of streams in the northeastern USA that are acidified primarily by acidic deposition. The ecological differences among the eastern Tennessee streams can be used to infer ecological responses of streams to acidification resulting from atmospheric deposition.

In a study of periphyton biomass and productivity conducted by Arnold and others (1981) in Pennsylvania streams with mean pHs of 5.2 (range 4.5 to 5.9) and

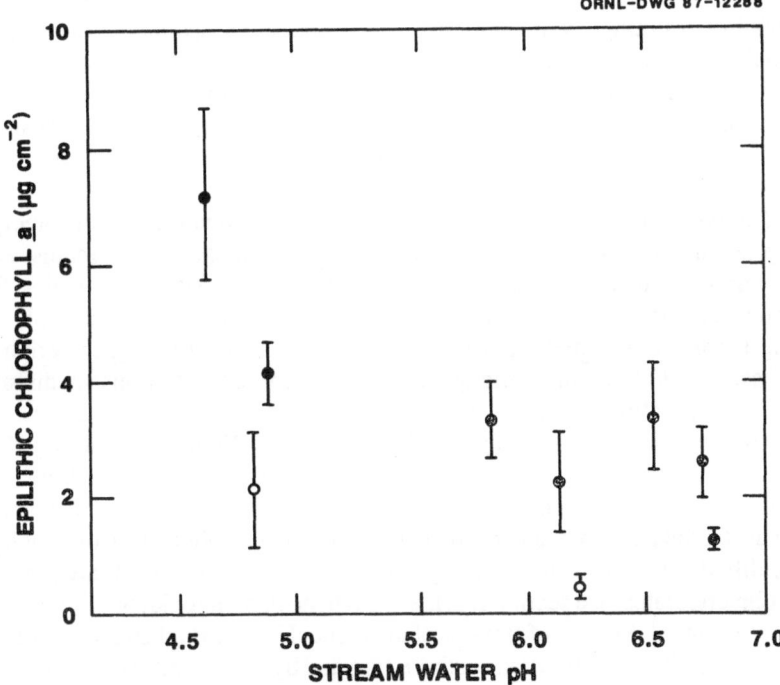

**Figure 4-2.** Epilithic chlorophyll *a* levels measured during October 1984 in southern Appalachian streams ranging in pH from 4.6 to 6.8 (mean and one standard error indicated, $n = 5$). The open points denote samples from upstream (pH 4.85) and downstream (pH 6.20) sites on a stream (Hemlock Creek) with substantial amorphous $Al_2O_3$ coatings on streambed materials (as determined from x-ray defraction). The headwaters of this stream are highly acidic (pH 4.5) due to highway excavation that exposed a pyritic phyllite.

6.8, differences in stream pH were attributed to acid precipitation and differential buffering in the study watersheds. In apparent contrast with findings of the above studies, Arnold and co-workers reported no significant differences in chlorophyll *a* on acetate slides submerged in the streams for 4-week periods. However, they did report significantly greater ash-free dry mass on slides in the acidic stream on all seven of the sampling dates and significantly greater primary productivity per unit area on three of the seven dates. It is unclear why the chlorophyll *a* levels are not also higher in the more acidic stream, although the lack of differences could have resulted from the relatively short duration of exposure of the slides.

In a study of mountain streams in the English Lake district with pH values ranging from 5 to >7, Sutcliffe and Carrick (1973) noted that streams with pH values <5.7 appeared to have lower periphytic biomass compared to streams with pH >6.0; however, this conclusion was based on observations only, and chlorophyll or biomass measurements were not reported.

In an experimental acidification study conducted in artificial, recirculating streams, Maurice and others (1987) also reported lower periphyton biomass in the acidified channel (pH 4) compared to the control channel (pH 7.5). They speculated that the reduced periphyton growth was due to reduced nutrient availability resulting from increased P adsorption by sediments and reduced microbial nutrient regeneration in the acidified channel. However, this study probably does not provide a good analog for an acidic precipitation effect in natural streams because of the lack of nutrient inputs in the experimental channels. Acidified natural streams continue to receive nutrient inputs from the surrounding watershed.

Collectively, these studies indicate that streams acidified from acidic precipitation generally have greater periphyton biomass than nonacidified streams. However, this generalization is largely confined to streams with pH values ≤5 for extended periods of time (several months). Streams that exhibit low pH only during storm or snowmelt episodes do not necessarily have higher periphyton biomass. In addition, in acidic streams with high concentrations of dissolved metals, precipitation of hydrous metal oxides can coat rock surfaces and result in reduced algal biomass (McKnight and Feder, 1984; Winterbourn et al., 1985). However, this situation is not typical of streams acidified as a result of acidic precipitation and represents a situation more similar to acid mine drainage impact.

A substantial body of evidence indicates that increased abundance of periphyton in acidified streams results largely from reduced densities of algal-grazing macroinvertebrates (see section II,E), although much of the evidence is correlative or circumstantial. Studies of streams of different pH in the English Lake district by Sutcliffe and Carrick (1973), in southern England by Townsend and others (1983), and in south Sweden by Otto and Svensson (1983) have indicated that grazing invertebrates, particularly the Ephemeroptera, are reduced in taxa richness and total number in acidic streams (pH ≤5.6) compared to streams of higher pH (see section II,E). Lamberti and Resh (1983) and Mulholland and others (1983) have demonstrated experimentally that reduction in density of grazing invertebrates in natural and laboratory streams results in markedly increased periphyton biomass.

The investigations of Hendrey (1976), Arnold and others (1981), and Mulholland and others (1986) have indicated that numbers of grazing invertebrates were lower in acidic streams with relatively high periphyton biomass compared to higher-pH streams with lower periphyton biomass. In studies involving experimental acid additions to natural and experimental streams, Hall and others (1980) and Allard and Moreau (1985) observed that increased periphyton abundance coincided with reductions in grazing invertebrates.

## 2. Species Composition

Periphyton community composition is also altered in streams acidified by acidic precipitation, with measurable changes occurring in streams with pH values less than ~5.5 for extended periods. In one of the earliest experimental stream acidification studies, Patrick and others (1968) reported shifts in diatom species composition with pH reduction from near neutrality to 5.2 to 5.4 and heavy mortality of the entire diatom community with further reduction in pH to 4.0 to 4.5. In studies of acidic Norwegian streams (pH <5) with heavy accumulations of algal biomass, two algal taxa, the diatom *Tabellaria flocculosa* and the green alga *Mougeotia* sp., have been observed to dominate the periphyton community, together often comprising over 70% of total cell numbers (Leivestad et al., 1976; Hendrey, 1976). Although these authors do not indicate how important these two species are in higher-pH streams, apparently they are found less frequently than in acidic streams. Leivestad and others (1976) also report a change in the relative abundance of different diatoms in seven Norwegian streams known to be affected by acidic precipitation. They indicate that certain diatom species known to tolerate low pH increased in relative abundance from about 3% of the diatom community in 1949 to about 12% in 1975. In their study of acidified and circumneutral Pennsylvania streams, Arnold and others (1981) reported the occurrence of some species of diatoms (particularly *Achnanthes minutissima*) only in the higher-pH stream and others (particularly *Eunotia exigua* and *E. curvata*) only in the acidic stream. They also reported more taxa of green algae in the acidic stream. Ziemann (1975) also noted the presence of a distinct acidophilic diatom flora in a stream with pH <6 compared to one with pH of ~6.5 in Germany.

Studies conducted in southern Norway (Leivestad et al., 1976), Quebec (Duthie and Hamilton, 1982), and eastern Tennessee (Mulholland et al., 1986) have shown that diatoms and green algae comprise a much greater proportion of the periphyton community in acidic streams than they do in streams with pH >5.6. In the eastern Tennessee study, blue-green algae were very abundant in streams with pH >5.6 but were largely absent in streams with lower pH. In both the Quebec and eastern Tennessee studies, red algae (*Batrachospermum moniliforme* and *Rhodochorton* sp., respectively) were abundant in the most acidic streams but absent in streams with pH >5.

The observed changes in periphyton community composition in streams acidified from acidic precipitation may be a direct result of changes in water chemistry, particularly in $H^+$, dissolved Al, inorganic C, and nutrient concentrations, or may be the result of changes in the invertebrate grazer community. The decline in importance of blue-green algae as components of the periphyton community at low pH is probably a result of physiological limitations at high concentrations of $H^+$. Brock (1973) has documented the absence of blue-green algae in environments with pH <4 to 5 and suggests that the lack of blue-greens (cyanobacteria) at low pH may be the result of the absence of internal cell structures such as chloroplasts to protect chlorophyll from acidic degradation. (See section II,A for effects on other procaryotes.)

Gensemer and Kilham (1984) present data that indicate differential tolerance of five freshwater algae to increased concentrations of metals at pH values <6. Tolerance of different species of algae to high concentrations of Al may be inversely related to their Ca requirements (Foy and Gerloff, 1972). The toxicity of Al to a green alga was shown to be highest at pH values of 5.8 to 6.2 and lower at lower and higher pHs, indicating that the $Al(OH)_2^+$ species may be most toxic to this alga (Helliwell et al., 1983). Thus, changes in periphyton community composition with acidification may also result from differential tolerance of increased concentrations of metals or changes in metal speciation with pH (Campbell and Stokes, 1985).

Changes in the availability of phosphorus with acidification may also influence the species composition of the periphyton community. Planas and Moreau (1986) have presented evidence of increased phosphorus availability in acidified artificial stream channels (pH 4 to 4.5), presumably from resolubilization of sediment P at low pH. However, coprecipitation of $PO_4$ with Al and increased P limitation of the epilithic community (as determined by phosphatase enzyme activity) have been observed in natural streams as pH increases to values >5 downstream from more acidic headwater reaches (Mulholland et al., 1986). Strong longitudinal gradients in P availability resulting from headwater acidification could result in substantial shifts in species composition, depending on the P uptake kinetics characteristic of different species. However, we know of no studies that have explicitly addressed this issue. The multitude of interacting factors (pH, nutrient availability, metal concentrations, and biotic factors) makes such investigations difficult.

Decreased concentrations of total dissolved inorganic carbon and the shift in speciation to free $CO_2$ with acidification may favor algal species that have efficient inorganic carbon uptake kinetics (relatively high half-saturation coefficients for inorganic carbon uptake) (Moss, 1973; Goldman et al., 1974) or that use free $CO_2$ only. For example, Goldman and others (1974) demonstrated in laboratory chemostat experiments that *Scenedesmus quadricauda* outcompeted *S. capricornutum* at low values of inorganic carbon concentration due to the lower half-saturation constant for carbon uptake of the former. However, most studies of carbon uptake limitation have focused on the effect of reduced concentrations of dissolved free $CO_2$ in lakes at high pH values. Although total dissolved inorganic carbon is low in acidic streams, free $CO_2$ is generally not reduced because it is controlled by soil respiration rates and exchange with the atmosphere. Comparisons of the air-equilibrated and closed sample pH of water collected from streams in the mid-Atlantic and southeastern United States for the National Stream Survey showed that most streams are supersaturated with respect to atmospheric $CO_2$ as evidenced by the increase in pH when the samples were equilibrated with 300 ppm of $CO_2$ (Kaufmann et al., 1988). Obligate free $CO_2$ users may be favored at low pH because, unlike species that are able to use $HCO_3^-$, they do not experience a reduction in inorganic C supply at low pH. The increased importance of *Batrachospermum moniliforme* in the periphyton community of a stream with a pH of 4.5 compared to streams with pH >5.6 noted by Mulholland and others (1986) may be at least partially because this species is known to be an obligate $CO_2$ user (Raven and Beardall, 1981).

Differences in the intensity of grazing by macroinvertebrates may also result in differences in periphyton community composition between acidified and nonacidified streams. Patrick (1977) has shown that diatoms are frequently grazed preferentially by macroinvertebrates in mixed algal communities in streams. Mayfly larvae have been shown to be effective diatom grazers, and diatom abundance has been shown to be inversely related to grazing by mayflies (Lamberti et al., 1987; Steinman et al., 1987). Reduced populations of macroinvertebrate grazers in acidified streams may, therefore, explain the greater relative abundance of diatoms in the periphyton communities of acidified streams.

## C. Bryophytes

Very few studies have been conducted to examine the response of stream bryophytes to acidification, despite the increased abundance of benthic bryophyte mats, particularly of *Sphagnum*, that is commonly reported as a response to lake acidification (Grahn et al., 1974; Hendrey and Vertucci, 1980; Havas and Hutchinson, 1983; Roelofs, 1983). Hall and others (1980) observed increases in moss biomass in their experimental acidification of a natural stream to pH 4. H.L. Boston (Environmental Sciences Division, Oak Ridge National Laboratory, personal communication) studied bryophytes in several soft water streams in eastern Tennessee with pHs ranging from 4.5 to 6.4 and found significantly higher biomass and areal productivity and slightly higher chlorophyll-specific productivity of the dominant liverwort (*Scapania undulata*) at the most acidic sites (pH <5). Boston attributes the higher chlorophyll-specific production rate to higher concentrations of free $CO_2$ and higher tissue P concentrations at the acidic sites. However, the higher biomass and areal productivity at the more acidic sites appeared to be related to reduced loss rates of bryophyte stems at these sites. Individual stems were longer and more highly branched in the more acidic streams, indicating reduced rates of biomass loss. Although living bryophytes are usually not consumed in streams, there have been some reports of consumption by caddis flies, stone flies, and mayflies (Jones, 1949; Chapman and Demory, 1963). Reductions in densities of these invertebrates in acidified streams could explain the greater biomass of moss in acidified streams. Alternatively, the reduction in microbial activity that has been observed at acidified sites (see section II,A) may be responsible for lower biomass loss rates, allowing increased biomass and areal productivity at these sites. This hypothesis is supported by the studies of Satake and Miyasaka (1984) that indicated the presence of bacteria and numerous holes in the cell wall of an aquatic liverwort and suggested that bacteria may play an important role in the decomposition of living moss.

## D. Benthic Meiofauna

The role of meiofauna in streams has not been investigated, but based on studies in lakes and marine systems it seems likely that this group of organisms plays an important role in nutrient cycling and decomposition and provides a food source for large consumer organisms. We are not aware of any comparative surveys that have addressed the effects of acidification on the so-called meiofauna or micro-consumer taxa and communities (e.g., protozoa, zooflagellates, rotifers, gastro-trichs, and nematodes). Laboratory studies using ciliated protozoans show that the species richness of at least this group of meiofauna declines with decreasing pH, particularly when the pH falls below 5.0 (Bick and Drews, 1973). Findings of this study indicate that several species of ciliated protozoans have pH tolerance limits substantially <5.0 and presumably could tolerate relatively acidic conditions without any direct adverse affects from pH alone. These experiments, however, did not include additions of Al or other metals that have been shown to increase in acidic surface waters. These ciliated protozoans may be less tolerant to low pH in the presence of toxic metals because of the potential additive or synergistic toxic effect of some metals at low pH. Thus, it is likely that the species richness of this group of protozoans will decline with decreasing pH at levels <5.0.

In another laboratory study, Faucon and Humman (1976) examined the effect of pH on a common benthic gastrotrich from a stream. They reported that the gastrotrich, *Lepidodermella* sp., was almost totally eliminated below a pH of 6.0 under laboratory conditions. The pH gradient used in the bioassay, however, was achieved by mixing water from an unpolluted stream with water from an acidic stream receiving acidic mine drainage. Because of the potential confounding effects of high concentrations of toxic metals typically associated with acidic mine drainage, it is unclear whether these results can be extrapolated to streams acidified by acidic deposition.

Effects of acidic conditions on other microconsumer taxa, such as zooflagel-lates, rotifers, and some nematodes, are essentially unknown. Further, there appear to have been no studies of the indirect effects of acidification on streams meiofauna that could result from, for example, alterations in the food supply of bactivorous and carnivorous meiofaunal taxa. The effect of acidification on the microconsumer community in streams is clearly an area where additional research is needed.

## E. Benthic Macroinvertebrates

Benthic macroinvertebrate taxa and communities in streams have been widely used as pollution indicators and are, therefore, potentially useful biomonitors of water quality changes due to acidification. In addition, these organisms are an important source of food for many stream fishes and can play a significant role in the physical breakdown of organic matter (inter alia Cummins et al., 1973; Wallace et al., 1982; Mulholland et al., 1985) and nutrient regeneration in streams (Mulholland et al., 1983; Merritt et al., 1984). Hence, acidification-caused changes in the benthic macroinvertebrate populations and communities may not only influence nutrient and organic carbon dynamics in streams but may also alter the available food supply of stream fishes.

## 1. Comparative Surveys of Benthic Macroinvertebrates

Surveys of benthic macroinvertebrate populations, communities, and taxa among streams with gradients in pH have been conducted in several areas of Europe and North America. Common findings that emerge from these surveys are that species richness, species diversity, functional group diversity, and, in some cases, the density and biomass of benthic invertebrates decrease with declining pH of stream water. Some of the apparent acidification effects appear to arise directly from toxic conditions to certain taxa, whereas others may result from either indirect effects of acidification associated with changes in, for example, food supply, competition, or predation, or from other factors that may or may not be directly related to acidification.

Although the objective was not to examine the effects of acidification on stream communities, Jones (1948) was one of the first to document a possible effect of acidification on benthic macroinvertebrate communities based on a survey of four streams in the Black Mountain district of South Wales, United Kingdom. Three of the streams were poorly buffered with the pH of stream water ranging from 6.5 to <5, depending on discharge, whereas the other stream was calcareous with a pH seldom <7, except during full flood. The most acidic streams drained areas with organic-rich (peat) soils, which are a source of organic acids that can affect stream pH. Sulfate concentrations during base flow were also relatively high, exceeding 150 μeq/L, suggesting that atmospheric deposition of sulfuric acid was also an important source of acidity in these streams.

The species richness of macroinvertebrates was highest in the alkaline stream (pH 6.8 to 7.8) and lowest in two streams that were acidic (pH 2.4 to 4.4) during storms. These latter two streams had either species of mayfly or none at all, lacked Crustacea and Mollusca, and had impoverished stone fly and caddis fly faunas. The most acidic stream, however, did not have the lowest taxa richness and, except for the calcareous stream, had the most species of semiaquatic insects (Hemiptera). The most acidic stream was also the only one of the four surveyed that lacked fish, suggesting that the distribution of some macroinvertebrate species among the streams may have been indirectly affected by fish predation. Although there are other possible causes for the observed differences in benthic invertebrate taxa between the acidic and nonacidic streams, such as differences in discharge, substrate and food quality and quantity, and temperature, the observed differences in the macroinvertebrate community among streams are qualitatively consistent with those observed in later comparative studies of acidic and circumneutral streams, suggesting that acidification was an important causal factor.

A similar comparison involved a series of studies on the River Duddon drainage in England (Minshall and Kuehne, 1969; Sutcliffe and Carrick, 1973; Minshall and Minshall, 1978; Willoughby and Mappin, 1988). In the first study, Minshall and Kuehne (1969) reported that the upper and lower portions of the basin differed in the distribution of mayflies, certain caddis flies, the river limpet (*Ancylus* sp.), and an amphipod (*Gammarus pulex*). These taxa were rare or absent in streams in the upper basin where the stream water had a lower conductivity and pH. Minshall and Kuehne (1969) concluded that nutrients or other water quality factors were responsible for the observed discontinuities in the macroinvertebrate taxa, but that temperature may have also played a role.

Whether the source of acidity in the upper basin streams in the River Duddon drainage was from acidic deposition or from internal sources is unclear. Chloride was the dominant anion and Na was the dominant cation (Minshall and Kuehne, 1969), suggesting a sea salt effect on stream water chemistry. If so, the relatively high sulfate levels, which generally exceeded 100 $\mu$eq/L, may have been caused in part by sea salt deposition.

In a second survey of macroinvertebrates in the River Duddon drainage, Sutcliffe and Carrick (1973) sampled 26 sites, many of which were near to, or identical with, those sampled by Minshall and Kuehne (1969). They also observed the same discontinuities in the invertebrate fauna described by Minshall and Kuehne between the upper basin, where stream pH was generally <5.7, and the lower basin, where pH was generally >5.7 throughout most of the year. Sutcliffe and Carrick concluded that pH was the overriding factor limiting the qualitative distribution of benthic macroinvertebrate taxa in the River Duddon drainage. Further, because the affected taxa are generally herbivores, they suggested that the limiting effect of pH operated indirectly through changes in the food supply.

To investigate whether water quality, food quality, or the two acting together determined the distribution of invertebrate taxa in the River Duddon drainage, Minshall and Minshall (1978) conducted a series of field bioassays. In situ, bioassays were conducted, involving transplanting selected invertebrate taxa contained in trays from a lower basin site with higher pH to an upper basin site with lower pH. The findings in the bioassay showed a greater net loss of invertebrates from the chambers transplanted to the upper basin site than among those at the reference or control site. Whether this loss is a result of greater mortality caused by the lower pH, greater emigration of invertebrates from the trays (which had holes through which macroinvertebrates could escape) at the upper basin site, or some combination of both is unknown. The tendency of some invertebrate taxa to avoid areas of low pH through increased drift is well documented (e.g., Hall et al., 1980; Zischke et al., 1983; Ormerod et al., 1987) and could have accounted for the observed difference in loss of invertebrates from the chambers.

To isolate the effects on survivorship of food quality and water quality, Minshall and Minshall (1978) also conducted separate laboratory bioassays using two invertebrate taxa (the mayfly, *Baetis rhodani*, and the amphipod, *Gammarus pulex*), which are present in the lower basin but absent from the upper basin. The two species, both with and without food, were exposed to water from an upper basin site with low pH. Water from a lower basin site was used for the control. The mortality rate of both species was greater in water from the upper basin, although there was also significant mortality after 4 to 5 weeks in the control. The findings indicated that the presence of food had no effect on survivorship of either species in water from the upper basin but did enhance the survival of the control or reference organisms. However, because detritus was the only source of food offered in the feeding experiments, these results do not provide conclusive evidence that food had no effect on the survival and abundance of herbivorous taxa, such as *Baetis*, which graze on epilithic algae.

Although Minshall and Minshall (1978) concluded that chemical factors are responsible for the absence of mayflies, *Gammarus*, and other taxa from the upper basin sites of the River Duddon, they discounted pH as being the primary factor. Given that pH affects the osmoregulatory ability of aquatic invertebrates (Havas, 1981), it is possible that some species absent from the upper basin in the River Duddon drainage were near the limit of their tolerance with respect to the concentration of some salts. If so, then episodic increases in acidic cations (e.g., $H^+$, $Al^{n+}$) may have caused osmoregulatory stress, resulting in the elimination of these more acid-sensitive species from upper basin sites that have a lower buffering capacity.

In a more recent study, Willoughby and Mappin (1988) also conducted comparative bioassay and feeding experiments to assess whether the absence of some species of mayflies in the acidic streams in the River Duddon drainage was due directly to the water chemistry or to the absence of suitable food. They demonstrated that *Ephemerella ignita* was relatively tolerant of the low-ionic-strength, low-pH waters in the upper portions of the drainage. Based on an analysis of food resources in the streams and the results of feeding and growth experiments, they concluded that the absence of this species from the more acidic streams was most likely due to its inaccessibility to suitable food rather than to a direct toxic effect of water quality. In contrast, results for other mayflies, such as *Baetis muticus* and *B. rhodani*, indicated that these taxa were excluded from the acidic streams because of a direct toxic effect of water chemistry.

Ziemann (1975) reported the results of a more limited survey of benthic macroinvertebrates at acidic and nonacidic sites in two streams in the Thuringian Forest in Germany. The pH of stream water ranged from 4.5 at the most acidic site to 6.5 at one of the nonacidic sites. There was no indication as to the source of the acidity. Species richness of three aquatic insect orders (Ephemeroptera, Trichoptera, and Plecoptera), particularly the mayflies and stone flies, decreased with declining pH. Mayflies were completely absent from the most acidic site, whereas at sites with a pH <6.0 where mayflies were present, *Baetis* was the only genus found and the stone flies consisted almost entirely of two genera, *Leuctra* and *Nemurella*. These apparent discontinuities in the distribution of genera among stream sites do not prove that acidification was the cause of the differences in species richness. The results are generally consistent with an acidification effect based on the evidence that many of the taxa absent from the lower-pH streams are known to be acid sensitive. Both *Baetis* and *Leuctra* are known to contain acid-tolerant species (Engblom and Lingdell, 1984).

Matthias (1983) reported similar results from a 2-year comparative survey of five small streams in the Nieste drainage located near Kassel, FRG. He found a decline in species richness of macroinvertebrates with declining pH. Mean pH of the streams ranged from 4.08 to 7.03, whereas the minimum pH that occurred during storms and snowmelt ranged from 3.7 to 6.4. The author speculated that the primary cause of acidification of the streams was most likely acidic deposition of strong mineral acids. The decline in species richness occurred in nearly all taxonomic groups of macroinvertebrates, including mayflies, stone flies, caddis flies, Crustacea, Coleoptera, Hydracarina, and Mollusca. Water from the acidic streams was found to acutely toxic to *Baetis* spp. larvae and to *Gammarus fossarum* collected from a circumneutral stream, indicating that the absence of these taxa from the acidic streams was due to a direct toxic effect of water quality.

In a survey of 12 streams in Scotland that drain both forested and nonforested watersheds, Harriman and Morrison (1982) reported comparable densities and biomass of benthic macroinvertebrates among the streams, which varied in pH, ranging from <4.4 to 5.8. Streams draining the forested watersheds had higher sulfate concentrations than the nonforested streams, perhaps caused by greater deposition of acidic aerosols scavenged from the atmosphere by the forest canopy. The lack of difference in benthic macroinvertebrate densities and biomass between the acidic and circumneutral streams suggests that the density and biomass of macroinvertebrates is not adversely affected by acidification. However, these investigators used a kick net to sample invertebrates, a method that is not quantitative. Hence, the absence of significant differences in density and biomass of macroinvertebrates among streams may arise from sampling variation resulting from the method used. Nonetheless, the more acidic forested streams had lower species richness and contained no mayflies, a pattern similar to that found in other comparative studies of acidic and circumneutral streams.

In a similar survey of 13 streams in West Wales, UK, with a mean pH ranging from 4.7 to 6.8, Stoner and others (1984) reported lower species richness in the more acidic streams, with the fauna being dominated by stone flies and Diptera. Mayflies were completely absent from the most acidic streams, and the taxa richness of caddis flies and Coleoptera, particularly the Elmidae, declined with decreasing pH.

The streams sampled by Stoner et al. (1984) had relatively high sulfate concentrations, ranging from 77 to 155 $\mu$eq/L. These high levels could not be accounted for by bulk deposition after correcting for the concentrating effect of evapotranspiration. Thus, natural sources of sulfate acidity from sulfide oxidation cannot be ruled out as a major source of acidity in these watersheds. Organic acids from natural sources may also have contributed to the low pH of some of the streams as suggested by the high UV absorbance of some stream water samples from areas draining peaty soils.

In a more thorough analysis of both the taxonomic and functional (i.e., feeding group) composition of benthic invertebrate communities in streams that differ in pH, 34 riffle sites in the Ashdown Forest of southern England in tributaries of the Medway and Sussex Ouse Rivers were sampled (Townsend et al., 1983; Townsend and Hildrew, 1984; Hildrew et al., 1984a; Hildrew et al., 1984b). The streams have relatively soft water and longitudinal gradients in pH, with the upstream reaches tending to be acidic (pH <5.6). The mean annual pH of all sites sampled ranged from 4.8 to 6.9. Townsend and others (1983) mention that the watersheds contain ferrous carbonate deposits and extensive areas of *Sphagnum*, the latter being a source of organic acids. Thus, natural sources of acidity from Fe oxidation and organic acids may account for the low pH of the streams surveyed.

The structure of the macroinvertebrate communities in these streams was strongly related to pH, with species and functional group richness, diversity, density, and equitability all declining with decreasing pH. The abundance of grazers and scrapers decreased with declining pH and were completely absent from the most acidic sites, a pattern consistent with that found in other comparative surveys to be discussed in later sections. Isolating the effects of pH on benthic invertebrate communities from other factors in this study, however, was confounded by the fact that the most acidic sites are small headwater streams whereas the nonacidic sites are higher-order streams. Greater accessibility of taxa to downstream sites and differences in available food supply among sites could, therefore, be contributing causes of the observed differences in functional and taxonomic richness and diversity among sites with contrasting pH.

Unfortunately, evidence of the role of pH and food limitation in structuring the benthic invertebrate community in these streams is only circumstantial. For example, grazer density was directly correlated with pH and inversely correlated with tree cover in the stream channel. Because tree cover may affect the available light for autotrophic production, reductions in the density of grazers with declining pH may have been due to light limitation, which reduced the available supply of epilithic algae on which grazers feed. Differences in grazer density among sites could also have been due to a combination of food limitation and toxic level of acidic cations. Unfortunately, no information was presented on either the toxicity to representative grazer taxa of water from the different streams or the abundance of epilithic algae available as food to grazers in these streams.

Evidence of the effects of acidification on other functional groups was also reported in this same study of streams in the Ashdown Forest (Hildrew et al., 1984b; Townsend and Hildrew, 1984). The densities of collectors and filter feeders, for example, were positively correlated with pH, whereas shredder density was inversely related to pH. The density of invertebrate predators, however, was not significantly correlated with pH. Townsend and Hildrew (1984) suggest that the inverse relationship between pH and shredders may be a result of the slower decomposition of detritus at low pH, resulting in greater food abundance for shredders. As shredder abundance was also negatively related to fish abundance in the stream, an alternative explanation is that fish predation on shredders or a combination of predation and food abundance may control the density of shredders in their study sites (Townsend and Hildrew, 1984). Hildrew and others (1984b) suggest that the direct relationship between collectors and pH was a result of greater decomposition of detritus with increasing pH, resulting in greater production of fine particles of organic detritus on which collectors feed.

To evaluate the contribution of pH to variation in community parameters, these investigators analyzed their data using linear regression. They reported that annual mean pH alone explained 48% of the variation in species richness, 53% of the variation in species diversity, 12% of the variation in density, and 42% of the variation in species equitability. These findings suggest that other factors are as important as pH in explaining the taxonomic composition and abundance of benthic invertebrates in these acid-sensitive streams in England. However, the fact that pH alone did not explain a higher fraction of the variability in these parameters is not surprising because the minimum pH is likely to be more important that the annual mean in determining the composition and abundance of the benthic community in poorly buffered streams. Second, pH is probably not the only water quality variable influencing the abundance of benthic invertebrates in streams undergoing acidification. Further, food quality and quantity, which affect the functional group composition and abundance of macroinvertebrates and which may or may not be affected by acidification, were not measured and thus were not included in the regression analyses.

In a similar analysis of benthic invertebrate communities in acidic streams in Sweden, Otto and Svensson (1983) reported a significant positive correlation between species richness and pH. The number of shredder and grazer and scraper species were also positively correlated with pH, but the correlations between collector, filter feeder, and predator species numbers and pH were not statistically significant.

Based on a more detailed analysis of the benthic invertebrate communities from an acidic stream and a nonacidic stream, Otto and Svensson (1983) reported higher species richness and a more equitable distribution of species among functional groups in the nonacidic stream. In the acidic stream, shredders were the dominant functional group, with scrapers, collectors (deposit feeders), and filter feeders present in low abundance.

In terms of the taxa affected, mayflies, caddis flies, Coleoptera, and Diptera showed the largest decline in number of taxa from the nonacidic to the acidic stream. Because many mayfly and caddis fly species are grazers, the absence of taxa from these two families could explain the decline in grazer taxa and grazer abundance in acidic streams.

To assess the influence of different physical and chemical factors on the density and biomass of benthic macroinvertebrates in streams differing in pH and other major ions, Aston and others (1985) analyzed survey data on water chemistry and benthic invertebrates from 72 streams in England. Using principal components analysis, they found that the total density of invertebrates and the density of Plecoptera, Ephemeroptera, Coleoptera, and Nematoda were significantly related to pH, Al, and alkalinity. This suggests that the density of these taxa is affected either directly by the extent of acidification, as reflected by the concentration of acidic cations and alkalinity, or indirectly by other factors that are correlated with pH, Al, and alkalinity.

These same variables also accounted for a significant fraction of the variation in the density of mayfly taxa known to be sensitive to acidic conditions (e.g., *Baetis scambus, Ephemerella ignita, Heptagenia sulphurea*, and *H. lateralis*). The densities of mayflies and aquatic beetles (Coleoptera) were also related to the concentration of base cations (i.e., Ca, Mg, Na, and K), accounting for 51% and 64% of the variation in the density of these two taxa, respectively. This finding suggests that base cations may have a mitigating effect on the toxicity of low pH and high Al to macroinvertebrates, a result that is well documented for fish (e.g., Brown, 1982; Muniz and Leivestad, 1980). A similar effect on aquatic invertebrates could explain why pH alone generally explains only a fraction of the variation in the distribution and abundance of benthic macroinvertebrate taxa among streams with different levels of acidic cations and base cations.

Raddum and Fjellheim (1984) surveyed the benthic macroinvertebrate fauna in acidic and circumneutral streams and rivers in Norway, where surface waters are considered to be more heavily impacted by acidic deposition than in other areas of Europe. Although the source of acidity in the acidic streams was not specifically identified, based on evidence from other studies of the chemistry of acidic streams in Norway (for example, Overrein et al., 1980), it is most likely acidic deposition. Raddum and Fjellheim documented that half of the stone fly species collected were found at all locations, both acidic and nonacidic. *Leuctra fusca* and *L. diura* were widely distributed, but were less common at acidified locations. Hydropsychid caddis flies appeared to be restricted by acidic water, and seldom were more than a few species of the other caddis fly groups present in the most acidic localities. Mayflies appeared to be the most acid-sensitive taxon, with only two species found in the most acidic waters, both of which were in the genus *Leptophlebia*. *Baetis rhodani* was not present in streams with a pH <5.55, and other baetid mayflies were not present at a pH <6.0.

Raddum and Fjellheim (1984) also reported that the ratio of the relative number of baetid mayflies to the number of stone flies was >1.0 when the pH was >6.0 but ≤0.5 when the pH was <6.0. This ratio increased with increasing Ca concentration, again indicating that higher Ca levels in water can mitigate the effect of acidification on at least some invertebrate taxa at intermediate pH (5.5 to 6.0).

In a survey designed both to map the short-term episodic changes in pH in Swedish streams and rivers draining montane areas and to determine long-term changes, benthic macroinvertebrates were sampled in 1983 in three relatively pristine nature reserves in Sweden, two of which had been previously surveyed (Engblom and Lingdell, 1984). In the northernmost area sampled, where acidic deposition is generally the lowest among the three areas, the species composition was still dominated by acid-sensitive taxa of benthic macroinvertebrates such as *Baetis lapponicus* and *Philopotamus montanus*. The findings indicate that the pH of stream water in this area of Sweden had been above 6.0 and had remained unchanged since the period from 1961 to 1966, based on the known acid sensitivity of the species present and the high frequency of occurrence of acid-sensitive species. In contrast, macroinvertebrate fauna in the southernmost area sampled, where acidic deposition is generally greater, was dominated by acid-tolerant species such as *Leptophlebia vespertina, Ameletus inopinatus, Baetis rhodani, Ephermerella aurivillii* (Ephemeroptera), and *Nemoura cinerea* (Plecoptera). The absence of acid-sensitive species in this region indicated that the pH of stream water had often been below 5.0. In the third area, some acid-sensitive species such as *Baetis lapponicus* that were common in 1971 (which implies that pH was seldom <5.5 in the area) were absent in 1983. From these results, Engblom and Lingdell inferred that the pH minimum in streams in this third area had declined by more than 0.5 pH unit since the early 1970s and had been well below 5.0 in many streams in the area, causing the elimination of acid-sensitive taxa such as *Baetis lapponicus.*

Circumstantial evidence of acidification effects of benthic macroinvertebrate communities in North American streams has also been documented from survey data (Fiance, 1978; Arnold et al., 1981; Kimmel et al., submitted; Mackay and Kersey, 1985; Simpson et al., 1985; Smith et al., 1987; Hall and Ide, 1987; Rosemond, 1987). In a survey that was restricted to a single genus of mayflies, *Ephemerella*, Fiance (1978) sampled 22 stream sites in the Hubbard Brook Forest, New Hampshire. He reported that this genus declined in abundance with decreasing pH and organic matter. The species was absent from sites with a pH <5.5. Experimental acidification of a reach of stream to a pH of 4.0 had no effect on the emergence of this species, but it did affect growth rate and recruitment.

Arnold and others (1981) and Kimmel and others (1985) compared the invertebrate communities in acidic and nonacidic streams in Pennsylvania. Arnold and others (1981) reported that an acidic stream had a lower biomass and species richness of invertebrates compared to nonacidic streams in the same area. The lower biomass was caused primarily by a paucity of grazers and scrapers that feed on epilithic algae. Differences in pH among the study streams were attributed to acidic deposition and to differential buffering in the associated terrestrial watershed. Kimmel and others (1985) reported a lower density, diversity, and species richness in a stream with a pH of 4.6 to 6.0, compared to a nearby stream with a pH of 6.1 to 7.4. Two genera, *Ascellus* (Isopoda) and *Leuctra* (Plecoptera), both of which are known to contain acid-tolerant species, dominated the benthic invertebrate community in the acidic stream throughout most of the year.

In a survey of 12 streams in south-central Ontario, an area with relatively high deposition rates of strong acids, Mackay and Kersey (1985) found that the generic richness of aquatic insects decreased with declining pH, with the taxa richness of mayflies and stone flies exhibiting the largest difference between acidic and nonacidic streams. The functional group diversity of caddis flies also declined with decreasing pH, the most acidic streams containing predators and shredders but lacking grazing, gathering, and filtering caddis flies.

Based on historical comparisons of benthic macroinvertebrate data from three sites in two streams, Hall and Ide (1987) presented evidence of effects of acidification on aquatic insect communities in streams in Algonquin Park, Ontario. The insect communities were first sampled at the sites in the late 1930s and early 1940s and again in 1984 and 1985. At the site where current spring pH fluctuations are small (pH declines from 6.4 to 6.1), the same taxa present in the original survey were present during 1984 and 1985. However, at the two sites that currently exhibit large spring pH depressions (pH declines from 6.4 to 4.9), many mayfly and stone fly taxa present during the original survey and known to be intolerant of low pH ($<5.0$) were not found in 1984 and 1985. Further, several acid-tolerant mayfly and stone fly taxa not recorded at any of the three sites during the original surveys in 1937 to 1942 were found at the sites with the large spring depressions in pH, indicating that some species replacement (i.e., from acid-sensitive to acid-tolerant taxa) had occurred. These results suggest that some poorly buffered streams in this region of Canada have acidified sometime within the last five decades and that episodic acidification associated with storms and snowmelt can affect the macroinvertebrate communities in poorly buffered streams.

Simpson and others (1985) sampled four streams in the Adirondack Mountains, New York, with a pH range of 4.4 to 7.2. The source of acidity in these streams was not discussed, but atmospheric deposition is assumed to be at least one contributing source. Simpson and others observed lower species richness in the more acidic streams, with fewer than half as many taxa as the streams with a higher pH. The acidic streams also contained few mayflies and elmid beetles, and they were dominated by stone flies of the genus *Leuctra* and *Isoperla*, which comprised from 56% to 86% of the individuals.

In a similar survey of eight sites in three drainages in the Adirondacks, Smith and others (in press) found that total macroinvertebrate generic richness, generic diversity, mayfly density and richness, collector-gatherer richness, and scraper density and richness were positively correlated with stream pH. Atmospheric deposition was the primary source of acidity in these streams, as indicated by sulfate being the dominant anion and by the absence of an internal source of sulfate in the watersheds studied (Driscoll et al.,, 1987). Mayflies and elmid beetles were absent from the most acidic of the sampling sites.

This study was one of the few studies in which the effects of acidification on macroinvertebrate densities was examined. Smith and others (submitted) found no significant relationship between pH and total invertebrate density. Regression analysis, however, indicated that stream pH and benthic organic matter were the two most important parameters influencing the variation in the structure of invertebrate communities among the stream sites sampled.

In a study of the effects of acidification on macroinvertebrates, excluding the Chironomidae and Oligochaeta, in streams in the Great Smoky Mountains National Park, Tennesse, Rosemond (1987) found a significant negative effect of pH on density, species richness, species diversity, and functional group richness. All of the streams sampled, which ranged in base flow pH from 4.5 to 6.8, drain unlogged spruce-fir watersheds and contain relatively high nitrate concentrations ($>50$ $\mu$eq/L). In addition, at higher elevations, some of the watersheds contain a pyritic phyllite, which has been exposed to oxidation at outcrops and from natural landslides. Thus pyrite oxidation and nitrification appear to be important sources of acidity in the acidic streams. These are clear water streams, however, and unlike many acid mine drainage streams, there is no indication of Fe precipitation in the stream channel.

The density of macroinvertebrates in these streams was also significantly and positively correlated with the standing stock of benthic organic matter (BOM), indicating that both water quality and food availability were important in determining differences in the macroinvertebrate communities between these sites. Stream pH and organic matter together accounted for 62% of the variation in the density of macroinvertebrates at the low stream order sites.

The composition of functional feeding groups also appeared to be affected by pH, with the composition shifting from a dominance by shredders at low pH to fewer shredders, fewer collector/gatherers, and an increased number of grazers and predators at high pH sites. There were also shifts in the species composition and richness of major taxa. The number of species of mayflies, stone flies, and caddis flies was positively correlated with pH and negatively correlated with inorganic monomeric Al. At high pH, grazers were dominated by the mayfly *Epeorus pleuralis*, whereas at low pH this species was absent, and grazers, which were in low abundance, were dominated by *Ameletus lineatus*.

In summary, comparative surveys of benthic macroinvertebrate communities in acidic and circumneutral streams show a consistent pattern of decreasing species richness and diversity and decreasing functional group richness, with declining pH, particularly at pH <6.0. Decreases in the density and biomass of macroinvertebrates with declining pH have also been documented, indicating that the apparent loss of acid-sensitive species of macroinvertebrates may not be offset by increases in the abundance of acid-tolerant taxa.

The benthic macroinvertebrate taxa most sensitive to acidic conditions appear to be mayflies (Ephemeroptera), snails, and mussels (Mollusca). Stone fly (Plecoptera) and caddis fly (Trichoptera) taxa exhibit a wide range of responses to acidic conditions, with some species being acid sensitive and others being acid tolerant. The Chironomidae appear to be generally more acid tolerant than most taxa, perhaps because they live within the stream sediments where the surrounding water may be better buffered by ion exchange and weathering processes.

The decline in functional group richness with declining pH is primarily associated with lower densities of grazer and scraper taxa, most of which are mayflies. This apparent reduction in, or loss of, grazers appears to be a direct result of the toxic effect of low pH and high Al levels on these taxa rather than an indirect effect associated with alterations in their food supply. However, detailed studies of the response of grazers to alterations in the species composition of attached algae in streams have not been conducted.

Although the evidence that acidification was the cause of many if not most of the differences in benthic macroinvertebrate communities among the streams surveyed is circumstantial, the consistent patterns of change in taxonomic and functional group composition and abundance with declining pH suggest that acidification is one of the primary causes. None of these studies, however, demonstrates a causal link between acidic deposition and stream acidification, although acidic deposition appears to be the only likely cause of acidification in many cases. Additional studies are needed to clarify the importance of episodic versus chronic, long-term changes in water quality on the benthic macroinvertebrate communities in poorly buffered streams.

## 2. Experimental Studies of Acidification Effects on Benthic Macroinvertebrates

To determine more directly the effect of both episodic and chronic, long-term depressions in pH on stream organisms and communities, natural and artificial streams have been dosed with acid and/or aluminum. Hall and his colleagues (Hall et al., 1980; Pratt and Hall, 1981) describe the results of an experiment in which they dosed a third-order stream in New Hampshire with acid for several months and measured emergence and drift of macroinvertebrates. The pH was depressed from a range of 5.7 to 6.4 in the upstream reference area to 3.9 to 4.5 in the treated section, which represents a 100-fold increase in hydrogen ion concentration. According to Hall and others (1980), the stream reach dosed with acid had not been previously stressed by acidification.

Emergence of adult mayflies, some stone flies, and some true flies (Diptera) was significantly lower in the acidified reach compared to the reference section. The authors do not indicate whether this apparent reduction in emergence was due to mortality or increased drift of these taxa from the acidified reach. Some taxa also exhibited net increases in drift rates in the acidified reach, with the species affected, such as *Ephemerella* and *Epeorus*, generally being those known to be more acid sensitive based on presence-absence data obtained in surveys of acidic and nonacidic streams. More acid-tolerant taxa, such as the stone fly, *Leuctra* sp., exhibited no net increase in drift rate.

The sensitivity of different functional feeding groups of macroinvertebrates in the acidified reach was also generally consistent with that expected as indicated by survey data. Drift rates of collectors, scrapers and grazers, and predators exhibited large net increases during the first week in the acidified reach, whereas shredders showed no such response. There was no difference in drift rates between the treated and reference reaches after the first week of dosing, which suggests that all of the acid-sensitive organisms were eliminated from the acidified reach in the first week through mortality and/or drift. Alternatively, the remaining individuals could have become acclimated to the lowered pH and discontinued their drift. The density of macroinvertebrates measured after the dosing was terminated was 75% lower in the acidified reach than in the reference section. Hall and others (1987) inferred an effect of acidification on the density of macroinvertebrates. Benthic samples, however, were not collected from the treated and reference sections prior to the dosing to determine if they were similar in the two reaches before the pH was depressed.

In other experiments designed to simulate acidic snowmelt episodes, first- and third-order reaches of the same stream described above were dosed in separate experiments with hydrochloric acid of $AlCl_3$ for short periods (Hall et al., 1987). Drift rates of mayflies, blackflies (Simuliidae), and chironomid larvae were greater from both the acid- and Al-treated reaches than from the reference reaches. The drift rate, however, was greater during the Al additions than during acid additions, even when pH values produced by the dosing with acid and Al were comparable (pH of 5.2 to 5.4), implying an additive or synergistic stress effect of aluminum on invertebrates. Hall and others (1987) suggest that fluctuations in Al concentrations at pH values not toxic to biota may be an important factor regulating the distribution and abundance of aquatic macroinvertebrates in poorly buffered streams.

In similar experiments designed to create simultaneous episodes of low pH and low pH with increased Al, Ormerod and others (1987) dosed a soft water stream in Wales for 24 hours with $H_2SO_4$ and $Al_2(SO_4)_3$ at two successive points. The pH was depressed from 7.0 to 4.28 and 5.02, respectively, in the acid- and Al-treated reaches. Aluminum concentrations were 52 and 347 $\mu g/L$, respectively, in the two zones; the former concentration was not significantly different from that in the reference zone, indicating that aluminum was not mobilized from the stream sediments in the reach dosed only with $H_2SO_4$.

This study showed a greater effect of the combined mineral acid and Al dosing on macroinvertebrates than the acid dosing alone. Drift rates of blackflies were greater in both the acid and Al treatments than in the reference areas. The drift rate of several taxa, including *Baetis rhodani, Ephemerella ignita, Protonemura meyeri, Dixa puberula*, and *Dicronata* sp. increased in the Al treatment but not in the acid treatment, indicating an additive or synergistic effect of increased Al on macroinvertebrates at low pH.

Despite the greater drift rates from the treated reaches, *Baetis* was the only species to exhibit a decline in benthic density during the treatment, and the decline occurred only in the Al-treated reach. Total density of invertebrates on the surface of the benthos, however, was greater in the reference reach than in either the acid or Al reaches, and densities at greater depth were similar among the treated and reference reaches. Whether this difference would have persisted with chronic, long-term acidification is unknown. The result is nonetheless interesting and suggests that the so-called infaunal macroinvertebrates living in the stream sediments may be less susceptible to acidification than the epifaunal forms that live on or above the sediments. This apparent difference in the response to acidification between epifaunal and infaunal taxa may be due to (1) buffering of interstital water by groundwater infiltration and/or by ion exchange and weathering reactions by sediments, which renders infaunal forms less susceptible to acidification, (2) differences in habitat preferences between acid-sensitive and acid-tolerant taxa, or (3) a combination of both.

As discussed in section II,A, the biomass and productivity of bacteria associated with fine-grained sediments in depositional areas of streams were not significantly related to the pH of the overlying water, whereas the activity and biomass of bacteria on rocks and organic substrates that extend up into the water column above the sediments were significantly correlated with pH (Palumbo et al., 1987a and 1987b). This finding is consistent with the apparent difference in response to acidification between the macroinvertebrate infauna and epifauna observed by Ormerod and others (1987) and suggests either that the bacteria and macroinvertebrates in these two habitats are exposed to water of different quality despite being collected from nearby locations in the same stream, or that the infaunal and infloral communities are less sensitive than the epifaunal and epifloral communities to acidic conditions.

Comparisons of benthic macroinvertebrate infauna and epifaunal communities among acidic and circumneutral lakes have shown no difference in the population of infauna, but direct observations indicate that acidic lakes (pH <4.9) have lower densities of epifauna compared to circumneutral lakes (Raddum, 1980; Collins et al., 1981). That sediment buffering may account for the apparent lack of an effect of acidification on the infauna of lake sediments is supported by documented pH gradients in lake sediments, with pH increases of more than one-half of a unit occurring within the upper few centimeters of the sediment-water interface (Kelly and Rudd, 1984; Cook et al., 1986; Schiff and Anderson, 1986). Although we are unaware of documented pH gradients in stream sediments, $O_2$ gradients in stream sediments have been documented (e.g., Grimm and Fisher, 1984), indicating that the chemistry of pore water in streams can be very different from that of the water flowing over the sediments. It is likely, therefore, that similar gradients in the pH of pore water may occur in experimentally acidified streams, particularly where the sediment consists of reactive, fine-grained particles that impede the rapid exchange of the pore water with overlying stream water.

The fact that all taxa exhibiting increased drift rates in the study by Ormerod and others (1987) did not show corresponding declines in density in either of the treated reaches may be because these taxa seek refuge in the sediments. If so, the overall response of epifaunal invertebrates to episodic depressions of pH may depend on the availability of less acidic refugia resulting from groundwater inputs and/or buffering (e.g., ion exchange, sulfate and nitrate reduction) in the sediments where the infauna reside. Thus, although its role in modifying the acid-base chemistry of interstitial water in stream sediments is not yet known, buffering of the pore water may prove to be an important factor in reducing the effects of acidification on the benthic infauna of streams.

The issue of whether most epifaunal taxa are more sensitive to acidification than the infauna due to physiological, behavioral, and/or morphological differences is presently difficult to assess because of the lack of information on responses of representative species in each group exposed to the same acidic conditions. Many of the acid-sensitive epifaunal taxa (e.g., mayflies) in streams have filamentous gills that serve several important physiological functions, including acid-base balance of internal fluids (i.e., hemolymph), respiratory gas exchange, and osmoregulation. All of these functions are adversely affected by declining pH (Havas, 1981), which may cause species with filamentous gills to be relatively more sensitive to acidification compared to species without such gills. Some epifaunal species with filamentous gills are relatively acid tolerant; other species within the same genus (e.g., *Baetis*) are more acid sensitive, however, indicating that the filamentous gill is not the major determining factor of differences in acid sensitivity among benthic macroinvertebrate taxa. It is more likely that differences in sensitivity among taxa are due to a combination of morphological, behavioral, and/or physiological/biochemical characteristics. These may include factors such as the presence of hemoglobin that buffers the hemolymph from external changes in pH (Havas, 1981), the ability to alter the area of surfaces involved in osmoregulation (Wigglesworth, 1938), the permeability of the integument to water and hydrogen ions, and the ability to detect and avoid acidic conditions by burrowing in sediments or drifting downstream until less acidic or nonacidic refugia are encountered.

The fact that Ormerod and others (1987) found no significant differences in mortality of *Baetis* nymphs between the acid and Al treatments using in situ bioassays but did observe much greater drift rates of this species in the acid and Al treatments indicates that the response of this species to elevated Al was primarily a behavioral drift response. This finding was confirmed by calculations showing that most of the decline in the density of *Baetis* in the stream reach dosed with Al was accounted for by the increased drift rate.

Experimental streams, as opposed to experiments in natural streams, have also been used to determine the effects of acidification on stream organisms and communities. Zischke and others (1983) used a 122-m section in each of three outdoor, 520-m-long stream channels supplied with water from the Mississippi River. One channel served as the reference (pH 8.0), a second was acidified with $H_2SO_4$ to pH 6, and the third was acidified to pH 5 for an experimental period of 17 weeks.

The density of macroinvertebrates in the acidified channels was lower than in the control channel during most of the acidification study. Densities at the end of the acidification period, however, were higher in the treated channels than in the reference channel because of an increased abundance of the acid-tolerant taxa. The diversity and species richness of macroinvertebrates was lower in the acidified channels than in the reference channel at the end of the study and decreased with declining pH. The amphipod (*Hyalella azteca*) and the snail (*Physa gyrina*) were classified as the most acid sensitive, the latter species being eliminated from the pH 5 channel. Chironomids, some amphipods such as *Crangonyx*, and flatworms were classified as being intermediate in sensitivity, whereas damselflies, isopods, and leeches were classified as being the most acid tolerant.

Although the experimental acidification on the two channels did not result in significant increases in Al, the reductions in macroinvertebrate species richness and diversity with declining pH are generally consistent with the patterns reported in surveys of streams differing in pH and Al levels. This indicates that pH depressions to levels of <5 alone can cause significant changes in the benthic macroinvertebrate communities in streams. However, because of the lack of opportunity for acclimation to low pH, the use of both organisms from a well-buffered environment, and high dose rates of acid (i.e., 100- to 1,000-fold increases in H ion concentrations), the validity of extrapolating these results to more acid-sensitive environments where organisms may be acclimated or adapted to low pH is debatable. Nonetheless, the consistent community and taxonomic responses to declining pH between natural and experimental streams suggests that many macroinvertebrate taxa have only limited ability or no ability to acclimate to acidic conditions.

Allard and Moreau (1985, 1986) conducted a similar experiment involving the diversion of water from a small stream into three wooden channels; one of these was acidified with $H_2SO_4$, the second was dosed with acid plus $Al_2(SO_4)_3$; the third served as a control. The pH of water in both the acid and acid plus Al treatments was 4.0, whereas the total dissolved Al concentration was 80 and 190 $\mu l/L$, respectively. The control channel had a pH of 6.3 to 6.9 and an Al concentration of 70 $\mu g/L$.

After 73 days of acidification, macroinvertebrate densities in the treated channels were only one-third of that in the control. However, there was no difference in the biomass of invertebrates between the channel dosed with acid only and that dosed with acid plus Al. Grazers and scrapers disappeared from both of the acidified channels soon after the dosing began. The authors concluded that this occurred as a result of increases in acidity and not because of changes in the food supply of grazers, based on the observation that grazers left the acidified channels before any modifications in the food supply could have occurred.

Their finding that Al had no additive effect on macroinvertebrates beyond that caused by the low pH contrasts with that of Ormerod and others (1987) and Hall and others (1987). The apparent absence an Al additive effect may be caused by a difference in Al speciation or by differences between the studies in other water quality factors that can mediate the effects of Al. Allard and Moreau (1986) report a relatively high concentration of organic matter (6.0 $\mu g/L$) in their stream water. Organic matter has been shown to complex Al, rendering it less toxic to aquatic organisms (Driscoll et al., 1980; Baker and Schofield, 1982). Alternatively, Al may be less toxic at low pH ($<4.5$), where it is more soluble and hence less likely to precipitate. Schofield and Trojnar (1979) and Baker and Schofield (1982) found that Al was most toxic to fish in solutions of intermediate pH (5.2–5.4) that are oversaturated with respect to the solubility of readily forming mineral phases of Al. This suggests that Al is more toxic when it hydrolyzes in oversaturated solutions and forms particulate Al oxyhydroxide. Hutchinson and others (1987), however, obtained results that are contradictory to these and suggested that differences among the studies in the aging of oversaturated Al solution, which would affect the formation of polymeric and colloidal Al forms, may be responsible for the different responses observed.

Burton and others (1985) also conducted experiments to examine the effects of low pH on stream invertebrates by using recirculating wooden stream channels in a greenhouse. One of the channels was acidified to a pH of 4.0 with $H_2SO_4$ and the second untreated channel with a pH of 7.0 served as the control. The streams were stocked with invertebrates and detritus from a nearby stream; most of the taxa added to the channels were detritivores.

Over the 264-day exposure period, macroinvertebrate populations in the acidified channel declined, whereas those in the control channel remained unchanged or increased due to reproduction. Some taxa, such as the snail (*Physa heterostropha*), the amphipod (*Gammarus lacustris*), the isopod (*Ascellus intermedius*), and the caddis flies (*Lepidostoma liba* and *Pycnopsyche* sp.), exhibited significantly higher mortality rates in the acidified channel and/or lower reproduction rates in the case of the second two species compared to that in the control. Although some taxa considered acid sensitive, such as *Physa* and *Gammarus*, were still present in the acidified channel after 264 days of continuous exposure to a pH of 4.0, these species would probably have been eliminated from the channel over time due to the lack of reproduction in the case of *Physa* or low reproduction in the case of *Gammarus*.

In summary, experimental studies of the effects of acidification on stream benthic macroinvertebrates show that pH depressions alone reduce species richness, diversity, and density, patterns that are consistent with those observed in comparative surveys of acidic and circumneutral streams. The response of different functional feeding groups to experimental acidification is also similar to that observed in natural streams, with grazers, collectors, and predators declining with decreasing pH and shredders exhibiting no response. Aquatic insect emergence and reproduction of some noninsect taxa (e.g., snails, amphipods) also decrease or stop with declining pH in experimentally acidified streams. The decline in reproduction may be a sublethal effect of acidification that has important implications for the long-term persistence of species that appear to be acid tolerant based on short-term bioassays.

These experimental studies show that depression in pH without concomitant increases in Al can account for some of the differences between communities in acidic and nonacidic streams. Where the concentration of dissolved organic matter is low, however, the addition of Al has an additive or synergistic effect on benthic macroinvertebrates at low pH (i.e., the effect on macroinvertebrates is not accounted for only by the pH effect resulting from Al hydrolysis). Macroinvertebrates also exhibit increased drift rates in response to episodic and chronic increases in both $[H^+]$ and in $[H^+]$ plus Al. The apparent lack of an effect of pH depressions on macroinvertebrate biomass, however, indicates that the density and/or the growth rate of acid-tolerant taxa increases, thereby compensating for the loss of biomass of acid-sensitive species due to mortality or emigration.

Future experimental studies and surveys of benthic invertebrates in acidic and circumneutral streams would benefit from measurements of both pore water and overlying stream water to which the macroinvertebrate epifauna and infauna are exposed. In addition, comparative data are needed on the responses of infauna and epifauna to acidic conditions in terms of changes in their taxonomic and functional group composition and abundance.

## F. Fish

Concern over the acidification of surface waters has focused primarily on fish because of the clear economic and sociological importance of some species. In addition, fish depend on lower trophic levels for their food and, therefore, may serve as sensitive integrators of ecosystem function.

Although there have been fewer detailed studies of fish populations and communities in acid-sensitive streams compared to lakes, declines in abundance and losses of fish populations in acidified streams and rivers and fish kills in acidic streams have been documented. The acidification of Norwegian streams and rivers and attendant fish kills (e.g., Leivestad and Muniz, 1976), particularly during spring snowmelt, and declines in natural reproduction of salmon and trout (e.g., Jensen and Snekvik, 1972) were among the first reported cases of adverse biological effects of acidic deposition. Similar claims of adverse effects of acidification on fish populations in streams and rivers in North America have since been reported (e.g., Watt et al., 1983).

Various causes have been suggested for the effects of acidic deposition on fish, including (1) acutely or chronically toxic water quality (e.g., low-pH, high-Al concentrations), particularly during high-acid, short-term events, and (2) decreased food availability in acidic environments.

## 1. Surveys of Fish Populations and Communities

Arnold and others (1980) reported data on the fish species richness and pH and alkalinity of streams in Pennsylvania and reported temporal patterns in these parameters based on a comparison of the "earliest" and the "most recent" data for which there was at least a 1-year interval between sampling dates. Findings in this study showed pH for most recent data at lower levels than the earliest pH measurements in 34% of the streams (from a mean of 6.99 to 6.37). In streams for which there were fishery data, 56% had fewer species than reported earlier, suggesting a loss of fish species. Brown and Brocksen (1984) applied chi-square analysis to the relationship between changes in both pH and number of fish species for the 174 streams examined by Arnold and others (1980) for which there are data; however, no consistent relationship is indicated between increases or decreases in pH and changes in the number of fish species in the streams in Pennsylvania. The conclusion of Arnold and others, "a definite overall trend of many streams becoming more acidic and/or less alkaline, and losing some fish populations at the same time," does not appear to be supported by the data presented.

Fish communities and water chemistry in 49 small streams located in the Laurel Hill region in the Allegheny Plateau of southwestern Pennsylvania were also surveyed (Sharpe et al., 1984). Fish were sampled by electrofishing a 100-m section of each stream. The results for four of the streams were summarized by Sharpe and others (1984), and Haines (1986) summarized the results for all 49 streams. The mean base flow pH and alkalinity in the four streams ranged from 5.4 to 6.0 and 14 to 161 μeq/L, respectively. Only the stream with the highest base flow alkalinity supported reproducing populations of brook char (*Salvelinus fontinalis*), rainbow trout (*Salmo gairdneri*), and mottled sculpin (*Cottus bairdi*). Two of the streams were fishless during a fall survey, and only three brook char and one brown trout (*Salmo trutta*), all of which were of hatchery origin, were found in the fourth stream during the fall survey. Similar results were obtained in a spring survey the following year, except that the fourth stream where the four hatchery fish had been found was fishless, and a single brook char was captured in one of the streams that had been fishless during the fall survey.

Of the 49 streams sampled in the Laurel Hill region, Haines (1986) reported that 33 contained reproducing fish populations as shown by presence of young-of-the-year fish of nonhatchery origin. Four of the streams had remnant fish populations (defined as having only 2 year classes), all of which had a pH $\leq 6.0$ and Al concentrations $<100$ µg/L. Twelve of the 49 streams lacked reproducing fish populations, and at least two of these 12 were entirely fishless (Sharpe et al., 1984). Only one of the 12 streams without reproducing fish populations had a pH $>6.0$, and eight had a pH $<5.0$ and Al concentrations $\geq 400$ µg/L.

To assess the effects of water quality on fish survival, Sharpe and others (1983) conducted in situ bioassays in one of the streams with low alkalinity, using the stream with highest alkalinity as a control. The study demonstrated that water quality was acutely toxic to brook char, rainbow trout, brown trout, and mottled sculpin. Low pH and high Al levels were suggested as probably causes for fish mortality in the low-alkalinity stream on the basis of the symptoms of the dead and dying fish.

In a survey of 12 streams in Scotland draining both forested and nonforested watersheds, Harriman and Morrison (1982) found that brown trout were absent from most of the streams draining forested catchments, which tended to be acidic. Streams draining the nonforested catchments, which tended to be nonacidic, contained juvenile trout, indicating the presence of reproducing trout populations in these streams.

Results of in situ bioassays using fertilized eggs of Atlantic salmon (*Salmo salar*) showed that eggs placed in the most acidic stream died within two months, but those in the less acidic stream remained alive and hatched within 3 to 4 months. Although these findings are circumstantial, the implication is that acidification is a contributing factor to the absence of brown trout in the acidic forested streams.

In a more extensive survey, Harriman and others (1987) sampled fish in 27 streams in the Galloway area of southwest Scotland. The pH of the streams on the date fish were sampled ranged from 3.88 to 7.3. Five of the streams were fishless and an additional two streams contained no brown trout but did have eels (*Anguilla anguilla*). These seven streams had pH and Al levels in the range known to be toxic to fish.

Townsend and others (1983) surveyed the fish communities in the 34 stream sites in the Ashdown Forest in southern England where benthic invertebrates were also sampled. The most acidic headwater sites were fishless, and the first species to occur as acidic conditions improved downstream were brown trout, bullheads (*Cottus gobio*), and stone loach (*Noemacheilus barbatulus*). Although the evidence is circumstantial, this spatial pattern in fish distribution suggests that acidic conditions precluded the presence of these species from the upstream reaches of the drainage. The greatest fish species richness occurred at sites well downstream or below lakes. Species richness was correlated with July water temperature, which could explain 42% of the variability in fish species richness among the sample sites. Fish abundance was also significantly correlated with increasing maximum discharge, explaining nearly 42% of the variation among sites. The authors, however, did not indicate whether pH or other acidification-related water quality parameters (e.g., alkalinity) accounted for a significant fraction of the variation in fish species richness and abundance among streams.

Based on a survey of fish communities in streams in west Wales, Stoner and others (1984) reported that brown trout were absent from streams draining the afforested (planted forest) watersheds in which the pH ranged from <4.0 to >7.0. Of the streams draining unafforested watersheds (i.e., forest allowed to regrow naturally), where the pH ranged from <4.5 to >7.0, ten were fishless, and nine contained only sparse populations of brown trout. Whether these differences in the status of fish populations result from acidification or from other factors related to watershed and stream management is unclear.

To assess the effects of water quality on trout survival, Stoner and others (1984) conducted in situ bioassays in selected streams using age group 0 brown trout obtained from one of the nonacidic streams. The test fish were held in polyvinyl chloride (PVC) pipes covered at the ends with plastic mesh to allow water to flow through. Trout held in three of the afforested streams died in 7 to 10 days, whereas those placed in the streams draining the unafforested watersheds survived. The streams where fish died had a mean pH of <6.0 and a mean filterable Al concentration of >21 µeq/L. Mean filterable Al concentration accounted for 96% of the variation in mortality rate (expressed as time to 50% mortality) of brown trout. The mortality rate of fish was positively correlated with the gill Al concentration ($r^2 = 0.92$), a finding that is consistent with toxicological studies showing that Al accumulation on surfaces involved in respiratory gas exchange and osmoregulation is a primary mechanism by which elevated Al levels affect fish survival in acidic waters (Leivestad, 1982; Wood and McDonald, 1982).

In a more recent survey of fish populations in acid-sensitive streams in the UK, Sadler and Turnpenny (1986) sampled 61 upland streams in central and north Wales and the Peak district of England. Of the streams with a pH of <5.5, 80% were fishless compared to only 19% of the streams with a pH >5.5. Stream pH, labile monomeric Al, and heavy metal concentrations (Cu, Zn, and Pb) were highly correlated with the biomass of age class I+ brown trout in the streams. Sampling of the other age groups was not considered to be quantitative, so data for only the one age group were analyzed. The density and condition factor of fish were also lower in the more acidic streams, suggesting that growth was adversely affected by acidic conditions.

Anderson and Anderson (1984) surveyed fish populations at 23 locations in 21 streams in southern Sweden, all of which were known to have previously contained trout (*Salmo* spp.) or char (*Salvelinus* spp.). The abundance of brown trout declined with decreasing pH, and they were absent from streams having a summer baseflow pH of <6.0. One of the streams from which brown trout were absent contained brook char, a species known to be more acid-tolerant than brown trout (Leivestad et al., 1976; Muniz and Leivestad, 1980). Stream pH and alkalinity together could explain 50% of the variation in brown trout abundance among the streams. Aluminum was not included as a variable in the multiple-regression analysis.

Anderson and Anderson also reported that the average length of brown trout of a given age from the streams with a pH of 6.0 to 6.5 was greater than that from streams with a pH of 6.6 to 7.0. This finding suggests that the growth rate of brown trout increases with decreasing pH down to a pH of 6.0. Because brown trout abundance in these same streams also declined with decreasing pH, the apparent positive growth response of brown trout to declining pH may be an indirect result of decreased competition for available food and space among the remaining trout.

In a more recent study of fish populations in Swedish streams, Degerman and others (1986) surveyed the water chemistry and fish populations in 12 acid-sensitive streams on the west coast of Sweden. They found no significant correlations between the occurrence or abundance of either salmonids (brown trout and Atlantic salmon) or eels (*Anguilla anguilla*) and summer pH. The frequency of occurrence of streams with salmonid parr, however, was significantly greater in streams with summer alkalinities >250 μeq/L than for streams with an alkalinity of 0 to 90 or 100 to 250 μeq/L. The occurrence of eels, however, was not significantly correlated with the alkalinity of stream water.

These same investigators also found significant positive correlations between the abundance of both salmonid parr (age classes 0+ and I+ brown trout and Atlantic salmon combined) and eels and the alkalinity of stream water. The mean abundance of salmonid parr in seven large and small streams with a summer alkalinity >150 μeq/L showed no consistent trend over the 1956 to 1984 period. In contrast, parr abundance in small streams with a summer alkalinity of <250 μeq/L showed a consistent decline from the late 1950s onward. The authors suggest that streams with the lower summer alkalinity may undergo greater depressions in pH and increases in Al during spring runoff, resulting in the mortality of salmonid parr.

In summary, surveys of fish populations in acidic and circumneutral streams have focused almost exclusively on those with Salmonid (trout, char, salmon) species. These surveys indicate that the richness of fish species generally declines with decreasing pH, with many of the most acidic streams (pH <5.5) generally lacking fish entirely. Fish reproduction is impaired when pH falls to <6.0. Aluminum concentration appears to be a source of variation in the acute mortality of fish in acidic streams. Of the salmonid species that are present in streams during all or a portion of their life history, rainbow trout, brown trout, and Atlantic salmon are relatively acid sensitive, and brook char are relatively acid tolerant. In areas with spring snowmelt, brown trout and rainbow trout are generally absent where the summer base flow pH is <6.0, suggesting that spring pH depressions may be an important cause for the extirpation of these two acid-sensitive taxa.

## 2. Experimental Studies of Acidification Effects on Stream Fishes

Experimental studies involving alterations in the hydrogen ion and/or Al concentration in stream water or placement of fish in streams of different pH and Al content have also been conducted to determine more directly the effects of low pH and elevated Al levels on fish. Hall and others (1980) reported that no mortality of brook char was observed when the pH of Norris Brook, New Hampshire, was experimentally depressed from slightly <6 to ~ 4 for six months by adding $H_2SO_4$ directly to the stream. Downstream movement of char to areas of higher pH was, however, observed after the onset of the experimental acidification, which coincided with the period of high flow in the spring. This downstream movement of char may have resulted from an active avoidance response to low pH caused, in part, by the experimental addition of $H_2SO_4$, a physical displacement (washout) caused by the high flows, or some combination of both. The overall effect of this experimental acidification on the brook char population in Norris Brook, however, is unknown.

The only other reported study of the response of free-roaming stream fish to experimental acidification was that of Zischke and others (1983). They reported that depressing the pH of water flowing through experimental stream channels from an ambient value of 8 to 6 had no effect on spawning and embryo production of fathead minnows (*Pimephelas promelas*) stocked in the channels. Newly hatched larvae, however, did not survive to the juvenile stage at pH 6. In contrast, there was very little spawning at pH 5, and no eyed embryos were produced at this pH.

Because barriers at the downstream ends of the channels prevented fish from leaving during the acidification, it is impossible to draw any conclusions from this study about the avoidance behavior of adult fathead minnows to low pH. Nonetheless, the findings suggest that there would be no successful recruitment of fathead minnows in streams with a pH ≤6 because of the absence of spawning or an absence of embryo and/or juvenile survival.

In situ toxicity tests using confined fish have been conducted by several investigators to determine the acute toxicity of acidified stream water (Sharpe et al., 1983; Anderson and Nyberg, 1984; Stoner et al., 1984; Ormerod et al., 1987; Gagen and Sharpe, 1987). One shortcoming of using the findings in such tests to assess the potential effects of acidification on fish populations and communities is that avoidance behavior is precluded by confining the fish. Despite this limitation, such studies provide conclusive evidence that water quality in some acidic streams is acutely toxic to yearling and adult fish. Gagen and Sharpe (1987), for example, reported that the $LT_{50}$ for hatchery-reared rainbow and brown trout and brook trout char held in an acidic stream in Pennsylvania was < 3 days. The median pH and total soluble aluminum concentration during the exposure of three species, respectively, were 5.28, 4.80, and 4.94, and 480, 480, and 450 µg/L.

Anderson and Nyberg (1984) found that the mortality of both hatchery-reared and native yearling brown trout confined in high-altitude streams in Sweden during acid spring episodes was generally 100%, depending on the change in water quality and when fish were placed in the stream during the episode. The maximum mortality rate did not occur at the minimum pH, but rather when the pH was still relatively high (> 5.5) and declining. Further, when exposed to the same conditions, wild trout survived longer than hatchery-reared trout, suggesting that wild trout were either more acclimated to acid conditions or were inherently more resistant to such conditions due to natural selection.

In contrast to the widely held view of the importance of Al as a cause of fish mortality in acidic environments, Anderson and Nyberg (1984) suggest that Fe and Mn were more important than Al as a cause of the acute mortality of brown trout during the acid spring episodes. The basis for this observation was that most of the Al was organically complexed and hence was rendered less toxic to fish, and the concentrations of even total Al during spring snowmelt were generally less than that found to be acutely toxic to fish.

Evidence that Al is more toxic to fish than low pH alone is based primarily on controlled toxicity tests in which pH and Al concentrations are varied separately and together. The mortality rate of confined brown trout and Atlantic salmon in a reach of stream dosed with acid plus Al (pH 5.0) to simulate acid episodes was significantly greater than that of fish in a reach dosed only with acid (pH 4.28) (Ormerod et al., 1987). Although these results demonstrate the additive effect of Al on fish mortality and the potential acute toxicity to fish of acid episodes, it is not possible to conclude from these in situ toxicity tests that such episodes are responsible for the reported decline and loss of fish species in some acid-sensitive streams in Europe and North America.

In summary, in situ toxicity tests of ambient stream water and experimentally acidified water show that, when sufficiently acidic, the water can be acutely toxic to stream fishes. Low pH alone may be the cause for the lack of successful reproduction and recruitment of some fish species under chronically acidic conditions. The major uncertainties about the overall effect of acidification on stream fishes are the lethal and sublethal effects of acidic conditions on all life history stages and avoidance behavior exhibited toward low pH and elevated metal concentrations during acid episodes.

## III. Effects of Acidification on Biological Processes

### A. Primary Production

Experimental acidification studies and field surveys indicate that rates of primary production per unit area of stream generally increase with acidification resulting from acid precipitation (Hendrey, 1976; Arnold et al., 1981; Parent et al., 1986; Planas and Moreau, 1986; Mulholland et al., 1986). However, it is unclear whether periphyton production per unit biomass is increased at low pH or whether the observed increases in area-specific primary production result primarily from increased periphyton biomass. Parent and others (1986) reported greater periphyton productivity per unit chlorophyll in experimentally acidified channels (pH 4.0–4.6) compared with controls (pH 6.5–6.7); however, this was not observed until 33 days of acidification. Arnold and others (1981) also reported significantly higher primary productivity per unit chlorophyll on three of seven dates in an acidic stream (pH 5.2) compared with a circumneutral stream (pH 6.8) in

Pennsylvania. Conversely, Hendrey (1976) reported lower primary productivity per unit chlorophyll in an experimentally acidified channel (pH 4) compared with higher-pH stream channels, despite higher productivity per unit area in the acidified channel. Mulholland and others (1986) found no statistically significant differences in primary productivity per unit chlorophyll among soft water streams ranging in pH from 4.5 to 6.4 in eastern Tennessee, although generally the lowest chlorophyll-specific productivity was measured at the most acidic site.

Several mechanisms have been hypothesized to account for the increased area specific rates of primary production observed in acidified streams. They include (1) low pH preferences of the naturally occurring periphyton community in streams, (2) reduced grazing by macroinvertebrates at low pH, (3) reduced microbial decomposition at low pH, (4) reduced competition with nonperiphytic components of the epilithic community for space and nutrients at low pH, and (5) stimulation due to increased bioavailability of P or some micronutrient at low pH (Hendrey, 1976; Planas and Moreau, 1986; Stokes, 1986). The first of these mechanisms seems unlikely. As discussed in section II,B, shifts in species composition are generally observed with acidification, with reductions in some of the species dominant at circumneutral pH. Although some species may have higher rates of photosynthesis at low pH, others are clearly inhibited, and there is little reason to believe that entire stream periphyton communities are to some extent inhibited at pH values to which they are normally exposed (pH 6–8).

There is more evidence to support the other hypotheses. Several studies indicate a strong effect of reduced invertebrate grazing on periphyton productivity. As discussed in section II,E, studies show that grazing invertebrates as a group are particularly sensitive to acidification; as cited in section II,B, studies show that higher rates of areal primary productivity were correlated with reduced abundance of grazing invertebrates in experimentally acidified channels (Hendrey, 1976; Parent et al., 1986) and in acidic natural streams (Arnold et al., 1981; Mulholland et al., 1986) compared with higher-pH streams. Other experimental stream studies unrelated to the acidification issue have also indicated an inverse relationship between areal primary productivity and abundance of grazing invertebrates (Mulholland et al., 1983; Lamberti et al., 1987).

Reduction in microbial decomposition of periphyton and reduced competition with other components of the epilithic community for space and/or nutrients at low pH may also contribute to increased rates of primary productivity, although no studies have directly addressed these issues. As discussed in section II,A, bacterial activity (decomposition) and abundance (competition) is certainly reduced at low pH; as cited in section II,A, Planas and Moreau (1986) have presented evidence from an experimental stream acidification study of increased P availability at low pH, which could account for the increased rate of primary productivity measured. These authors also noted an increase in S uptake rate that they hypothesize could be the result of increased $SO_4$ concentrations and could have stimulated primary production. However, it is difficult to accept that periphyton are S limited under natural conditions.

The concentration and speciation of dissolved metals are often strongly pH dependent, and the responses of algae to such acidification-induced changes can be either positive or negative (Campbell and Stokes, 1985). Wuhrmann and Eichenberger (1975) experimentally demonstrated a stimulatory effect on primary production from small additions of a variety of trace metals to stream water in artificial stream studies. However, data from other studies indicate that it is doubtful that periphyton productivity is increased as a result of increased availability of some metal micronutrient with acid precipitation. A study involving in situ incubations of stream water and periphyton from an acidic stream (pH 4.8) showed higher rates of primary production when citrate was added to the water to complex trace metals compared with unamended controls, although the differences were not statistically significant because of high variability (Mulholland et al., 1986). The laboratory studies of Gensemer and Kilham (1984) showed that acidification of a standard algal culture medium to pH 6 resulted in reduced algal growth caused by increased concentrations of free trace metals.

Perhaps the most intriguing hypothesis concerning the effects of acidification on periphyton productivity is a reduction resulting from reduced availability of P due to coprecipitation of $PO_4$ and Al (Almer et al., 1978, Dickson, 1978, 1980). This oligotrophication of the stream ecosystem is likely to occur in stream reaches where pH is >5 but that are downstream from highly acidic headwaters with high dissolved Al concentrations. As discussed briefly in section II,B, Mulholland and others (1986) found much higher phosphatase activity in streams with pH >5.6 that are downstream from highly acidic reaches (pH <5), compared with highly acidic streams or streams without acidic headwaters. Studies of longitudinal trends in stream P concentration and experimental neutralization of highly acidic stream water indicated substantial coremoval of $PO_4$ and Al from solution as pH increased downstream (Mulholland and Elwood, unpublished data). Similarly, Nalewajko and Paul (1985) reported reduced algal growth when Al was added to lake water of pH 5.2 to 6.9 and little effect at pH 4.5; they attribute much of the reduction to coprecipitation of Al and $PO_4$ at the higher-pH range. The effect of acid precipitation on streams is likely to be most severe in the headwaters, where total acidic deposition is higher and soil and stream buffering processes are less effective. However, precipitation of amorphous aluminum oxides as pH increases downstream may result in $PO_4$ removed via coprecipitation or adsorption and reduced rates of primary productivity well downstream of acidic stream reaches, thereby extending the effects of stream acidification longitudinally.

## B. Decomposition of Organic Matter

There is considerable evidence that rates of leaf mass loss in streams are reduced as a result of acidification and that this reduction begins to occur when pH drops below 6. Studies comparing acidic streams (pH <5.6) with streams of higher pH in the River Duddon drainage, Great Britain (Minshall and Minshall, 1978), the English Lake district (Chamier, 1987), southern Sweden (Otto and Svensson, 1983), southwestern Pennsylvania (Kimmel et al., 1985), and south-central Ontario (Mackay and Kersey, 1985) have all indicated lower decomposition rates in the more acidic streams. Chamier (1987) also reported that the differences in mass loss rate between acidic and circumneutral streams were much greater for a slowly degrading leaf species (oak) than for a more rapidly degraded species (alder). In a comparison of four soft water streams in eastern Tennessee, ranging in pH from 4.5 to 6.4, Mulholland and others (1987) found that leaf mass loss rate varied directly with stream pH, with significantly lower mass loss rate in streams with pH ≤5.7, compared to one with a pH of 6.4. In a similar comparison of leaf mass loss rates in five Adirondack Mountains streams ranging in pH from 4.75 to 6.14, Osgood (1987) also reported lower rates at the sites with lower pH. Finally, after experimentally acidifying artificial streams from a high pH ranging from 7.0 to 7.4 to a low pH of 4, Burton and others (1985) reported decreased leaf mass loss rates compared to a reference channel.

There is considerable evidence that the reduction in mass loss rate of leaves in acidified streams is the result of reduced microbial activity rather than reduced macroinvertebrate leaf shredding. In a survey of streams in southeastern England, Hildrew and others (1984b) reported that microbial cellulase activity was directly related to pH, with substantial inhibition in streams with a mean annual pH below about 5.6 to 5.8. In her study of English Lake district streams, Chamier (1987) reported that high levels of fungal and bacterial colonization were associated with high rates of leaf mass loss. Mackay and Kersey (1985) incubated leaves in both fine mesh bags (200 $\mu$m) and coarse mesh bags (5 mm) and found that mass loss rates in both types of bags were lower in the acidic streams, despite similar shredder abundance in the two streams. Mulholland and others (1987) found that microbial ATP and respiration rates and bacterial production rates associated with leaf material in streams ranging in pH from 4.6 to 6.4 followed the same pattern as that for leaf mass loss. Further, the number and biomass of macroinvertebrate shredders found in the leaf bags were lowest at the site with the highest pH (6.4) and highest rate of leaf mass loss. However, estimates based on measurements of microbial respiration indicated that microbial oxidation could account for only about one-half of the differences in mass loss rate measured between the highest pH site and the more acidic sites (Mulholland et al., 1987).

As discussed in section II,A, the low rates of microbial decomposition of leaves in acidic streams may be the result of water chemistry factors, such as high concentrations of $H^+$ and Al, or precipitation of $Al(OH)_3$ coatings on leaf surfaces. Reduced microbial decomposition in acidic streams may also be a result of reduced abundance of microinvertebrates. Protozoan grazing of bacteria has been shown to stimulate bacterial metabolism and the oxidation of organic matter in aquatic ecosystems (Javornicky and Prokesova, 1963; Fenchel and Harrison, 1976). Laboratory and field acidification studies conducted in Norway indicated a reduction of protozoans with increasing acidity and showed that lower rates of leaf mass loss may have been related to reduced numbers of microinvertebrates (Leivestad et al., 1976).

We know of no studies of the effect of acidification on microbial decomposition of detritus buried in stream sediments; however, decomposition in lake sediments has been shown to be unaffected by acidification of the overlying water column

(Gahnstrom et al., 1980; Gahnstrom and Fleischer, 1985; Kelly et al., 1984). The insensitivity of the sediment community to water column acidification appears to be the result of microbially mediated alkalinity-generating processes (e.g., denitrification, sulfate reduction) that maintain interstitial water at pH $\geqslant 6$ (Rudd et al., 1986). Microbial decomposition of detritus in stream sediments that are sufficiently organic rich or that have sufficiently low rates of pore water exchange would be expected to be affected very little by stream water acidification.

## C. Secondary Production

### 1. Benthic Macroinvertebrates

We are not aware of any direct comparisons of the estimated production rate of benthic macroinvertebrates among comparable streams that differ in pH. Indirect evidence, however, suggests that the secondary production rate should decrease as pH declines to values $<6.0$. The bases for this are that (1) species and functional group richness decrease with declining pH, as discussed in section IIE,1; (2) the density and biomass of macroinvertebrates have been shown to decrease with declining pH in some streams (e.g., Hall et al., 1980; Arnold et al., 1981; Townsend et al., 1983; Zischke et al., 1983; Kimmel et al., 1985; Allard and Moreau, 1986; Rosemond, 1987); (3) the growth rate of some invertebrate taxa decreases at relatively low pH ($<4.5$) in experimentally acidified streams (Fiance, 1978; Allan and Burton, 1986); and (4) the production rate of bacteria that may provide a source of food for some functional groups of invertebrates decreases at low pH (Palumbo et al., 1987a, 1987b).

These findings together suggest at least a potential for a decline in the production rate of macroinvertebrates in streams with declining pH. Compensating factors such as increases in the density, biomass, and growth rate of acid-tolerant taxa with declining pH, however, could result in little or no change in the secondary production rate of benthic invertebrates despite the loss of acid-sensitive taxa. The fact that some studies show no significant difference in the total density and/or biomass of macroinvertebrates between streams differing in pH (e.g., Harriman and Morrison, 1982; Simpson et al., 1985; Smith et al., submitted) suggests that such compensatory changes occur with declining pH, that other factors are more important than hydrogen ion activity in regulating the total density and biomass of invertebrates in some streams with low pH, or that sampling in these studies was inadequate to detect differences in density and biomass.

Allard and Moreau (1986) reported that invertebrate density in stream channels experimentally treated with acid and with acid plus Al were lower than in the control, whereas invertebrate biomass was not significantly different between the treated and control channels. This lack of a difference in biomass between the acidified and control channels was caused by the presence of a large acid-tolerant species of Chironomidae in the treated channels, which constituted a large fraction of the biomass, essentially accounting for as much biomass as numerous smaller larvae in the control channel. Because small chironomids may have a higher biomass turnover rate than the larger species (Benke et al., 1979; Benke 1984), secondary productivity may have been greater in the control than in the treated channel.

Uncertainty concerning sublethal effects of low pH and Al on the growth and biomass turnover rate of macroinvertebrates precludes drawing any firm conclusion, although results of an experimental study by Hermann and Anderson (1986) suggest that the biomass turnover rate of some acidophilic taxa is more likely to decrease than to increase as pH declines and Al concentration increases. These investigators reported that the respiration rate of three species of acidophilic mayflies increased when exposed to low pH and Al, resulting in a greater expenditure of energy for respiration. This finding implies that the mayflies would have less energy for growth and reproduction. Unfortunately, estimates of biomass turnover in these and other studies of streams differing in pH are lacking. Thus, it is impossible to draw any conclusions about the overall effect of acidification on the production rate of benthic invertebrates in stream ecosystems with declining pH. Future experimental and comparative studies of benthic invertebrates in lotic ecosystems would clearly benefit from inclusion of such measurements.

## D. Nutrient and Carbon Cycling and Availability

Reductions in the input, retention, cycling, and availability of limiting nutrients, such as P, and changes in the available supply of organic C are some of the potential effects of stream acidification that may reduce system productivity. Such alterations could occur as a result of several different factors resulting from acidic deposition. These include (1) reductions in the microbially mediated mineralization of nutrients due to acidification-caused declines in bacterial productivity; (2) increased immobilization of nutrients and dissolved organic C in soil and sediments through adsorption, precipitation, and/or flocculation processes associated with the protonation of exchange sites and the dissolution and subsequent precipitation of oxyhydroxides resulting from the hydrolysis of Al, Fe, and Mn; and (3) alterations in biochemical reactions associated with the enzymatic hydrolysis or organically bound nutrients such as organophosphates. Although there is limited documented evidence for such effects on P and organic C supplies in lakes (Dickson, 1978, 1980; Jansson, 1981), there is even less evidence for such effects in streams.

Omernik (1977) reported that in large areas of the northeastern USA, the concentration of both total P and orthophosphorus in stream water was overpredicted by regionally based empirical models of the relationship between land use in the surrounding watershed and levels of these two P fractions in stream water. The areal distribution of streams with negative residual errors (i.e., those for which the observed concentration was less than the predicted concentration) for the predictive model shows that many are in areas of the eastern USA known to be sensitive to acidic deposition. Omernik also reported significant correlations between both total and orthophosphorus concentration in stream water and pH of the surface soil in the watershed. Both of these results are generally consistent with that expected if the pH-dependent adsorption of P by acidic soils were the primary mechanism controlling P mobility and transport in drainage waters leading to streams, and if coprecipitation and adsorption of P by hydrous oxides of Al, Fe, and Mn mobilized under acid conditions were occurring. It must be emphasized, however, that this is circumstantial evidence and does not prove that a causal relationship exists between acidic deposition, soil acidity, and P supplies in stream water.

In a study of acidic and circumneutral streams in the Great Smoky Mountains National Park, Tennessee, low concentrations ($\leq 2$ μg/L of soluble reactive phosphorus (SRP) were found at all sites, and invididual samples were frequently below the analytical detection limit of 0.5 μg/L (Elwood and Mulholland, in preparation). Phosphorus was likely to have been the limiting nutrient at all of the sites based on the high N-to-P ratio in stream water. Soluble reactive phosphorus concentrations were generally lowest at the sites with intermediate pH levels (5–6) and highest at pH levels <5 and >6 (Figure 4–3). A similar pattern was observed for DOC cconcentration (Figure 4–4), suggesting that there was a common, pH-related mechanism controlling the concentration of both SRP and DOC at

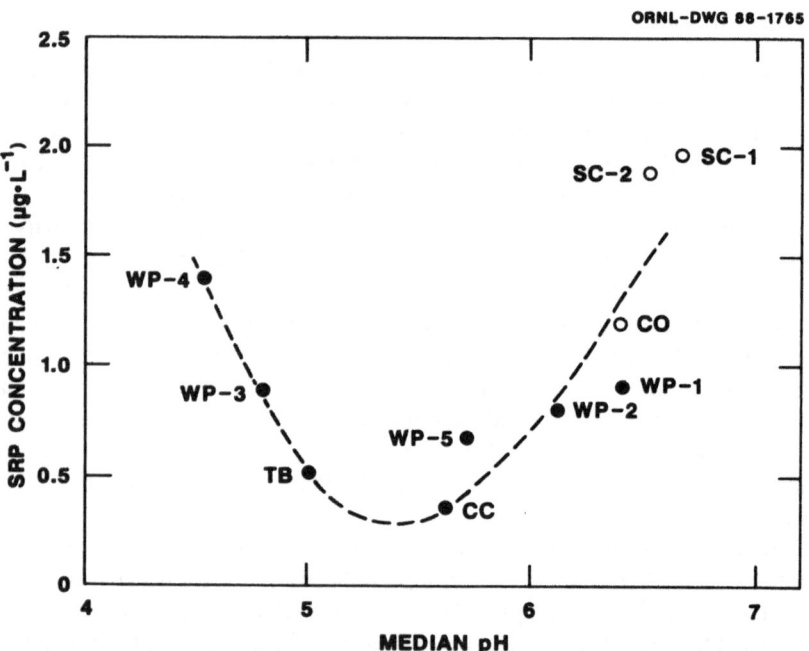

**Figure 4-3.** Relationship between the median SRP concentration and the median pH of stream water at sites in the Southern Blue Ridge Province of Tennessee and North Carolina. Values plotted as 0.5mg/L SRP, which is the analytical detection limit, were actually <0.5 μg/L.

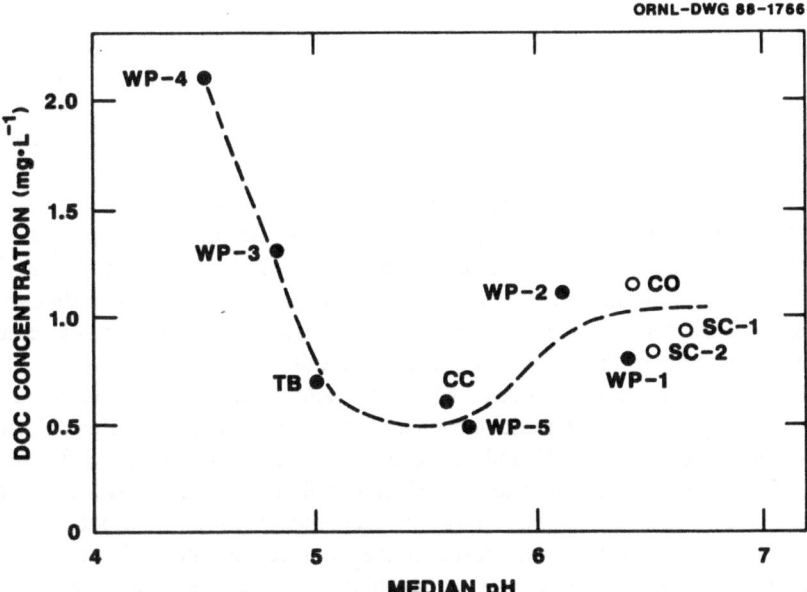

ORNL-DWG 88-1766

**Figure 4-4.** Relationship between the median DOC concentration and the median pH of stream water at sites in the Southern Blue Ridge Province, Tennessee and North Carolina.

these stream sites. Sites with a pH >6 and SRP concentrations >1 μg/L either lacked acidic upstream reaches with high inorganic monomeric Al concentrations or were located downstream of tributary inputs that are nonacidic. These patterns in SRP, DOC, and Al concentrations with pH suggest that Al precipitation via coprecipitation or adsorption reactions downstream from acidic reaches reduce SRP and DOC concentrations.

To determine if these differences in SRP concentrations reflected differences in the available supply of P to epilithic microbes growing on rocks on the stream bottom, we conducted assays of phosphatase activity (using nitrophenyl phosphate as the substrate). The turnover of orthophosphorus by epilithic microbes on rock from several sites was also estimated from measurements of the uptake rate of carrier-free $^{32}PO_4$ from stream water by the attached microbes. At the sites where the SRP concentration was generally the lowest, ATP-specific phosphatase activity was greatest, indicating that P limitation of the epilithic community was greater at the sites with the low orthophosphate concentration than at the sites with a higher SRP concentration.

The relative turnover of orthophosphorus was also greater at the intermediate pH (5–6) sites than at sites with either a higher (>6.0) or lower (<5.0) pH. This pattern is thus consistent with the SRP concentrations in water and the ATP-specific phosphatase activity of epilithic microbes and supports the hypothesis that P limitation is greatest at sites located downstream of acidic reaches in the drainage. The role of Al in scavenging P and DOC from the water column is suggested by the fact that there was a downstream decrease in the concentration of Al with increasing pH that could not be accounted for by dilution alone. This decline in Al suggests that Al is precipitating in the water column and/or adsorbing to surfaces on the stream bottom. Based on the known adsorption efficiency of Al trihydroxides for P (Kee Kwong and Huang, 1978; Hsu and Rennie, 1962; Sims and Ellis, 1983) and DOC (Davis, 1982), it seems likely that the scavenging of P and DOC from the water column by adsorption to particulate aluminum oxyhydroxide is at least partly responsible for the lower concentration of these materials and the greater P limitation at sites of intermediate pH. These results imply that the effect of acidification on the available supply of P and DOC in streams can extend well downstream of the most acidic reaches of a drainage.

In a somewhat similar study, McKnight and Feder (1984) compared the concentration of "filtrable ionic phosphate" in two upstream tributaries, of which one was acidic (pH of 3.75) and the second was circumneutral (pH 7.3), and at sites downstream of the confluence of these tributaries where Al and Fe was precipitating. The P concentration was lowest in the two upstream tributaries (both 2 μg/L) and increased downstream of the confluence where Fe and Al was precipitating. This increase in P concentration in the presence of Fe and Al precipitation is contrary to expectations and suggests either that colloids containing P passed through the filters and were included in the analysis of "filtrable ionic phosphate" at the downstream sites or that there was an additional, unaccounted-for source of P input to the stream between the upstream tributaries and the downstream sampling point.

In experiments designed to simulate acidic snowmelt episodes, Hall and others (1987) examined the effects of short-term (several hours) inputs of $H^+$ and Al on the chemistry and biology of a poorly buffered mountain stream in New Hampshire. Hydrochloric acid and AlCl, were added in separate experiments to first- to third-order reaches of the stream. Total P, Ca, Mg, and Al concentrations increased in the stream during the addition of HCl, apparently a result of dissolution of minerals and exchange reactions in the stream sediments. Samples for total P analysis were not filtered, so the contribution of suspended particles to the net increase in total P concentration is unknown.

Total P, Ca, and Mg were also mobilized from sediments during the addition of Al, presumably due to weathering and exchange reactions associated with the depression in pH caused by the hydrolysis of the added Al. These results indicate that episodic pH depression may mobilize P associated with stream sediments as a result of weathering and ion exchange reactions. They found no evidence that P was removed from the water column by the increased concentrations of Al associated with the additions of HCl and $AlCl_3$.

This apparent lack of scavenging of P with increasing Al concentrations could be a result of insufficient Al precipitation, the inclusion of particle-bound P in the analysis for total P, or a combination of both. If the concentration of soluble Al in stream water was less than the solubility product of aluminum trihydroxide phases, Al precipitation and the resultant scavenging of P from the water column due to coprecipitation of $AlPO_4$ and adsorption of P to Al hydroxides would not be expected. If Al precipitation did occur and sorbed P was bound to aluminum oxyhydroxide particles, the P associated with these or other particles may have been resolubilized during the oxidation of samples for total P analysis. In either case, whether a significant fraction of the P mobilized from sediments during episodic pH depressions would be available to stream organisms during these high flow periods is problematic given that the biomass of the epilithic community is frequently reduced during periods of high discharge due to scouring by suspended sediments.

In summary, data on P and DOC concentrations in acidic and circumneutral streams indicate that both SRP and DOC concentrations tend to be the lowest in acidic streams with intermediate pH (5.0–5.5) and to increase at pH <5 and >6. Phosphorus limitation is generally inversely related to orthophosphate concentration.

## IV. Future Research Needs

Despite a substantial body of literature on the biological and ecological effects of stream acidification, there is a need for additional comparative and experimental research on streams to define more clearly and quantify the direct and indirect effects of acidification on certain taxonomic groups and processes. For heterotrophic microorganisms, research is needed on the direct effects of water quality changes associated with acidification on various microbially mediated processes and on the biomass and biochemical and physiological activity of fungi. Research is also needed on the direct and indirect effects of acidification on the species composition, abundance, and productivity of benthic algae and bryophytes (mosses and liverworts) in streams.

In terms of the invertebrate fauna in streams, research is needed, in particular, on the functional role of various taxa of the so-called meiofauna and how these taxa are directly and indirectly affected by stream acidification. Additional research is also needed on the direct and indirect effects of acidification on representative taxa of stream macroinvertebrates that live on or above the sediment-water interface (so-called epifauna) and that live in the sediments (so-called infauna). Studies are also needed to determine the direct toxic effects on benthic meiofauna and macroinvertebrates of episodic changes in stream chemistry associated with storms and snowmelt events. Because of the importance of macroinvertebrates as a food resource for some stream fishes, experimental and comparative studies should be conducted to determine the overall effects of acidification on the growth, biomass turnover rate, and production rate of the dominant invertebrate taxa in streams.

Similarly, additional studies are needed to define better the chemical dose and acute effects of episodic acidification on various life history stages of stream fishes. Research is also needed to determine the sublethal effects on stream fish of changes in water chemistry associated with acidification, including the effects on growth, reproduction, and behavior.

Finally, experimental and comparative studies are needed to more clearly define the effects of acidification on the cycling, transport, and availability of nutrients in streams. Research is also needed on the effects of acidification on both the concentration of soluble organic carbon substrata and the microbial oxidation rate of particulate organic carbon in or above stream sediments.

Some of the above research needs can best be accomplished with carefully designed comparative studies, but others will require detailed experimental studies under tighly controlled conditions. An important, overriding consideration in the design of the studies must be the inclusion of environmental condition, including chemical doses, that duplicate as close as possible those likely to be encountered by test organisms in natural streams that have been acidified or are undergoing acidification.

## Acknowledgments

We wish to extend our thanks to S.W. Christensen, C.C. Coutant, H.L. Boston, A.V. Palumbo, and D.C. Adriano for their constructive reviews of this manuscript.

This work was sponsored in part by the Electric Power Research Institute under contract RP2326-1 with Martin Marietta Energy Systems, Inc., under contract DE-AC05-84OR21400 with the U. S. Department of Energy. This chapter is Publication No. 3162, Environmental Sciences Division, ORNL.

## References

Allan, J.W. and T.M. Burton. 1986. *In* B.G. Isom, S.D. Dennis and J.M. Bates, eds. *Impact of acid rain and deposition on aquatic biological systems*, 54–66. American Society of Testing and Materials, Philadelphia.

Allard, M. and G. Moreau. 1985. Can J Fish Aquat Sci 42:1676–1680.

Allard, M. and G. Moreau. 1986. Air Water Soil Pollut 31:673–679.

Almer, B., W. Dickson, C. Ekstrom, and E. Hornstrom. 1978. *In* J.O. Nriagu, ed. *Sulfur in the environment: Part II, ecological impacts*. John Wiley and Sons, New York.

Anderson, B., and P. Anderson. 1984. Rept Inst Freshw Res Drottningholm 61:28–33.

Anderson, P., and P. Nyberg. 1984. Rept. Inst. Freshw. Res., Drottningholm, 61:34–47.

Arnold, D.E., P.M. Bender, A.B. Hale, and R.W. Light. 1981. *In* R. Singer, ed. *Effects of acidic precipitation on benthos*, 15–33. North American Benthological Society, Springfield, Ill.

Arnold, D.E., R.W. Light, and V.J. Dymond. 1980. *Probable effects of acid precipitation on Pennsylvania waters*. EPA-600/3-880-012, Corvallis Environmental Research Laboratory, U.S. Environmental Protection Agency, Corvallis, Oregon. 20 p.

Aston, R.J., K. Sadler, A.G.P. Milner, and S. Lynam. 1985. *The effects of pH and related factors on streams invertebrates*. Central Electric Generating Board, CERL No. TPRD/L/2792/N84, United Kingdom. 24 p.

Baker, J.P., and C.L. Schofield. 1982. Water Air soil Pollut 18:289–309.

Barlocher, F. 1980. Oecologia 47:303–306.

Barlocher, F. 1987. Can J Bot 65:76–79.

Barlocher, F. and B. Kendrick. 1974. J Ecol 62:761–791.

Barlocher, F. and J. Rosset. 1981. Trans Brit Mycol Soc 76:479–483.

Benke, A.C. 1984. *In* V.H. Resh and D.M. Rosenberg, eds. *The Ecology of aquatic insects*, 289–322. Praeger Scientific, New York.

Benke, A.C., D.M. Gillespie, F.K. Parrish, T.C. Van Arsdall, Jr., R.J. Hunter, and R.L. Henry III. 1979. *Biological basis for assessing impacts of channel modification: Invertebrate production, drift, and fish feeding in a southeastern blackwater river.* Environmental Resource Center, Georgia Institute of Technology, Atlanta.

Bick, H., and E.F. Drews. 1973. Hydrobiologia 42:393–402.

Brock, T.D. 1973. Science 179:480–483.

Brown, D. 1982. Water Air Soil Pollut 18:343–351.

Brown, D., and R.W. Brocksen. 1984. *In A specialty conference on atmospheric deposition, November, 1982*, 342–351. Air Pollution Control Association, Detroit.

Burton, T.M. and J.W. Allan. 1986. Can J Fish Aquat Sci 43:1285–1289.

Burton, T.M., R.M. Stanford and J.W. Allan. 1982. *In* F.M. D'Itri, ed. *Acid precipitation, effects on ecological systems*, 209–235. Ann Arbor Science, Ann Arbor, Michigan.

Burton, T.M., R.M. Stanford, and J.W. Allan. 1985. Can J Fish Aquat Sci 42:669–675.

Campbell, P.G.C., and P.M. Stokes. 1985. Can J Fish Aquat Sci 42:2034–2049.

Chamier, A.-C. 1985. Bot J Linn Soc 91:67–81.

Chamier, A.-C. 1987. Oecologia 71:491–500.

Chamier, A.-C. and P.A. Dixon. 1982. J Gen Microbiol 128:2469–2483.

Chapman, D.W., and R. Demory. 1963. Ecology 44:140–146.

Collins, N.C., A.P. Zimmerman, and R. Knoechel. 1981. *In* R. Singer, ed. *Effects of acidic precipitation on benthos*, 35–48. North American Benthological Society, Springfield, Ill.

Cook, R.B., C.A. Kelley, D.W. Schindler, and M.A. Turner. 1986. Limnol Oceanogr 31:134–148.

Cummins, K.W., R.C. Petersen, F.O. Howard, J.C. Wuycheck, and V.I. Holt. 1973. Ecology 54:336–345.

Davis, J.A., 1982. Geochim Cosmochim Acta 46:2381–2393.

Degerman, E., F.-E. Fogelgren, B. Tengelin, and E. Thornelof. 1986. Air Water Soil Pollut 30:665–671.

Dickson, W. 1978. Verh Internat Verein Limnol 20:851–856.

Dickson, W. 1980. *In* D. Drablos and A. Tollan, eds. *Proceedings International Conference on Ecological Impact of Acid Precipitation*, SNSF Project, 75–83. Norwegian Ministry of Environment, Oslo.

Driscoll, C.T., J.P. Baker, J.J. Bisogni, Jr., and C.L. Schofield 1980. Nature 284: 161–164.

Driscoll, C.T., B.J. Wyskowski, C.C. Cosentini, and M.E. Smith. 1987. Biogeochemistry 3:225–241.

Duthie, H.C., P.B. Hamilton. 1982. *In* R.G. Wetzel, ed. *Periphyton of Freshwater Ecosystems: Proceedings First International Workshop on Periphyton of Freshwater Ecosystems*, 185–190, Dr. W. Junk, The Hague, Netherlands.

Engblom, E., and P.E. Lingdell. 1984. Rept Inst Freshw Res Drottningholm 61:60–68.

Faucon, A.S., and W.D. Hummon. 1976. Hydrobiologia 50:205–209.

Fenchel, T., and P. Harrison. 1975. *In* J.M. Anderson and A. Macfadyen, eds. *The role of terrestrial and aquatic organisms in decomposition processes*, 285–299. Blackwell Scientific Publishers, Oxford, U.K.

Fiance, S.B. 1978. Oikos 31:332–339.

Foy, C.D., and G.C. Gerloff. 1972. J Phycol 8:268–271.

Gagen, C.J., and W.E. Sharpe. 1987. Bull Environ Contam Toxicol 39:7–14.

Gahnstrom, G., G. Andersson and S. Fleischer. 1980. *In* D. Drablos and A. Tollan, eds. *Proceedings International Conference on Ecological Impact of Acid Precipitation*, SNSF Project, 306–307. Norwegian Ministry of Environment, Oslo.

Gahnstrom, G., and S. Fleischer. 1985. Ecological Bulletins 37:287–292.

Gensemer, R.W., and S.S. Kilham. 1984. Can J Fish Aquat Sci 41:1240–1243.

Goldman, J.C., W.J. Oswald, D. Jenkins. 1974. J Water Poll Control Fed 46:554–574.

Grahn, O., H. Hultberg, L. Landner. 1974. Ambio 3:93–94.

Grimm, N.B., and S.G. Fisher. 1984. Hydrobiologia 111:219–228.

Haines, T. 1986. *In Acid deposition: Long-term trends*, 300–304. National Academy Press, Washington, D.C.

Hall, R.J., C.T. Driscoll, and G.E. Likens. 1987. Freshw Biol 17: (in press).

Hall, R.J., and F.P. Ide. 1987. Can J Fish Aquat Sci 44:1652–1657.

Hall, R.J., G.E. Likens, S.B. Fiance, and G.R. Hendrey. 1980. Ecology 61:976–989.

Hanlon, R.D.G., and J.M. Anderson. 1979. Oecologia 38:93–99.

Harriman, R., and B.R.S. Morrison. 1982. Hydrobiologia 88:251–263.

Harrison, R., B.R.S. Morrison, L.A. Caines, P. Collen, and A.W. Watt. 1987. Water Air Soil Pollut 32:89–112.

Havas, M. 1981. *In* R. Singer, ed. *Effects of acidic precipitation on benthos*, 49–55. North American Benthological Society, Springfield, Ill.

Havas, M., and T.C. Hutchinson. 1983. Nature 301:23–27.

Helliwell, S., G.E. Batley, T.M. Florence and B.G. Lumsden. 1983. Environ Technol Lett 4:141–144.

Hendrey, G.R. 1976. *Effects of pH on the growth of periphytic algae in artificial stream channels*. Sur Nedbors Virkning pa Skog og Fisk Project IR 25/76.

Hendrey, G.R., K. Baalsrud, T.S. Traaen, M. Laake, and G. Raddum. 1976. Ambio 5:224–227.

Hendrey, G.R, and F.A. Vertucci. 1980. *In* D. Drablos and A. Tollan, eds. *Proceedings of the International Conference on Ecological Impact of Acid Precipitation*, SNSF Project, 314–315. Norwegian Ministry of Environment, Oslo.

Herrmann, J., and K.G. Anderson. 1986. Water Air Soil Pollut 30:703–709.

Hildrew, A.G., C.R. Townsend, and J. Francis. 1984a. Freshw Biol 14:297–310.

Hildrew, A.G., C.R. Townsend, J. Francis, and K. Finch. 1984b. Freshw Biol 14: 323–328.

Hsu, P.H., and D.A. Rennie. 1962. Can J Soil Sci 28:48–61.

Hutchinson, N., K. Holtze, J. Munro, and T. Pawson. 1987. *In* H. Witters and O. Vanderborght, eds. *Ecophysiology of acid stress in aquatic organisms*, Annals of the Royal Zoological Society of Belgium, vol. 117, 201–218, Supplement 1. Brussels.

Iqbal, S.H., and I. Webster. 1973. Trans Brit Mycol Soc 61:331–346.

Iqbal, S.H., and J. Webster. 1977. Trans Brit Mycol Soc 69:233–241.

Jansson, M. 1981. Arch Hydrobiol 93:32–44.

Javornicky, P., and V. Prokesova. 1963. Int Revue Ges Hydrobiol 48:335–350.

Jensen, K.W., and E. Snekvik. 1972. Ambio 1:223–225.

Jones, J.R.E. 1948. J Anim Ecology 17:51–65.

Jones, J.R.E. 1949. J Animal Ecol 18:67–88.

Kaufmann, P.R., A.T. Herlihy, J.W. Elwood, M.E. Mitch, W.S. Overton, M.J. Sale, J.J. Messer, K.A. Cougan, D.V. Peck, K.H. Reckhow, A.J. Kinney, S.J. Christie, D.D. Brown, C.A. Hagley, and H.I. Jager. 1988. *Chemical characteristics of streams in the Mid-Atlantic and Southeastern United States. Volume I: Population Descriptions and*

*Physico-Chemical Relationships*. EPA/600/3-88/21a. U.S. Environmental Protection Agency, Washington, D.C.

Kaushik, N.K., and H.B.N. Hynes. 1971. Arch Hydrobiol 68:1465–1515.

Kee Kwong, K.E.N., and P.M. Huang. 1978. Nature 271:336–338.

Kelly, C.A., and J.W.M. Rudd. 1984. Biogeochemistry 1:63–77.

Kelly, C.A., J.W.M. Rudd, A. Furutani, and D.W. Schindler. 1984. Limnol Oceanogr 29:687–694.

Kimmel, W.G., D.J. Murphey, W.E. Sharpe, and D.R. DeWalle. 1985. Hydrobiologia 124:97–102.

Krulwich, T.A., and A.A. Guffanti. 1983. Ad in Microb Physiol 24:173–214.

Lamberti, G.A., L.R. Ashkenas, S.V. Gregory, and A.D. Steinman. 1987. J North Amer Benthological Society 6:92–104.

Lamberti, G.A., and V.H. Resh. 1983. Ecology 64:1124–1135.

Leivestad, H. 1982. *In* R.E. Johnson, ed. *Acid Rain/Fisheries*, 157–164. American Fisheries Society, Bethesda, Md.

Leivestad, H., G. Hendrey, I.P. Muniz, and E. Snekvik. 1976. *In* F.H. Braekke, ed. *Impact of acid precipitation on forest and freshwater ecosystems in Norway*, SNSF Project Research Report FR6, 87–111. Norwegian Ministry of Environment, Oslo.

Leivestad, H., and I.P. Muniz. 1976. Nature 259:391–392.

Mackay, R.J., and K.E. Kersey. 1985. Hydrobiologia 122:3–11.

McKnight, D.M., and G.L. Feder. 1984. Hydrobiologia 119:129–138.

Matthias, U. 1983. Archiv Hydrobiol Suppl 65:407–483.

Maurice, C.G., R.L. Lowe, T.M. Burton, and R.M. Stanford. 1987. Water Air Soil Pollution 33:165–177.

Merritt, R.W., K.W. Cummins, and T.M. Burton. 1984. *In* V.H. Resh and D.M. Rosenberg, eds. *The ecology of aquatic insects*, 134–163. Praeger Scientific, New York.

Minshall, G.W., and R.A. Kuehne. 1969. Arch Hydrobiol 66:169–191.

Minshall, G.W., and J.N. Minshall. 1978. Arch Hydrobiol 83:324–355.

Morrison, S.J., and D.C. White. 1980. Appl Environ Microbiol 40:659–671.

Moss, B. 1973. J Ecology 61:157–177.

Mulholland, P.J., J.W. Elwood, J.D. Newbold, and L.A. Ferren. 1985. Oecologia 66:199–206.

Mulholland, P.J., J.W. Elwood, A.V. Palumbo, and R.J Stevenson. 1986. Can J Fish Aquat Sci 43:1846–1858.

Mulholland, P.J., J.D. Newbold, J.W. Elwood, and C.L. Hom. 1983. Oecologia 58:388–366.

Mulholland, P.J., A.V. Palumbo, J.W. Elwood, and A.D. Rosemond. 1987. J North American Benthological Society 6:147–158.

Muniz, I.P., and H.P. Leivestad. 1980. *In* D. Drablos and A. Tollan, eds. *Ecological Impact of Acid Precipitation*, Proc. Int. Conf. SNSF Project, 320–321. Oslo.

Nalewajko, C., and B. Paul. 1985. Can J Fish Aquat Sci 42:1946–1953.

Omernik, J. 1977. *Nonpoint source—stream nutrient level relationships: A nationwide survey*. U.S. Environmental Protection Agency, EPA-600/3-77-105, Corvallis Environmental Research Laboratory, Corvallis, Oregon.

Ormerod, S.J., P. Boole, C.P. McCahon, N.S. Weatherley, D. Pascoe, and R.W. Edwards. 1987. Freshw Biol 17:341–356.

Osgood, M.P. 1987. Microbiological studies of three Adirondack streams exhibiting pH gradients. Ph.D. Thesis, Rensselaer Polytechnic Institute, Troy, N.Y.

Otto, C., and B.S. Svensson. 1983. Arch Hydrobiol 99:15–36.

Overrein, L.N., H.M. Seip, and A. Tollan. 1980. *Acid precipitation: Effects on forest and fish*. SNSF Project Final Report FR19, Norwegian Ministry of Environment, Oslo.

Palumbo, A.V., M.A. Bogle, R.R. Turner, J.W. Elwood, and P.J. Mulholland. 1987a. Appl Environ Microbiol 53:337–344.

Palumbo, A.V., P.J. Mulholland, and J.W. Elwood. 1987b. Can J Fish Aquatic Sci 44:1064–1070.

Palumbo, A.V., P.J. Mulholland, and J.W. Elwood. 1988. *In* S.S. Rao, ed. *Microbial interactions in acid stressed aquatic ecosystems*, 69–90, CRC Press, Boca Raton, Fla.

Parent L., M. Allard, D. Planas, and G. Moreau. 1986. *In* B.G. Isom, S.D. Dennis, and J.M. Bates, eds. *Impact of acid rain and deposition on aquatic biological systems*, 28–41, ASTM STP 928, American Society for Testing and Materials, Philadelphia.

Patrick, R. 1977. Amer Scientist 66:185–191.

Patrick, R., N.A. Roberts, and B. Davis. 1968. *The effect of changes in pH on the structure of diatom communities*. Notulae Naturae of Academy Natural Sciences of Philadelphia, Number 416.

Planas, D., and Moreau. 1986. Water Air Soil Pollution 30:681–686.

Pratt, J.M., and R.J. Hall. 1981. *In* R. Singer, ed. *Effects of acid precipitation on benthos*, 77–95. North American Benthological Society, Hamilton, N.Y.

Raddum, G.G. 1980. *In* D. Drablos and A. Tollan, eds. *Proceedings of an International Conference on the Ecological Impact of Acid Precipitation*, SNSF Project, 330–331. As, Norway.

Raddum, G.G., and A. Fjellheim. 1984. Verh Internat Verein Limnol 22:1973–1980.

Raven, J.A., and J. Beardall. 1981. British Phycol J 16:165–175.

Rimes, C.A., and R. Goulder. 1986. Freshwater Biol 16:633–651.

Roelofs, J.G.M. 1983. Aquatic Botany 17:139–155.

Rosemond, A.D. 1987. *The effects of acidification on the macroinvetebrate communities of streams*. Masters Thesis, University of North Carolina, Chapel Hill.

Rosset, J., and F. Barlocher. 1985. Trans Br Mycol Soc 84:137–145.

Rudd, J.W.M., C.A. Kelly, V. St. Louis, R.H. Hesslein, A. Furutani, and M.H. Holoka. 1986. Limnol Oceanogr 31:1267–1280.

Sadler, K., and A.W.H. Turnpenny. 1986. Water Air Soil Pollut 30:593–599.

Satake, K., and K. Miyassaka. 1984. J Bryol 13:277–280.

Schiff, S.L., and R.F. Anderson. 1986. Water Air Soil Pollut 30:941–948.

Schofield, C.L., and J.R. Trojner. 1979. *In* T. Toribara, ed. *Proceedings 12th Rochester International Conference on Environmental Toxicology*, 341–366, University of Rochester, Rochester, N.Y.

Sharpe, W., W. Kimmel, E. Young, and D. DeWalle. 1983. Northeast Environ Sci 2:171–178.

Sharpe, W.E., D.R. De Walle, R.T. Leibfried, R.S. Dinicola, W.G. Kimmel, and L.S. Sherwin. J Environ Qual 13:619–631.

Shearer, C.A., and J. Webster. 1985. Trans Brit Mycol Soc 84:489–501.

Simpson, K.W., R.W. Bode, and J.R. Colquhoun. 1985. Freshw Biol 15:671–681.

Sims, J.T., and B.G. Ellis. 1983. Soil Sci Soc Am J 47:912–916.

Smith, M.E., B.J. Wyskowski, C.M. Brooks, C.T. Driscoll, and C.C. Cosentini. Can J Fish Aquat Sci (submitted).

Steinman, A.D., C.D. McIntire, S.V. Gregory, G.A. Lamberti, and L. Ashkenas. 1987. Journal North American Benthological Society 6:175–188.

Stokes, P.M. 1986. Water Air Soil Pollut 30:421–438.

Stoner, J.H., A.S. Gee, and K.R. Wade. 1984. Environ Pollut (Series A) 35:125–157.

Suberkropp, K.F., and M.J. Klug. 1976. Ecology 57:707–719.

Sutcliffe, D.W., and T.R. Carrick. 1973. Freshw Biol 3:437–462.

Townsend, C.R., and A.G. Hildrew. 1984. Verh Internat Verein Limnol 22:1953–1958.

Townsend, C.R., A.G. Hildrew, and J. Francis. 1983. Freshw Biol 13:521–544.

Wallace, J.B., J.R. Webster, and T.F. Cuffney. 1982. Oecologia 53:197–200.

Watt, W.D., C.D. Scott, and W.J. White. 1983. Can J Fish Aquat Sci 40:462–473.

Wigglesworth, V.B. 1938. J Exptl Biol 15:235–247.

Willoughby, L.G., and R.G. Mappin. 1988. Freshw Biol 19:145–155.

Winterbourn, M.J., A.G. Hildrew, and A. Box. 1985. Fresh Biol 15:363–374.

Wood, C.M., and D.G. McDonald. 1982. *In* R.E. Johnson, ed. *Acid rain/fisheries*. American Fisheries Society, Bethesda, Md.

Wood-Eggenschwiler, S., and F. Barlocher. 1983. Trans Brit Mycol Soc 81:371–379.

Wuhrmann, K., and E. Eichenberger. 1975. Verh Internat Verein Limnol 19:2028–2034.

Ziemann, H. 1975. Int Rev Ges Hydrobiol 60:523–555 (German abstract in English).

Zischke, J.A., F.W. Arthur, K.J. Nordlie, R.O. Hermanutz, D.A. Standen, and T.P. Henry. 1983. Water Res 17:47–63.

# Effects of Acidic Precipitation on Lake Ecosystems

Harold H. Harvey*

## Abstract

The acidification of surface waters has been demonstrated by direct observation of declining pH, by reconstructions of historic pH values by means of sediment diatoms, and by declining alkalinity. The direct result of acidic deposition is cation depletion from sensitive watersheds. Anion composition shifts from bicarbonate to sulfate and nitrate, with the associated loss of buffering and increasing pH instability. At lower pH, metal recruitment from the watershed and mobilization from sediments is increased, resulting in higher concentrations in the water column.

In acidifying waters, aquatic organisms are exposed to the toxicity of hydrogen ion and metals, especially monomeric aluminum. Both $H^+$ and Al ion species interfere with Na-K ATPase in gills and thus regulation of plasma sodium and chloride. Episodic pH depressions induced by an acidic rain or snowmelt can bring about the death of fishes, both larval and adult. Such losses result in the failure to recruit new age classes into the population or the juvenilization of the population. Amphibians are intolerant of low pH and are susceptible to acidic stress as larvae in meltwater ponds and stream pools. Fish-eating birds appear to suffer as acidification lessens their food supply, but birds feeding on invertebrates may benefit via reduced competition from fishes.

Many invertebrates—snails, clams, some insects, and crayfish—are sensitive to hydrogen ion and are lost from lakes during acidification. This loss of invertebrate and fish diversity results in much simpler ecosystems, with more nutrients cycled through filamentous algae, moss, and fungi. These adverse effects of acidification can be reversed at least in part by lake neutralization: a clear demonstration of the cause–effects relationship between acidic deposition and loss of the biota.

*Department of Zoology, University of Toronto, Toronto M5S 1A1, Canada.

# I. Introduction

That atmospheric pollutants could result in acid deposition has been known for a long time. In his book *Air and Rain*, Angus Smith (1852) described the acidic nature of precipitation falling on Manchester, England. It has also been known for many years that man-made pollutants could be carried long distances. In *Brand*, published in 1866, Henrik Ibsen wrote: "A sickening fog of smoke from British coal drops in a grimy pool upon the land. . . ." Ibsen's home was in southeastern Norway. The linkage among long-range transport of pollutants, acidic deposition, and status of the biota in fresh waters was slow in coming. In 1922 Professor Dahl of the University of Oslo published a letter that he had received from a Dr. Høyer in Arendahl (southeastern Norway). In free translation, Dr. Høyer reported that in 1915 there were no small brown trout in Holmevatten, a lake. They were, however, in Myovatten, the next lake downstream, and he determined to catch some, but when he went back the following year, there were none to be found. This may have been the earliest description of the loss of brown trout from the lakes of southern Norway. In 1927, Dahl noted "a small change in direction to acid may have fatal consequences" for trout. Unfortunately, the significance of these observations went unrecognized for another half century. Dannevig (1959, 1966) identified the impact of precipitation on the acidity of surface waters, plus the effect on fish populations: reduced population density, faster growth, and larger size.

In the 1960s researchers in Sweden, most notably Svante Odén, were documenting trends of declining pH in surface waters (e.g., Odén and Ahl, 1970). Dickson (1970) reported the distribution of acidified lakes in western Sweden, and Hultberg and Stensson (1970) identified the effects of this acidification on the fish fauna in some of the lakes in southwestern Sweden. This surge in research into acidification became the basis of Sweden's case study for the United Nations Conference on the Human Environment (Royal Ministry, 1971).

In North America early studies of acidification were in close proximity to large point sources of pollution. Gorham and Gordon (1960) reported the very low pH of ponds south of the Sudbury smelter and northeast of the sintering plant at Wawa, Ontario (Gordon and Gorham, 1963). The acidification of the La Cloche Mountain lakes was documented by Beamish and Harvey (1971, 1972) as was the large-scale loss of fish populations (Beamish et al., 1975; Harvey, 1975). The difficulties encountered in public perception of the acidification phenomenon at this time have been reviewed by Brydges (1988). The acidification of the Adirondack Mountain lakes was described by Schofield (1976), based in part on a comparison of pH and alkalinity with data from the 1930s on the same water bodies. Brook trout and/or other fishes now are known to have disappeared from 49 Adirondack lakes, and in as many as 200 to 400 lakes fish populations may have been lost (Haines and Baker, 1986). The acidification of the Adirondack lakes plus the timely publications of Likens and others (1972, 1979) and Likens (1976) shifted the problem focus away from point sources and toward the long-range transport of acid-forming emissions. Interest in acidic precipitation was increasing rapidly during the mid-1970s, culminating in the First International Symposium on Acid Precipitation and the Forest Ecosystem, in Columbus, Ohio, in 1975 (Dochinger and Seliga, 1976).

By the mid to late 1970s, there was sufficient widespread evidence for the acidification of surface waters to stimulate federal, provincial and state governments to fund large-scale and long-term investigations into the causes and effects of acidic rain. With respect to the lacustrine environment, the objectives of this research were to:

1. Establish the sensitivity of surface waters, as defined by alkalinity, pH, and changing composition of cations and anions
2. Establish the susceptibility of sensitive waters to acidification by superimposing annual acidic deposition per unit area onto the sensitive areas, as defined above
3. Establish the rates of change in chemical water quality for surface waters, both by direct measurement, as in the calibrated watersheds, and by means of mathematical modeling
4. Assess the extent of acidification effects on the biota, most especially in terms of physiological stress on individuals, changing species abundance, and extinction of populations

The report of Leivestad and Muniz (1976) was the first linking an episodic pH depression, resulting from the melt of acidic snow, to a large-scale fish kill. Since then surprisingly little has been published of the stress on and mortality to the biota resulting from rain- or snow-induced pH depressions (Harvey and Whelpdale, 1986).

The major study of acidification in Norway was launched in early 1972 as the interdisciplinary research program Acid Precipitation: Effects on Forest and Fish (the SNSF project). This massive effort culminated in the international conference Ecological Impact of Acid Precipitation (Drabløs and Tolan, 1980) held at Sandefjord, Norway.

The large Canadian research programs of the late 1970s and early 1980s sparked the conference International Symposium on Acidic Precipitation at Muskoka, Ontario, in 1985 (Martin, 1986). All of these major conferences drew attention to the acidification problem and helped to define the direction of future research. The next such milestone conference is scheduled for 1990 in Glasgow, Scotland. A historical review of investigations into the acidic precipitation phenomenon may be found in Cowling (1982).

## II. Discussion

### A. Acidification of Surface Waters

Acidification of surface waters implies loss of alkalinity; that is, more alkalinity is consumed in the neutralization of acid inputs to lake surface and watershed than can be generated from the combined watershed plus within-lake processes. Where the initial supply of alkalinity was appreciable, such loss of alkalinity may take place with little change in pH. If the concentration of bicarbonate was small, pH will decline rapidly as the last of the bicarbonate buffering is lost. Early attempts at measuring both pH and alkalinity created problems in low ionic waters. In the case of pH, measurement often was by colorimetric comparator, and the indicator (dye) could change significantly the pH of the water sample. Alkalinity determination by colorimetric endpoint also tended to exaggerate the amount of alkalinity present, due to the buffering in the indicator itself and the tendency to overshoot in perception of the color change. In addition to the above, low ionic waters in general received sparse attention from limnologists. The result of these problems is that there has been a paucity of reliable historic data from which to interpret trends in chemical water quality.

## 1. Trends in pH

In Sweden, the downward trend in pH of rivers and lakes, such as Vänern, was apparent in 1965 to 1970 (Figure 5–1) with pH declines of about 0.4 units over one

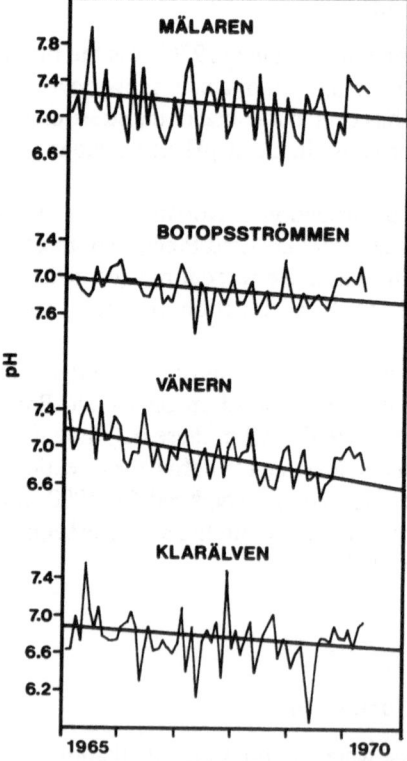

**Figure 5–1.** Five-year pH trends of lakes (outlets) and rivers in southern Sweden. (Source: Royal Ministry for Foreign Affairs, Sweden, 1971.)

to two decades (Dickson, 1970; Odén and Ahl, 1970; Royal Ministry for Foreign Affairs, 1971). In Norway, lakes showed a pH decline of about 0.8 units over 22 to 23 years, and rivers such as the Otra had fallen 0.5 pH units (Snekvik, 1972; Seip and Tollan, 1978).

In North America, Beamish and Harvey (1972) reported the rapid acidification of a set of lakes for which earlier pH measurements were available. Schofield (1976) was able to compare the current pHs of 217 Adirondack lakes above 610-m elevation with values from the 1930s and found a dramatic shift downward in pH. The pH of rivers in southern Nova Scotia has also declined, based on the comparisons of Thompson and others (1980), with the Tusket, Medway, and St. Mary's dropping about three-quarters of a pH unit from 1955 to 73.

## 2. Reconstructions of pH

Lake acidification may be hypothesized on the basis of the changing diatom compositions of the sediments and the known pH tolerances of these organisms. Such reconstructions have been attempted for lakes in Europe and North America (Battarbee and Charles, 1986). Lake Gårdsjön, in southwestern Sweden, showed a decline from ~pH 6.0 in the 1950s to pH 4.5 by 1979 (Renberg and Hedberg, 1982). Changes in chrysophytes in sediment cores also may be useful in establishing trends in acidification (Norton, 1986; Charles et al., 1986). Other atmospheric pollutants may be transported along with acidic precursors, deposited concurrently, and thus serve as indicators of conditions past and present. Acidification of the water columns will alter both deposition and mobilization of these materials, for example, zinc (Evans et al., 1983; Schindler et al., 1980a), resulting in a lower concentration in surface sediments. However, Carignan and Tessier (1985) cautioned that accumulation rates in sediments of acidic lakes cannot be inferred simply from rates of atmospheric deposition of, for example, zinc, because of the postdepositional diffusive transport of the metal below the sediment-water interface.

## 3. Decreasing Alkalinity

This has been demonstrated many times, for example, Dickson's (1980) illustration of alkalinity declining in Lake Unden, prior to liming in 1977. In south-central Ontario, Plastic Lake has been losing alkalinity at a steady rate (Figure 5–2) (Dillon et al., 1987). As the last of the alkalinity is exhausted in the next decade, pH should decline rapidly, resulting in fish kills greater than those reported to date (Harvey et al., 1982). Such examples are sparse in that declines in alkalinity are best demonstrated in calibrated watersheds or through long-term monitoring, both of which are costly.

## 4. Lake Neutralization

The magnitude of the acidification process can be measured by attempting to arrest the process via application of base sufficient to react with the incoming acid—in

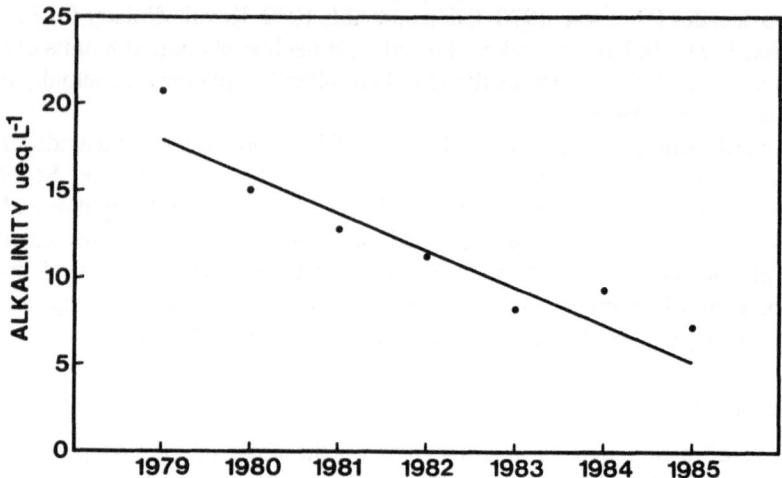

**Figure 5–2.** Decline in alkalinity in Plastic Lake, Ontario. (Source: Dillon et al., 1987. Reprinted with permission. Copyright © 1987 Macmillan Magazine Publishing.)

effect, an ecosystem-scale titration. This has been done many times; one example will suffice. Lake Lysevatten in southwestern Sweden was treated to restore pH and alkalinity (Figure 5–3) (Hultberg and Andersson, 1981). Some of the powdered limestone was lost to the sediment, some alkalinity was exported in the discharge, and the remainder was consumed. Within 4 years the lake was again back to zero alkalinity and the pH was dropping rapidly.

## 5. Cation Denudation

Acidic deposition yields a surplus of $H^+$, which reacts with carbonate and bicarbonate insofar as they are available, and contact time permits. Unreacted $H^+$ will react slowly with silicates and displace Ca, Mg, Na, Al, Mn, and Fe. Thus there is the possibility of Ca and Mg concentrations being elevated during the acidification process; that is, water hardness could increase (Henriksen, 1979, 1982; Almer et al., 1978; Dillon et al., 1979). The opportunity for $H^+$ to displace other cations may be limited, as in snowmelt or heavy precipitation, resulting in temporary pH depression as $H^+$ enters stream and lake waters. The ability of each watershed to neutralize the incoming acid is unique and is the most important variable in determining rates of acidification. The cation supply of a watershed may be sufficiently small that over the long-term Ca concentration in surface waters may decline. The data of Thompson and others (1980) show the Ca concentration having declined over the period 1954 and 1955 to 1973 for some eastern Canadian rivers.

## 6. Shifting Anion Composition

Acidic deposition is dominated by two mineral acids, sulfuric and nitric, varying in proportions as a function of sources, oxidation processes, and season. Reaction

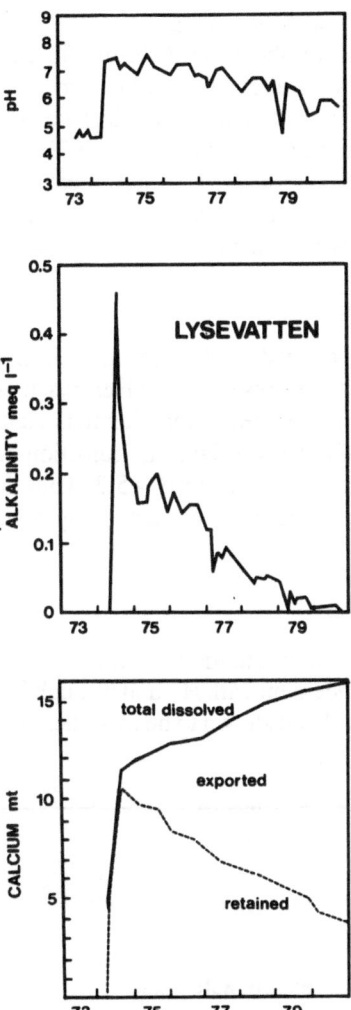

**Figure 5–3.** Decline in pH, alkalinity, and calcium in (lake) Lysevatten, southwestern Sweden, following neutralization with powdered limestone. (Source: Hultberg and Andersson, 1981.)

of these acids with bicarbonate yields ($CaSO_4$ and $Ca(NO_3)_2$, neutral salts with little buffering capacity. In areas subject to acidic deposition, $SO_4^{2-}$ concentration rises and $HCO_3^-$ declines. Wright (1983) reported a near-linear relationship between the excess sulfate in precipitation and the concentration of sulfate in lakes; similarly Kramer and others (1986) related sulfate input via precipitation to lake sulfate output. In south-central Ontario and Quebec, $SO_4^{2-}$ concentration is 4 to 8 ppm versus 0.5 to 1 ppm for background concentration. Low bicarbonate waters may become sulfate dominated, and whereas $SO_4^{2-}$ is not a very toxic ion, it does not provide the biota with the pH stability of $HCO_3^-$.

In areas where $HNO_3$ deposition is high, $NO_3^-$ may become one of the dominant ions in surface waters. In the river Mörrumsån in southern Sweden, $NO_3^-$ concentration has increased about threefold from 1965 to 1985, reaching 300 $\mu g/L^{-1}$ (Dickson, 1986). In the oligotrophic lakes of southern Sweden, $NO_3^-$ concentration has been increasing since the 1970s (Hörnström and Ekström, 1986), with values of 200–500 $\mu g/L^{-1}$ now being reported. In eastern North America, nitrate concentration in stream water at Hubbard Brook Experimental Forest increased about twofold from 1965 to 1974 (Likens et al., 1977).

## 7. Acidification Models

Various attempts have been made to describe the acidification process in empirical models. The approach of Henriksen (1979; Henriksen et al., 1986) has proved to be popular and useful in interpreting the chemical status of lakes (Figure 5–4). Acidic lakes are those of pH <4.7; lakes in transition have pHs between 4.7 and 5.3, and bicarbonate lakes are those of pH >5.3. Thus during acidification, a lake could track towards *a* or *b* (Figure 5–4), depending on rates of cation recruitment (Henriksen, 1980).

## 8. Mobilization of Metals

The effect of acidic deposition is to mobilize metals from watershed soils and lake sediments, both by displacement with $H^+$ and through increased metal solubility at lower pH. Aluminum has been shown to be in higher concentration in lakes of low

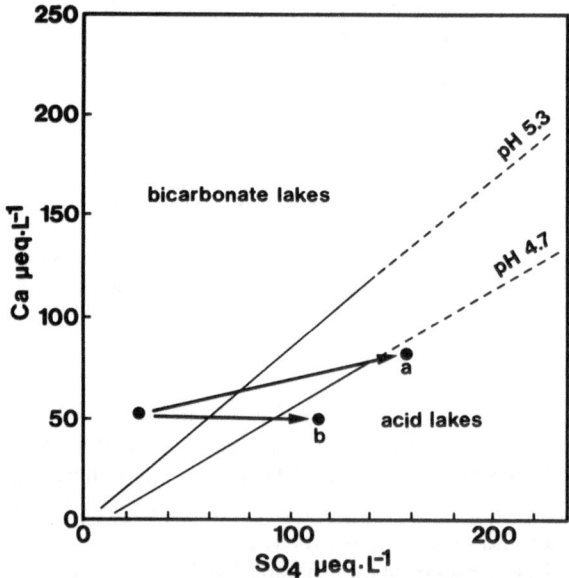

**Figure 5–4.** Bicarbonate, transition, and acidic lakes, with two hypothetical tracks during acidification, depending on rates of cation recruitment. (Source: Henriksen, 1980.)

pH (Dickson, 1975; Schofield, 1976) (Figure 5–5) and to increase in concentration in surface waters in response to acidic rain events or snowmelts (Muniz and Leivestad, 1979). The ionic form of aluminum is pH dependent (Figure 5–6), and the aluminum species differ in toxicity. Aluminum forms complexes with fluoride, phosphate, and organics, and thus the bioavailability of aluminum varies with concentration of these ions, which vary by site and season (LaZerte, 1984).

Some of the heavy metals have been shown to be present at higher concentrations in more acidic waters (Dickson, 1977). Whereas this is a static concept, the dynamics of metal concentrations during acidification have been documented also. Manganese is especially pH sensitive (Figure 5–7) and proportionately shows the greatest increase in concentration during acidification (Somers and Harvey, 1984). Thus the Mn concentration in the lake Stora Skarsjon rose tenfold as pH declined from 1957 to 1975 (Figure 5–8) and returned to initial concentration after the lake was limed in 1975. The Mn concentration of Lake 223 increased during experimental acidification of the lake (not the watershed), leading Schindler and others (1980) to conclude that the metal was recruited from the sediments. This mobilization of metal from sediments in acidic lakes may result, albeit briefly, in the lake becoming a net exporter of metal, with sediment-surface samples showing lower concentrations than samples from deeper strata (Dillon et al., 1978).

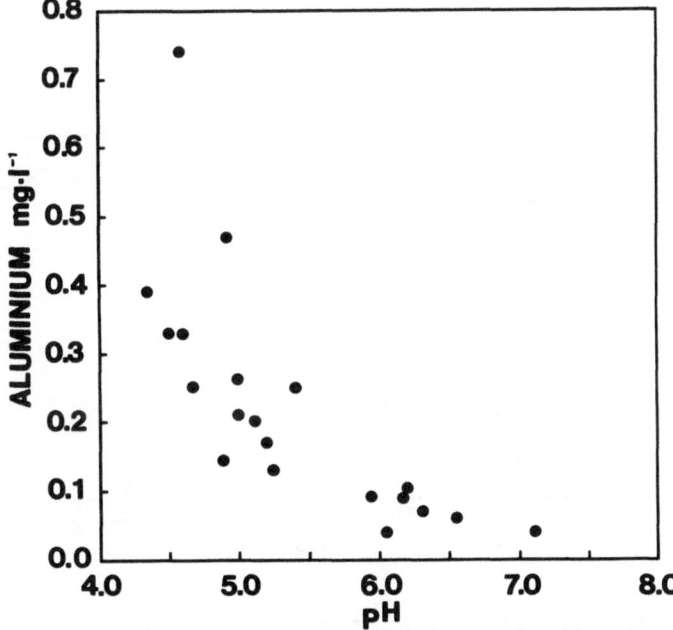

**Figure 5–5.** Aluminum concentration in relation to pH for 20 La Cloche Mountain lakes. (Source: Harvey, unpublished.)

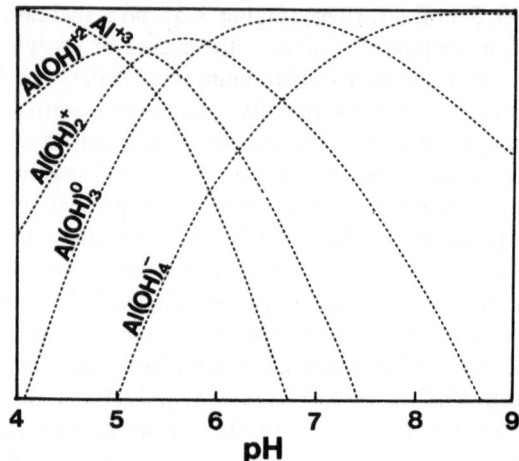

**Figure 5–6.** The proportions of the five ionic forms of aluminum in relation to pH. (Source: Harvey, unpublished.)

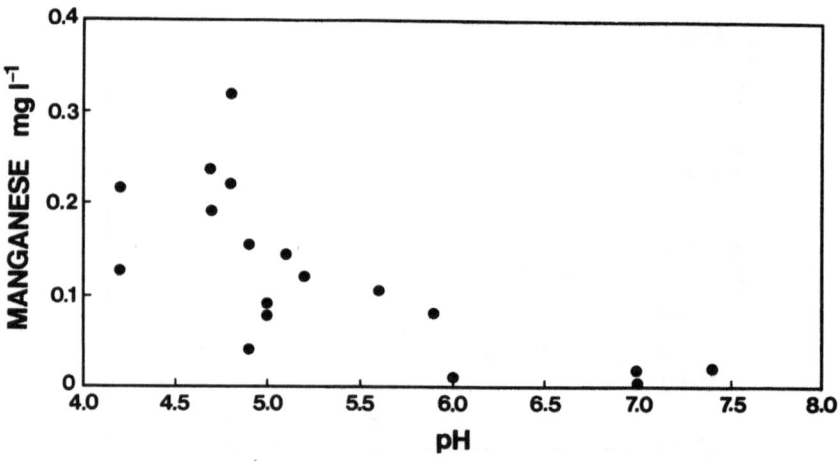

**Figure 5–7.** Manganese concentration in relation to pH for 18 La Cloche Mountain lakes. (Source: Harvey, unpublished.)

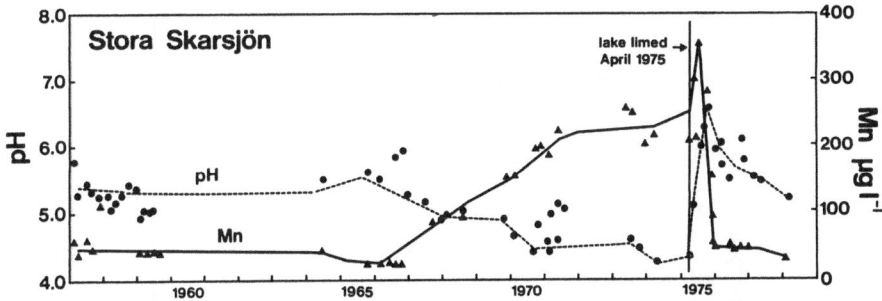

**Figure 5–8.** The pH decline and increase in manganese concentration during the acidification of (lake) Stora Skarsjön, southwest Sweden, until lake neutralization in 1975. (Source: Kalkning av sjöar och vattendrag, 1977–1981.)

## B. Effects on Fishes

### 1. Fish Kills

Massive fish kills have been reported from trout and salmon hatcheries, for example, the 10,000 rainbow trout reported killed in a Scottish hatchery (Galloway News, 1982), and from trout and salmon rivers, such as the kill of Atlantic salmon in the River Esk in northwestern England. Unfortunately, the analysis of chemical water quality sometimes lagged the observed death of fishes in such reports. Even the classical description of the 1975 Tovdal River (southern Norway) brown trout kill was complicated by the presence of ice on the river. Nonetheless, Leivestad and Muniz (1976) were able to show that the acidic stress resulted in a reduction in plasma $Na^+$ and $Cl^-$ concentrations of brown trout, and this was the probable cause of death. Fish kills, principally of pumpkinseeds, were reported annually for Plastic Lake (Harvey et al., 1982), with moribund fish showing reduced total $Na^+$, $Cl^-$, and $K^+$ concentrations.

The loss of fish populations may go largely unnoticed: at least 54 populations disappeared from the lakes of the Chickanishing River basin (north shore of Georgian Bay, Ontario) without the report of a dead or dying fish (Harvey and Lee, 1982). The result of such losses is a reduction in species diversity (Harvey, 1975; Somers and Harvey, 1984), which of itself may have some value in the diagnosis of the stress. In fish communities not affected by acid stress, the number of fish species is correlated strongly with lake size (Harvey, 1978, 1981) but in acid-stressed lakes, the diversity of species may correlate more strongly with lake pH (Figure 5–9; Harvey, 1975). However, lake area and pH are not independent variables, as a consequence of small lakes acidifying more readily than large lakes.

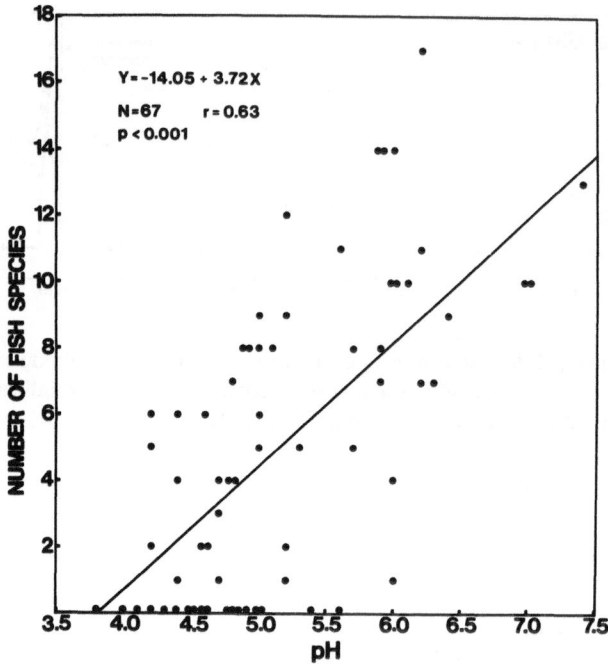

**Figure 5–9.** Relationship between lake pH and the number of fish species present, La Cloche Mountain lakes. (Source: Harvey, 1975.)

The acute toxicity of temporary pH depressions in streams and lakes, due to acidic rain events or snowmelts, has been demonstrated (Harvey and McArdle, 1986a; Harvey and Whelpdale, 1986) by holding healthy fish in surface waters during such acidic stress events. Heney Lake in south-central Ontario, for instance, routinely has a pH of about 6, but surface waters, and hence the outlet stream, decline to pH 4.7 at snowmelt. Such waters are rapidly lethal to caged rainbow trout.

## 2. Failed Recruitment

"In most detective stories the body has been discovered, and the problem is to find the murderer. In this case, however, salmon and trout are being killed *en masse* in the acid water courses of southern Norway, and their bodies are usually never found" (Jensen and Snekvik, 1972). Acid-stressed fish populations commonly show missing age classes resulting from the failure to spawn or the early loss of new year classes. At pH 5.0, Skidway Lake is at the pH threshold for recruitment failure for the white sucker (*Catostomus commersoni*) and appears not to have recruited a new year class in the previous decade (Figure 5–10). In contrast, Bentshoe Lake, at pH 5.9, had 13 age classes represented. Such recruitment failure has been reported for numerous fish species, including: pike (*Esox lucius*)

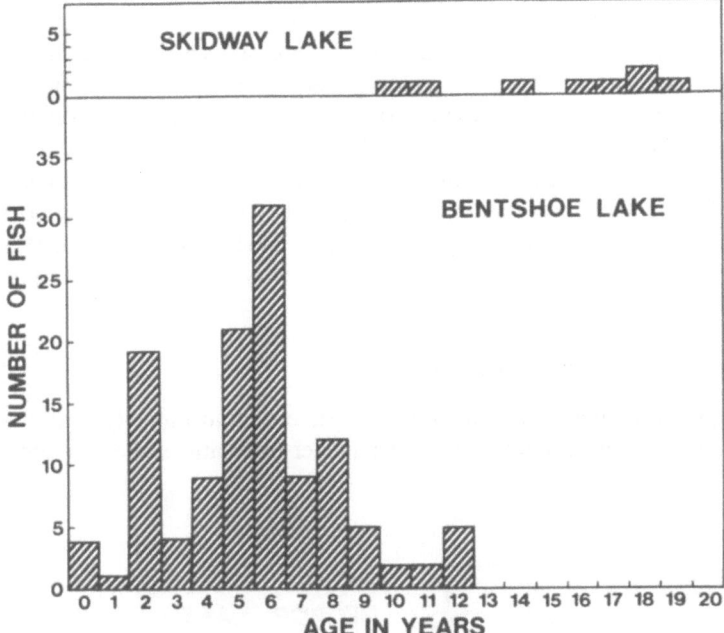

**Figure 5–10.** The age-class composition of samples of the white sucker populations in lakes Skidway (pH 5.0) and Bentshoe (pH 5.9). (Source: Harvey, unpublished.)

(Hultberg and Stensson, 1970), perch (*Perca fluviatilis*) (Almer, 1972; Rosseland et al., 1980), brown trout (*Salmon trutta*) (Jensen and Snekvik, 1972), rock bass (*Ambloplites rupestris*) (Ryan and Harvey, 1977), and yellow perch (*P. flavescens*) (Ryan and Harvey, 1980). Jensen and Snekvik (1972) noted that the population is destroyed via failed recruitment before the pH is depressed sufficiently to kill the larger fish. Baker and Schofield (1980) found white sucker and brook trout (*Salvelinus fontinalis*) sensitivity to low pH decreased with increasing age, whereas aluminum sensitivity increased with increasing age (eggs > sac fry > swim-up fry).

The hypothesis that low pH per se was the cause of the missing year classes is readily testable. Neutralization of acidic waters in which a small number of old fish were still present has resulted in the recruitment of new year classes into the population (Kalkning av sjöar och vattendrag, 1977–1981). More than 2,500 lakes in Sweden have been "limed" to raise the pH. Many of these lakes had only remnant fish populations but yielded new age classes following lake neutralization.

## 3. Juvenilization of Fish Populations

Fish species may spawn in winter, spring, summer, or fall, and fish may be shallow or deep lake spawners or inlet or outlet stream spawners. The pH depressions in streams and lakes (especially littoral or surface waters) coincide with snowmelts or acidic rain events. Thus some species through chronology or distribution may escape the stress of low pH at spawning. Fish also may be able to seek out chemical refuges of higher pH, but very little documentation exists. For some fish populations, stress appears to be greater later in life, yielding age-class compositions dominated by young animals (Harvey, 1980; Rosseland et al., 1980). During the acidification of George Lake, the age-class composition of white suckers shifted to immature fish (Figure 5–11), and the life history of the species shifted to a single spawning effort with very high postspawning mortality (Trippel and Harvey, 1987).

### 4. Population Density and Condition Factor

Acid-stressed populations tend to be in a state of transition, typically with the population growing smaller. Five white sucker populations showed large differ-

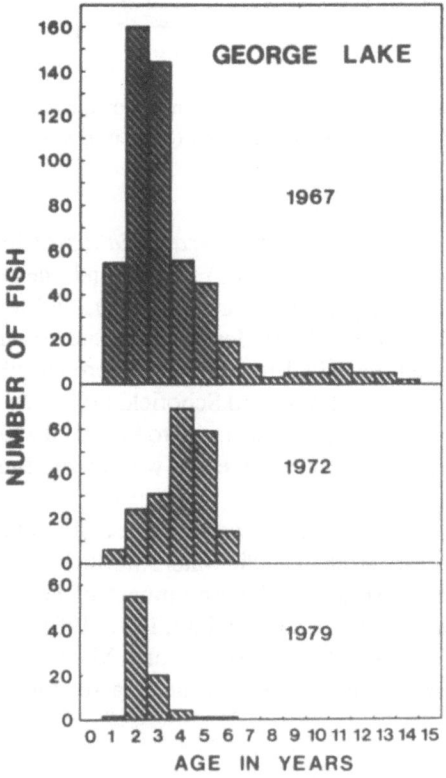

**Figure 5–11.** Age-class composition of samples of white sucker from George Lake during acidification. (Source: Harvey, 1982.)

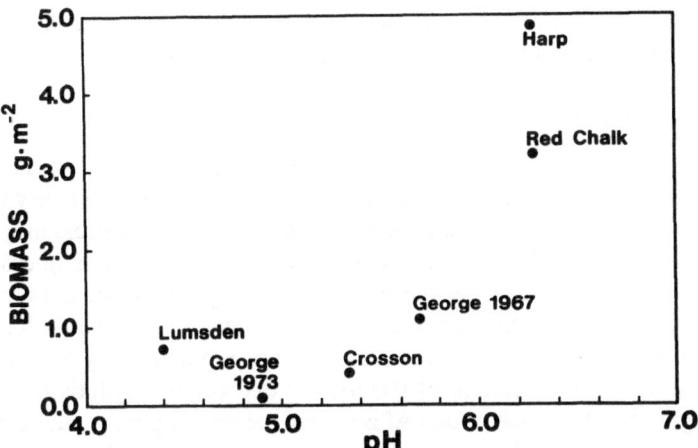

**Figure 5–12.** Biomass of white sucker, measured by mark-recapture, in relation to lake pH; the Lumsden population became extinct within a year of the estimate. (Source: Lumsden Lake—Beamish and Harvey, 1972; George Lake—Beamish et al., 1975; Crosson, Red Chalk, and Harp—Harvey, 1980.)

ences in biomass, based on mark-recapture population estimates (Figure 5–12); the critical pH range appears to be between pH 5.0 and 5.3. Low-density populations at low pH may or may not be food constrained. Chironomids in general tend to be acid tolerant; thus fishes feeding on chironomids may show good growth in acidic waters. The shift to acid-tolerant organisms in acidifying lakes has altered the diet of fishes. Perch, for example, have been observed to be feeding exclusively on corixids, normally a sparse item in their diet. Pike in an acidic lake obliged to feed on insects showed very poor growth (Hultberg and Stensson, 1970) and achieved about one-half the size at age of other populations (Harvey, 1982).

## 5. Species Presence or Absence

Several attempts have been made to derive the lower pH tolerance of fishes from large data sets of species presence or absence and lake chemistry (Harvey, 1979; Rahel and Magnuson, 1983; Wales and Beggs, 1986). Such an approach assumes equal opportunity of invasion, that other important variables such as lake size are independent of pH, and that fish communities are modified by pH or chance. The above notwithstanding, for frequently occurring species, some insight may be gained of the lowest pH tolerated, especially where the populations show evidence of stress (Harvey, 1979). The lethal pHs for any fish species derived from direct observations during extinction or from presence-absence data typically are appreciably higher than lethal pH values obtained in short-term toxicity tests.

Muniz and Leivestad (1980) suggested that the complication of Al toxicity in natural waters may account for much of these differences.

## 6. Aluminum Uptake

Aluminum is one of the most common elements in the earth's crust, but Al solubility is pH dependent, and it forms sparingly soluble compounds with many other substances. Thus in the natural environment acidification stress may take the form of $H^+$ or $H^+$ plus Al ionic species. Both $H^+$ and Al ions interfere with ionoregulation, for example, at the fish gill, where Na-K ATPase promotes the uptake of $Na^+$ in exchange for $NH_4^+$, and $Cl^-$ uptake in exchange for $HCO_3^-$. Aluminum uptake in acidic waters may be very rapid (Harvey and McArdle, 1986a), yielding concentrations of 100 $\mu g/g^{-1}$ dry weight of gill tissue within 24 hours (Figure 5–13). Much of this Al is within the mucus per se, with concentrations of Al up to 1,600 ppb, dry weight of mucus. In Al-exposed rainbow trout, Al was found within gill, esophagus, and stomach tissues, but little elsewhere (Lee and Harvey, 1986).

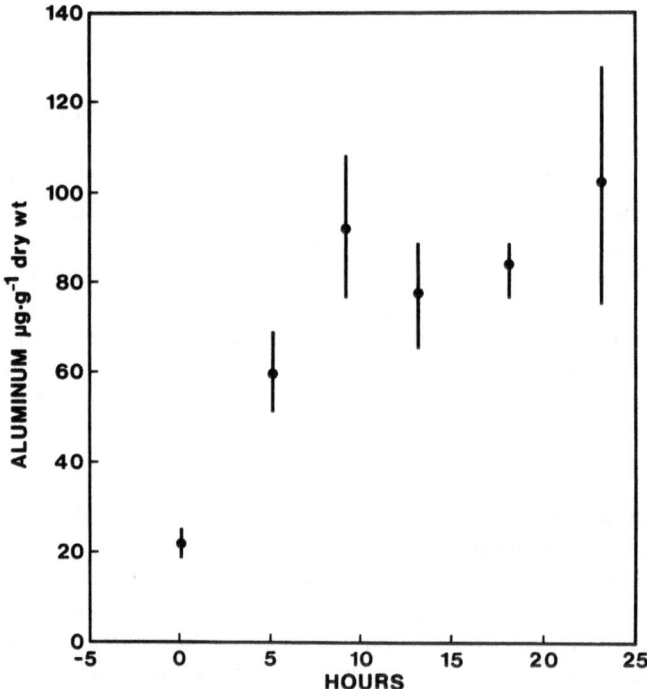

**Figure 5–13.** Aluminum concentrations of gill tissue following the exposure of rainbow trout to Plastic Lake, inlet stream No. 1; means plus standard errors. (Source: Harvey and McArdle, 1986a. Copyright © 1986 by D. Reidel Publishing Company. Reprinted with permission of Kluwer Academic Publ.)

## 7. Metal Concentrations in Fishes

As most metals are more soluble in water at low pH, it is to be expected that higher concentration of metals may be found in fishes from lakes and streams at low pH. Thus, Suns et al. (1980) reported mercury concentrations in walleye (*Stizostedion vitreum*) from low-calcium waters were approximately twice the concentrations in walleye taken from high-calcium lakes. Mercury concentration in muscle tissue has long been known to be a function also of size and growth (Scott, 1974). Bone manganese was two to five times higher in fish taken from two acid lakes (Fraser and Harvey, 1982), whereas cadmium concentration depended more on rate of fish growth than on Cd concentration in food or water (Bendell-Young et al., 1986). Within an acidifying lake, metal concentrations may differ greatly between fish species (Figure 5–14).

## C. Effects on Amphibians

Pough (1976) in the USA and Hagström (1977) in Sweden identified the linkage between acidic precipitation and amphibians. Pough (1976) reported that ponds in the region of Ithaca, New York, were acidic, precipitation in this area had an average pH of 3.98 (Likens and Bormann, 1974), and spotted salamanders (*Ambystoma maculatum*) embryos showed a high incidence of mortality. In a study of six stream pools in south-central Ontario, pH was 4.0 to 4.5 in two, 5.5 to 5.6 in three, and 6.3 in one, and total aluminum was 7 to 210 $\mu g/L^{-1}$ (Clark and

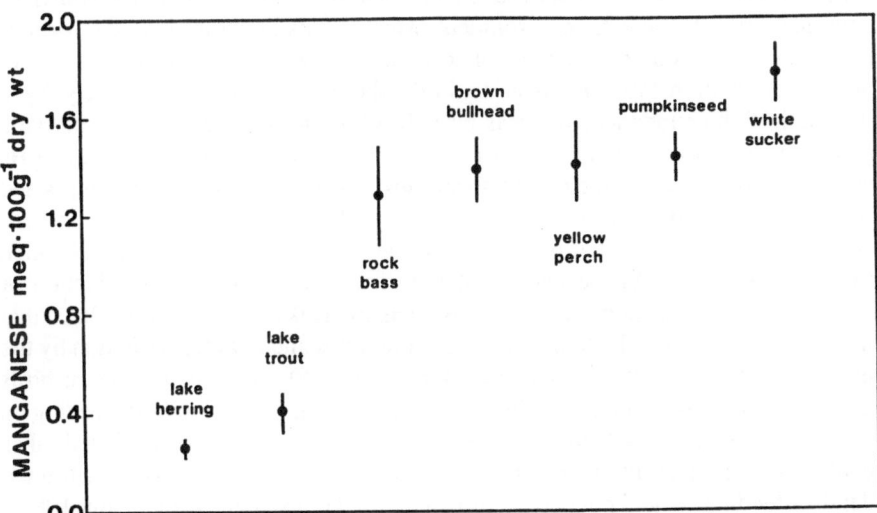

**Figure 5–14.** Manganese concentration of vertebrae of seven fish species in George Lake; means plus standard errors. (Source: Harvey et al., 1986. Copyright © 1986 by D. Reidel Publishing Company. Reprinted with permission of Kluwer Academic Publ.)

Hall, 1985). Hagström (1977) reported the extinction of frogs from an acidified lake. Since these reports, attempts have been made to relate the distribution of various species of amphibians to environmental pH and to determine under laboratory conditions the pH tolerance of frogs and salamanders. This latter work has focused on the early life history, in that half the terrestrial species of salamanders and 90% of frog species deposit eggs in water (Pough and Wilson, 1977). For almost all North American frogs and toads, fertilization is external, thus exposing sperm and ova to the pH of the pond or lake; in salamanders fertilization is internal, and only fertilized eggs are exposed to environmental pH. At low pH, ionoregulation in amphibians is altered, resulting in loss of sodium and chloride (Freda and Dunson, 1985). The embryonic stage appears to be the most sensitive.

Anuran species appear to vary in pH tolerance. Both the pine barrens tree frog (*Hyla andersoni*) and wood frog (*Rana sylvatica*) in the New Jersey Pine Barrens were breeding in all available ponds, with critical pHs about 3.8 (Freda and Dunson, 1986). In general, critical pH is in the range 4.0 to 4.5 for most species (Freda, 1986). Growth rates were poorer among frogs reared at sublethal pH 4.4 (Freda and Dunson, 1985). Acidic tolerance also varies within species, due possibly to genetic variation between populations (Pierce and Sikand, 1985). Other aspects of water quality also may be involved: Clark and LaZerte (1985) found the presence of monomeric aluminum at low pH further reduced hatching success in *R. sylvatica* and *Bufo americanus eggs*.

## D. Effects on Birds

There is now a modest literature on the benefits and disbenefits to birds of acidification of the aquatic environment. Most of this literature is in the form of field surveys yielding environmental correlates but weak in demonstrated cause-and-effect relationships. Fish-eating birds should be affected adversely by a decline in fish abundance, whereas birds feeding on invertebrates may derive a benefit from reduced competition for food, where this resource is limited. At pHs so low as to eliminate some or all fish species, invertebrates such as molluscs and some insects also disappear. Nilsson and Nilsson (1978) found a positive correlation between number of bird species and the pH of the lakes around which they were breeding. In the black-throated diver (*Gavia arctica*), production of young was correlated negatively with fish density; Eriksson (1986) postulated that this could be due to high abundance of invertebrates and reduced predation by the pike (*Esox lucius*). DesGranges and Darveau (1985) reported fish-eating birds chose large and less acidic lakes in southern Quebec in contrast to birds feeding on insects. Eriksson (1984) postulated that the increased clarity of acidified lakes could assist the "pursuit divers," but that "surface plungers," which have a limited depth of feeding, would not benefit from increased transparency (Eriksson, 1985).

Dippers feed on both invertebrates and fishes. In mid and north Wales the distribution of breeding dippers was related to environmental variables, and the birds were less common along acidic streams, which coincided with higher concentrations of Al and lower densities of ephemeropteran nymphs and trichopteran larvae (Ormerod et al., 1986).

A connection between Al and breeding success of pied flycatchers (*Ficedula hypoleuca*) was postulated by Nyholm and Myhrberg (1977) in that eggs with defective shells were from females with elevated concentrations of aluminum in medullary bones. Nyholm (1981) attributed the Al to emerging insects. The growth of ringed turtledoves (*Streptopelia risoria*) was not affected by Al in the diet (Carrière et al., 1986). Thus, the Al connection remains unclear. Water-quality parameters, including $SO_4$, $NO_3/NO_2$, pH alkalinity, Al, Mn, and Fe, influenced by acidification, had a small but significant effect on the growth of eastern kingbirds (*Tyrannus tyrannus*) (Glooschenko et al., 1986).

## E. Effects on Lake Benthos

It has been proposed that less taxa of benthic organisms are present in acidic versus circumneutral lakes (Mossberg and Nyberg, 1979; Ravera, 1986; Økland and Økland, 1986; Harvey and McArdle, 1986b). Gastropods and pelecypods were reported as being not found at pH 5.7 to 6.2 (Hagen and Langeland, 1973), pH 3.7 to 4.6 (Wiederholm and Eriksson, 1977), and pH <5.8 (Harvey and McArdle, 1986b) and reported sparse at pH 4.4 to 4.6 (J. Økland, 1969; Grahn et al., 1974; Raddum, 1980). Among the crustaceans *Lepidurus arcticus* was not found in lakes with pH <6.1 (Borgstrom and Hendrey, 1976), and *Gammarus lacustris* was never found in Norwegian lakes with pH <6.0 (K.A. Økland, 1969). *Mysis*, the large malacostracan, was reported as acid sensitive (Nero and Schindler, 1983). Harvey and McArdle (1986b) found no amphipods in three acidic lakes and daphnids absent or sparse in five acidic lakes; bosminids were absent from only the most acidic lake (pH 4.4) and were otherwise more abundant in the next six most acidic lakes.

In general, lake benthos was dominated by chironomids (Harvey and McArdle, 1986b), but chironomid density was independent of pH in 11 lakes. Mossberg and Nyberg (1979) found chironomid density to increase (up to 86% of the benthos of the most acid lake), whereas *Phaenopsectra* abundance declined. Ephemeropterans have been reported to be acid sensitive, based on not being found at pH 5.7 to 6.2 (Hagen and Langeland, 1973); fewer species present at pH 4 to 4.5 versus >6.5 (Hendrey et al., 1976); being lower in abundance (Grahn et al., 1974); and being absent from three acidic lakes (Harvey and McArdle, 1986b). Odonates, trichopterans, neuropterans, and plecopterans have been variously reported as pH density dependent (Hendrey et al., 1976; Grahn et al., 1974; Hagen and Langeland, 1973) or independent of pH (Harvey and McArdle, 1986b). At low pH, bacterial breakdown of detritus is inhibited and thus the food supply for shredders may be extended seasonally. The caddis fly *Clistornia magnifica* showed improved growth at low pH, apparently in response to increased fungal biomass on leaf litter (van Frankenhuyzen et al., 1985). Shredders in general appear to be acid tolerant (Hall et al., 1980), especially plecopteran and trichopteran shredders (Mackay and Kersey, 1985).

Some fish species such as perch and brook trout are more acid tolerant than the more acid sensitive invertebrate prey species such as mayflies, gastropods, and pelecypods. The result of this may be a dramatic shift in diet to acid-tolerant prey, most notably corixids, notonectids, and gyrinids, among fishes in acidifying lakes (Rosseland et al., 1980).

Crayfish function as animal predators, as plant grazers, and as scavengers on detritus (Crocker and Barr, 1968) and are important in the energy flow to vertebrates (Neill, 1951). Thus, alteration of the crayfish community can be expected to lead to perturbation of the littoral ecosystems. Published observations on the susceptibility of crayfish to acidification differ widely. Crayfish were reported (Morgan and McMahon, 1982; McMahon and Morgan, 1981) as being acid tolerant in acute toxicity experiments. *Orconectes virilis* in water of low calcium concentration were stressed at pH <5.5 (Malley, 1980), and recalcification of the exoskeleton was inhibited following molting. *Orconectes virilis* declined in abundance during the acidification of Lake 223 to pH 5.37 (Schindler and Turner, 1982). France (1981) found crayfish sensitive to chronic acid exposure with just-independent juveniles most susceptible to acid stress (France, 1984). Acidic stress interferes with ion exchange and with the reproductive cycle of female crayfish (Lockwood, 1968; Appelberg, 1981), in that uptake of $Na^+$ and $Cl^-$ is dependent on $H^+$ and $HCO_3^-$ concentrations in the hemolymph. Aluminum ion at pH 5.0 also acted to reduce $Na^+$ concentration in the hemolymph (Appelberg, 1985). Acidosis results in increased hemolymph Ca concentration, perhaps due to loss of exoskeleton carbonate (Defur et al., 1980; Morgan and McMahon, 1982). From the results of Berrill and others (1985), it is clear that *Cambarus robustus* is much more acid tolerant than *O. propinquus* or *O. rusticus*, but in a ten-lake survey, *C. bartoni* occurred in the least acidic (pH 6.4) and most acidic (pH 4.6) lakes.

The microbial community is pH dependent, with bacteria being inhibited at low pH and fungi replacing bacteria in the decomposition process (Hendrey et al., 1976). Detritus has been reported to accumulate in acidic waters (Leivestad et al., 1976; Hultberg and Grahn, 1976). This process of detrital accumulation is reversed rapidly following lake neutralization (Scheider et al., 1975; Hultberg and Andersson, 1981), resulting in the lake effervescing for 1 to 2 years postneutralization, as methane and carbon dioxide stream off. This is in contrast to the gases emitted (CO, $SO_2$, $C_2H_2$, $CS_2$) from beneath the so-called fungal mat, more correctly a mat of detritus plus algae plus fungi (Hultberg and Grahn, 1976; Grahn et al., 1974). In contrast, Schindler (1980) found rates of decomposition did not change during the experimental acidification of Lake 223.

In some acidified lakes, most especially in Scandinavia, *Sphagnum* has expanded in distribution from the shoreline to a depth of 10 m (Grahn et al., 1974). In water *Sphagnum* grew more rapidly, for instance, 10 cm annually versus 1 cm on land, and broke down only slowly, yielding strands up to 1 m in length. In addition *Sphagnum* is known to take up $Ca^{2+}$ in exchange for $H^+$, perhaps contributing to the acidification process. Such massive *Sphagnum* intrusions and rapid growth have not been reported in North America. In acidified lakes of the La

Cloche, Parry Sound, and Muskoka, Ontario, it has been more common to observe large quantities of the green alga, *Mougeotia*. Schindler (1980) reported the sudden appearance of *Mougeotia* during the experimental acidification of Lake 223. Leivestad and others (1976) postulated that the dense beds of *Sphagnum* could be blocking oxygen flow to the sediments in displacing the acidophilic angiosperms and thus altering the microbial community in the sediments.

## F. Effects on Plankton

During acidification many species of zooplankton disappear. The number of zooplankton species may be correlated with lake pH but also with lake size and with the number of fish species present (Figure 5–15). Because these are not independent variables, it is difficult to extract cause-and-effect relationships from such synoptic survey data. The possibility exists that a diverse zooplankton community is maintained by fish predation and that in the absence of such predation a relatively few species dominate the community. Sprules (1975)

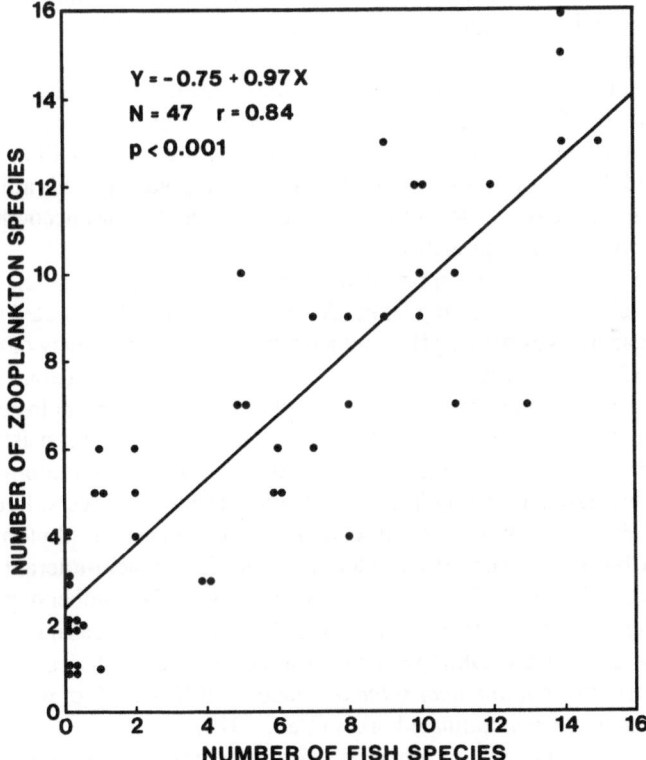

**Figure 5–15.** Number of zooplankton species in relation to number of fish species in a large set of lakes, most of which were acidifying. (Source: Harvey, 1975.)

determined the crustacean zooplankton species in 47 La Cloche lakes. Of all species, 64% rarely or never were found in lakes of pH <5.0; a few species, most notably *Diaptomus minutus*, were found in the most acidic lakes. Because only the same few species were present in acidic lakes, acid tolerance per se must be an important determinant of zooplankton communities in acidifying lakes. This conclusion is supported by the evidence for direct toxic action of $H^+$ and $Al^{+3}$ on zooplankters; for example, *Daphnia magna* lost $Na^+$ much more rapidly at pH 3 versus pH 7 (Potts and Fryer, 1979).

Eriksson and others (1980) postulated that, with the loss of fish predation during acidification, zooplankton grazing would predominate, and this would result in zooplankton predators of larger size. This is consistent with the observations of Hrbáček (1958) and Brooks and Dodson (1965). Eriksson and others (1980) proposed that dinoflagellates would dominate the phytoplankton, reducing primary productivity in acidic lakes. Dillon and others (1984) tested this hypothesis with three data sets. In two such lake pairs, dinoflagellates made up half the biomass of the two acidic lakes versus less than a quarter in the less acidic lakes. Productivity measured as phytoplankton biomass in acidic lakes was equal to or greater than that in the near-neutral lakes. In the third data set dinoflagellates did not dominate in the acidic lakes. Dillon et al. (1984) concluded that the data did not show dinoflagellate dominance inducing a reduction in primary production in acidifying lakes.

In acidic lakes small herbivores tended to dominate the zooplankton community (Yan and Strus, 1981). This would be expected if predation by large zooplankton did not increase concurrently with decreased fish predation. Dillon and others (1984) noted that the normally abundant, large, cyclopoid copepods are reduced in abundance in acidic lakes, despite the food supply, and this may account in part for the reduced invertebrate predation.

There are now numerous reports in the literature that phytoplankton diversity is reduced in acidic and acidifying lakes. Almer and others (1974) noted the paucity of green algae in lakes below pH 5, but a much increased frequency of occurrence at pH 5.0 to 5.8 for the diatoms, blue-green algae, golden-brown algae, and green algae. Thirteen species common in oligotrophic lakes were absent in acidic lakes, and for lakes with pH <5 phytoplankton was homogeneous and consisted of about ten species. In recently acidified lakes, the frequency of diatom values in sediments decreased markedly from older to recent sediment layers. Leivestad and others (1976) also reported significantly fewer species of phytoplankton in Norwegian lakes of lower pH and identified the shift to acid-tolerant periphytic diatom species. Yan (1979) also reported a reduction in the number of phytoplankton taxa, Clearwater Lake (pH 4.3) having half as many genera as morphologically comparable Blue Chalk Lake (pH 6.5). Whereas the chrysophytes dominated the nonacid lakes, *Peridinium inconspicuum* made up 30% to 55% of the phytoplankton of acidic, metal-contaminated lakes (Yan, 1979).

Despite the reduction in phytoplankton diversity, the available evidence indicates that both production and biomass are comparable in acidic and nonacidic lakes, if phosphorus concentrations are similar. Phytoplankton biomass was correlated with phosphorus concentration, with the latter lower in acidic lakes (Almer et al., 1974). Precipitation of phosphate by aluminum has been postulated as a cause of the low phosphorus concentrations in acidic lakes. Dillon and others (1979) added phosphate to acidified lakes, and the phytoplankton biomass increased. Both phosphorus and nitrogen cycling were unaffected in acidified lakes (Schindler and Turner, 1982). In the experimental acidification of Lake 223, biomass and productivity increased slightly at the lower pH, due apparently to natural variation. These apparent differences in productivity and nutrient cycling need to be explained.

## III. Summary Recommendations

The effects of acidic deposition were first diagnosed in freshwater environments and research was most vigorous there. With growing recognition of the acidification phenomenon during the 1970s and 1980s, there emerged a social-political need-to-know: (1) the geographical distribution of acidification of surface waters; (2) the rates of change of water quality due to acidic deposition; (3) the magnitude of the loss of biota from fresh waters; (4) the amount of the freshwater resource at risk; (5) that there were demonstrable cause-and-effect relationships between acidic deposition and chemical and biological response; and (6) that the course of acidification was predictable for increasing and decreasing emissions of $SO_2$ and $NO_x$.

The extent to which these objectives have been met in the many hundreds of published investigations has been the subject of much public debate and governmental assessment, most typically in the context of how much proof is enough, prior to implementation of reductions in emissions. Institutionalization of the acidic rain problems within government has resulted in new criteria, new research directions, and longer time scales for future work.

In the absence of a reduction in acidic deposition below effects thresholds, there remains a need for more or continuing information of some kinds. For example, the calibrated watersheds yield data on both rates of change and the processes involved. There remains the need for monitoring low-alkalinity lakes for trends. Episodic event monitoring of pH depressions at snowmelt and acidic rains in terms of stress on the biota has received scant attention. This is due largely to spring problems of access, fragile ice cover, and the tradition of summer limnology. Chemical refuges, that is, places where the biota can escape from pH depressions, have been postulated but never documented. Although pH tolerances have been reported for many organisms, the approach may be too simple; for example, species disappearing at pH 6 may not only be reacting to the 1 ppb $H^+$ at that pH. Little has been reported about the effects on the biota of acidification-induced changes in cations such as Ca and Mg and anions $HCO_3$, $SO_4$, and $NO_3$. The role of Al in $PO_4$ precipitation from the water column in acidic lakes is still largely speculative. There are many reports of increased concentrations of metal in acidic lakes but relatively few reports of increased uptake and the biological consequences of such metals.

Much more can be done using existing data sets and by adding to historic data sets with respect to the changing chemical composition and changing biota in surface waters in relation to acidic deposition.

# References

Almer, B. 1972. *Forsurningens inverkan på fiskbeståndet i västkustsjoar.* Information från sötvattenslaboratoriet, Drottningholm, Sweden. Nr 12. 47 p.

Almer, B., W. Dickson, C. Ekström, and E. Hörnström. 1978. *In* J.O. Nriagu, ed. *Sulphur in the environment*, 271–311, Wiley, New York.

Almer, B., W. Dickson, C. Ekström, E. Hörnström, and U. Miller. 1974. Ambio 3:30–36.

Appelberg, M. 1985. Hydrobiologia 121:19–25.

Appelberg, M.P. 1981. *In Proceeding of the 5th International Symposium on freshwater crayfish*, 59–70.

Baker, J.P., and C.L. Schofield. 1980. *In* D. Drabløs and A. Tollan, eds. *Proceedings of the International Conference on ecological impact of acid precipitation*, 292–293, Sandefjord. SNSF project, Ås, Norway.

Battarbee, R.W., and D.F. Charles. 1986. Water Air Soil Pollut 30:347–354.

Beamish, R.J. 1970. *Factors affecting the age and size of the white sucker Catostomus commersoni at maturity.* Ph.D. Thesis, University of Toronto. 170 p.

Beamish, R.J., and H.H. Harvey. 1971. The Globe and Mail, Toronto, July 26, 1971.

Beamish, R.J., and H.H. Harvey. 1972. J Fish Res Board Can 29:1131–1143.

Beamish, R.J., W.L. Lockhard, J.C. Van Loon, and H.H. Harvey. 1975. Ambio 4:95–102.

Bendell-Young, L.I., H.H. Harvey and J.F. Young. 1986. Can J Fish Aquat Sci 43:806–811.

Berrill, M., L. Hollett, A. Margosian, and J. Hudson. 1985. Can J Zool 63:2586–2589.

Borgström, R., and G.R. Hendrey. 1976. *pH tolerance of the first larval stages of Lepidurus arcticus (Pallas) and adult Gammarus lacustris G.O. Sars.* IR 22/76. 37 p. SNSF project, Ås, Norway.

Brooks, J.L., and S.I. Dodson. 1965. Science 150:28–35.

Brydges, T.G. 1989. Proceedings, Svante Oden Commemorative Symposium Skokloster, Sweden, 1987.

Carignan, R., and A. Tessier. 1985. Science 228:1524–1526.

Carrière, D., K.L. Fischer, D.B. Peakall, and P. Anghern. 1986. Can J Zool 64:1500–1505.

Charles, D.F., D.R. Whitehead, D.S. Anderson, R. Bienert, K.E. Camburn, R.B. Cook, T.L. Crisman, R.B. Davis, J. Ford, D.B. Fry, R.A. Hites, J.S. Kahl, J.C. Kingston, R.G. Kreis, M.J. Mitchell, S.A. Norton, L.A. Roll, J.P. Smol, P.R. Sweets, A.J. Uutala, J.R. White, M.C. Whiting, and R.J. Wise. 1986. Water Air Soil Pollut 30:355–365.

Clark, K.L., and R.J. Hall. 1985. Can J Zool 63:116–123.

Clark, K.L., and B.D. LaZerte. 1985. Can J Fish Aquat Sci 42:1544–1551.

Cowling, E.B. 1982. *In* F.M. D'Itri, ed. *Acid precipitation: Effects on ecological systems.* Ann Arbor Sci Publ, Ann Arbor, Michigan. 506 p.

Crocker, D.W., and D.W. Barr. 1968. *Handbook of the crayfishes of Ontario.* Royal Ontario Museum, Toronto.

Dahl, K. 1922. Supplerende bemerkninger til undersokelser over orretens utdoen i det sydvestlige Norges fjeldvand. Norsk Jaeger og Fiskerforenings Tidsskrift, Meddelelsernes side Aargang. 64 p.

Dahl, K. 1927. Salmon and Trout Magazine 46:35–43.

Dannevig, G. 1959. Jeger og Fisker 3:116–118.

Dannevig, G. 1966. Jakt, fiske og friluftsliv 9:388–393.

Defur, P.L., P.R. Wilkes, and B.R. McMahon. 1980. Resp Physiol 42:247–261.

DesGranges, J.-L., and M. Darveau. 1985. Holarct Ecol 8:181–190.

Dickson, W. 1970. *pH-Conditions in lakes in the western part of Sweden in November-December 1970.* Report of the National Swedish Environmental Protection Board, Solna, Sweden.

Dickson, W. 1975. Rept Inst Freshw Res, Drottningholm 54:8–20.

Dickson, W. 1977. Verh Int Verein Limnol 20:851–856.

Dickson, W. 1980. *In* D. Drabløs and A. Tollan, eds. *Proceedings of the international conference on ecological impact of acid precipitation,* 75–83. Sandefjord, Norway. SNSF project, Ås, Norway.

Dickson, W. 1986. *In* J. Nilsson, ed. *Critical loads for sulphur and nitrogen,* 199–210. National Environmental Protection Board, Solna, Sweden.

Dillon, P.J., D.S. Jeffries, W. Snyder, R. Reid, N.Y. Yan, D. Evans, J. Moss, and W.A. Scheider. 1978. J Fish Res Board Can 35:809–815.

Dillon, P.J., R.A. Reid, and E. de Grosbois. 1987. Nature 329:45–48.

Dillon, P.J., N.D. Yan, and H.H. Harvey. 1984. Crit Rev 13:167–194.

Dillon, P.J., N.D. Yan, W.A. Scheider, and N. Conroy. 1979. Arch Hydrobiol 13:317–336.

Dochinger, L.S., and T.A. Seliga, eds. 1976. *Proceedings of the First International Symposium on Acid Precipitation and the Forest Ecosystem, Columbus, Ohio, May 12–15, 1975.* U.S.D.A. Forest Service General Technical Report NE-23, 1074 p.

Drabløs, D., and A. Tollan, eds. 1980. *Ecological impact of acid precipitation. Proceedings of an international conference, Sandefjord, Norway, March 11–14, 1980.* 383 p. SNSF project, Ås, Norway.

Eriksson, M.O.G. 1984. Ambio 13:20–22.

Eriksson, M.O.G. 1985. Ornis Scand 16:1–7.

Eriksson, M.O.G. 1986. Ornis Scand 17:245–248.

Eriksson, M.O.G., L. Henrikson, B.I. Nilsson, G. Nyman, H.G. Oscarson, J.A.E. Stenson, and K. Larsson. 1980. Ambio 9:248–249.

Evans, H.E., P.J. Smith, and P.J. Dillon. 1983. Can J Fish Aquat Sci 40:570–579.

France, R.L. 1981. *In* Goldman, C.R. (ed.) *Proceeding of the 5th International Symposium on freshwater crayfish,* 98–111. AVI Publ. Co. Westport, Connecticut.

France, R.L. 1984. Can J Zool 62:2360–2363.

Fraser, G.A. and H.H. Harvey. 1982. Can J Fish Aquat Sci 39:1289–1296.

Fraser, G.A., and H.H. Harvey. 1984. Can J Zool 62:249–259.

Freda, J. 1986. Water Air Soil Pollut 30:439–450.

Freda, J. and W.A. Dunson. 1985. Copeia 1985:415–423.

Freda, J., and W.A. Dunson. 1986. Copeia 1986 (2):454–466.

*Galloway News*, October 28, 1982. P.I. Galloway, Scotland.

Glooschenko, V.,.P. Blancher, J. Herskowitz, R. Fulthorpe, and S. Rang. 1986. Water Air Soil Pollut 30:553–567.

Gordon, A.G., and E. Gorham. 1963. Can J Bot 41:1063–1078.

Gorham, E., and A.G. Gordon. 1960. Can J Bot 41:371–378.

Grahn, O., H. Hultberg, and L. Lander. 1974. Ambio 3:93–94.

Hagen, A., and A. Langeland. 1973. Environ Pollut 5:45–59.

Hagström, T. 1977. Sver Nat 11:367–369.

Haines, T.A., and J.P. Baker. 1986. Water Air Soil Pollut 31:605–629.

Hall, R.J., G.E. Likens, S.B. Fiance, and G.R. Hendrey. 1980. Ecology 61:976–989.

Harvey, H.H. 1975. Proc Int Assoc Theoret Appl Limnol 19:2406–2417.

Harvey, H.H. 1978. Proc Int Assoc Theoret Appl Limnol 20:2406–2417.

Harvey, H.H. 1979. Fish Mar Serv Tech Rep 862:115–128.

Harvey, H.H. 1980. *In* Drabløs, D, Tollan A (eds) *Proceedings of the international conference on ecological impact of acid precipitation*, 93–98. Sandefjord, Norway. SNSF project, Ås, Norway.

Harvey, H.H. 1981. Proc Int Assoc Theoret Appl Limnol 21:1222–1230.

Harvey, H.H. 1982. *In* R.E. Johnson, ed. *Acid rain: Fisheries*, 227–242. American Fisheries Society, Bethesda, Md.

Harvey, H.H., P.J. Dillon, G.A. Fraser, K.M. Somers, P.E. Fraser, and C. Lee. 1982. Abs Am Chem Soc, Div Environ Chem 22:438–441.

Harvey, H.H., G.A. Fraser, and J.M. McArdle. 1986. Water Air Soil Pollut 30:515–521.

Harvey, H.H., and C. Lee. 1982. *In* R.E. Johnson, ed. *Acid rain: Fisheries*, 45–55. American Fisheries Society, Bethesda, Md.

Harvey, H.H., and J.M. McArdle. 1986b. Water Air Soil Pollut 30:529–536.

Harvey, H.H., and J.M. McArdle. 1986a. Water Air Soil Pollut 30:687–694.

Harvey, H.H. and D.M. Whelpdale. 1986. Water Air Soil Pollut 30:579–586.

Hendrey, G.R., K. Baalsrud, T. Traacn, M. Laake, and G.G. Raddum. 1976. Ambio 5:224–227.

Henriksen, A. 1979. Nature 278:542–545.

Henriksen, A. 1980. *In* D. Drabløs and A. Tollan, eds. *Proceedings of the international conference on ecological impact of acid precipitation*, 67–74, Sandefjord. SNSF project, Ås, Norway.

Henriksen, A. 1982. *In* R.E. Johnson, ed. *Acid rain: Fisheries*, 103–121. American Fisheries Society, Bethesda, Md.

Henriksen, A., W. Dickson, and D.F. Brakke. 1986. *In* J. Nilsson, ed. *Critical loads for sulphur and nitrogen*, 87–120. National Environmental Protection Board, Solna, Sweden.

Hörnström, E., and C. Ekström. 1986. *Acidification and liming effects on phyto- and zooplankton in some Swedish West Coast lakes*. National Environmental Protection Board, Solna, Sweden. 110 p.

Hrbáček, J. 1958. Proc Int Assoc Theoret Appl Limnol 13:394–399.

Hultberg, H., and I.B. Andersson. 1981. *Liming of acidified lakes and streams: Perspectives on induced physical-chemical and biological changes*. Swedish Water and Air Pollution Research Institute. Göleborg. 61 p.

Hultberg, H. and O. Grahn. 1976. J Grt Lakes Res Suppl 2:208–217.

Hultberg, H., and J. Stensson. 1970. Fauna Flora 1:11–19.

Jensen, K.W., and E. Snekvik. 1972. Ambio 1:223–225.

Kalkning av sjöar och vattendrag 1977–1981. 1981. *In Information från sötvattenslaboratoriet, Drottningholm*, Sweden. Nr 4. 201 p.

Kramer, J.R., A.W. Andren, R.A. Smith, A.M. Johnson, R.B. Alexander, and G. Oehlert. 1986. *In Acid deposition long term trends*. National Academy Press, Washington, D.C. 506 p.

LaZerte, B.D. 1984. Can J Fish Aquat Sci 41:766–776.

Lee, C., and H.H. Harvey. 1986. Water Air Soil Pollut 30:649–655.

Leivestad, H., G. Hendrey., I. Muniz, and E. Snekvik. 1976. *In* F.H. Braekke, ed. *Impact of acid precipitation ecosystems in Norway*, 86–111. Research Report 6, SNSF project, Ås, Norway.

Leivestad, H., and I.P. Muniz. 1976. Nature 259:391–392.

Likens, G.E. 1976. Chem Eng News 54(48):29–44.

Likens, G.E., and F.H. Bormann. 1974. Science 184:1176–1179.

Likens, G.E., F.H. Bormann, and N.M. Johnson. 1972. Environment 14:33–40.

Likens, G.E., F.H. Bormann, R.S. Pierce, J.S. Eaton, and N.M. Johnson. 1977. *Biogeochemistry of a forested ecosystem*. Springer-Verlag, New York. 146 p.

Likens, G.E., R.F. Wright, J.N. Galloway, and T.I. Butler. 1979. Sci Am 241(4):43–51.

Lockwood, A.P. 1968. *Aspects of the physiology of crustacea*. Oliver and Boyd, London.

Mackay, R.J., and K. Kersey. 1985. Hydrobiologia 122:3–11.

McMahon, B.R., and D.O. Morgan. 1981. *In* Goldman, C.R. (ed.) *Proceedings of the 5th international symposium on freshwater crayfish*, 71–85. AVI Publ. Co., Westport, Connecticut.

Malley, D.F. 1980. Can J Fish Aquat Sci 37:364–372.

Martin, H.C., ed. 1986. *Acidic precipitation*. D. Reidel, Dordrecht, Holland. Part 1, 1053 p; part 2, 1118 p.

Morgan, D.O., and B.R. McMahon. 1982. J Exp Biol 97:241–252.

Mossberg, P., and P. Nyberg. 1979. Rep Inst Freshw Res, Drottningholm 58:77–87.

Muniz, I.P., and H. Leivestad. 1979. *Langtidseksponering av fish til surt vann. Forsøk med bekkerøye Salvelinus fontinalis Mitchell*. IR 44/79. SNSF project, Ås, Norway. 32 p.

Muniz, I.P., and H. Leivestad. 1980. *In* D. Drabløs and A. Tollan, eds. *Proceedings of the international conference on ecological impact of acid precipitation*, 84–92, Sandefjord, Norway. SNSF project. Ås, Norway.

Neill, W.T. 1951. Ecology 32:764–766.

Nero, R.W., and D.W. Schindler. 1983. Can J Fish Aquat Sci 40:1905–1911.

Nilsson, S.G., and I.N. Nilsson. 1978. Oikos 31:214–221.

Norton, S.A. 1986. Water Air Soil Pollut 30:331–345.

Nyholm, N.E.I. 1981. Environ Res 26:363–371.

Nyholm, N.E.I., and H.E. Myhrberg. 1977. Oikos 29:336–341.

Odén, S., and T. Ahl. 1970. *The acidification of Scandinavian lakes and rivers*, 103–122. Ymer, Årsbok.

Økland, J. 1969. Malacologia 9:143–151.

Økland, J., and K.A. Økland. 1986. Experientia 42:471–486.

Økland, K.A. 1969. Norw J Zool 17:111–152.

Ormerod, S.J., N. Allinson, D. Hudson, and S.J. Tyler. 1986. Freshwater Biol 16: 501–507.

Pierce, B.A., and N. Sikand. 1985. Can J Zool 63:1647–1651.

Potts, W.T.W., and G. Fryer. 1979. J Comp Physiol B 129:289–294.

Pough, F.H. 1976. Science 192:68–70.

Pough, F.H., and R.E. Wilson. 1977. Water Air Soil Pollut 7:307–316.

Raddum, G.G. 1980. *In* D. Drabløs and A. Tollan, eds. *Proceeings of the international conference on ecological impact of acid precipitation*, 330–331. Sandefjord, Norway. SNSF project, Ås, Norway.

Rahel, F.J., and J.J. Magnuson. 1983. Can J Fish Aquat Sci 40:3–9.

Ravera, O. 1986. Experientia 42:507–516.

Renberg, I., and T. Hedberg. 1982. Ambio 1:30–33.

Rosseland, B.O., I. Sevaldrud, D. Svalastog, and I.P. Muniz. 1980. *In* D. Drabløs and A. Tollan, eds. *Proceedings of the international conference on ecological impact of acid precipitation*, 336–337, Sandefjord, Norway. SNSF project, Ås, Norway.

Royal Ministry for Foreign Affairs and Royal Ministry of Agriculture. 1971. *Air pollution across national boundaries: the impact on the environment of sulfur in air and precipitation*. Author. 96 p.

Ryan, P.M., and H.H. Harvey. 1977. J Fish Res Board Can 34:2079–2088.

Ryan, P.M., and H.H. Harvey. 1980. Environ Biol Fishes 5:97–108.

Scheider, W.A., J. Adamski, and M. Paylor. 1975. Ontario Ministry of Environment. Toronto. 129 p.

Schindler, D.W. 1980. *In* D. Drabløs and A. Tollan, eds. *Proceedings of the international conference on ecological impact of acid precipitation*, 370–374. Sandefjord, Norway. SNSF project, Ås, Norway.

Schindler, D.W., R.H. Hesslein, R. Wagemann, and W.S. Broeker. 1980a. Can J Fish Aquat Sci 37:373–377.

Schindler, D.W., K.H. Mills, D.F. Malley, D.L. Findlay, J.A. Shearer, I.J. Davies, M.A. Turner, G.A. Linsey, and D.R. Cruikshank. 1985a. Science 228:1395–1401.

Schindler, D.W., and M.A. Turner. 1982. Water Air Soil Pollut 18:259–271.

Schindler, D.W., R. Wagemann, R.B. Cook, T. Ruszczynski, and J. Prokopowich. 1980b. Can J Fish Aquat Sci 37:342–354.

Schofield, C.L. 1976. Ambio 5:228–230.

Scott, D.P. 1974. J Fish Res Board Can 31:1723–1729.

Seip, H.M., and A. Tollan. 1978. Sci Total Environ 10:253–270.

Smith, R.A. 1852. *Air and rain: The beginnings of a chemical climatology*. Longmans, Green, London.

Snekvik, E. 1972. Vann 7:59–67.

Somers, K.M., and H.H. Harvey. 1984. Can J Fish Aquat Sci 41:20–29.

Sprules, W.G. 1975. J Fish Res Board Can 32:389–395.

Suns, K., C. Cury, and D. Russel. 1980. *The effects of water quality and morphometric parameters on mercury uptake by yearling perch*. Ontario Ministry of Environment, Toronto. Technical Report LTS 80-1. 16 p.

Thompson, M.E., M.C. Elder, A.R. Davis, and S. Whitlow. 1980. *In* D. Drabløs and A. Tollan, eds. *Proceedings of the international conference on ecological impact of acid precipitation*, 244–245. Sandefjord, Norway. SNSF, Ås, Norway.

Trippel, E.A., and H.H. Harvey. 1987. Can J Fish Aquat Sci 44:1018–1023.

van Frankenhuyzen, K., G.H. Geen, and C. Koiisto. 1985. Can J Zool 63:2298–2304.

Wales, D.L., and G.L. Beggs. 1986. Water Air Soil Pollut 30:601–609.

Wiederholm, T., and L. Eriksson. 1977. Oikos 29:261–267.

Wright, R.F. 1983. *Predicting acidification of North American lakes*. Report No. 4, Norwegian Institute Water Research, Oslo. 165 p.

Yan, N.D. 1979. Water Air Soil Pollut 11:43–55.

Yan, N.D., and R. Strus. 1981. Can J Fish Aquat Sci 37:2282–2293.

# Effects of Acidic Precipitation on Forest Ecosystems in North America

William H. Smith*

## Abstract

Forests are variable in species, topography, elevation, soils, and management. Air pollution deposition and influences are also variable and poorly documented in the field. Monitoring of species dynamics and productivity, necessary to detect effects of regional and global air pollutants or any other environmental stress, are presently deficient.

Regional air contaminants may influence reproductive processes, nutrient uptake or retention, metabolic rates (especially photosynthesis and respiration), and insect pest and pathogen interactions of individual trees. At the ecosystem level, regional air pollutants may influence nutrient cycling, population dynamics of arthropod or microbial species, succession, species composition, and biomass production. In the case of low-dose regional-scale pollution, symptoms are typically not visible (at least initially), undramatic, and not easily measured. The integration of regional pollutant stresses can be slower growth, altered competitive abilities, and changed susceptibility to pests. Ecosystem symptoms may include altered rates of succession, changed species composition, and biomass production. Visible symptom expression is not a useful assessment strategy.

The potential stresses imposed by regional-scale air pollutants must be viewed against the background of uncertainty associated with the changes in radiation balances associated with global-scale air contaminants. If global warming does accelerate in the near term, and especially if large forest regions are subjected to reduced moisture availability, the expansion of forest tree decline could be very great. Such an occurrence would pale the impact of regional-scale pollutant deposition.

*School of Forestry and Environmental Studies, Yale University, New Haven, CT 06511, USA.

# I. Introduction

*Acidic precipitation* is defined as rain, snow, fog, or other forms of precipitation having a pH less than 5.6. A more inclusive and appropriate term for acidic atmospheric pollution is *acidic deposition*. This term includes all forms of acidic or acidifying materials transferred from the atmosphere to the earth in the dry (gas, particulate) or wet (precipitation) form.

Acidic deposition is appropriately designated a regional-scale air pollutant. Air contaminants may be classified in terms of the distances they travel from point of release in the troposphere to deposition back to the earth. Local-scale pollutants travel a few to tens of km. Regional-scale pollutants travel hundreds or even thousands of km from point of precursor release, and global-scale pollutants may be distributed throughout the atmosphere of the earth. The regional-scale pollutant designation includes several contaminants in addition to acidic deposition.

Regional air pollutants of greatest documented or potential influence for forests include: oxidants, most importantly ozone; trace metals, most importantly heavy metals such as cadmium, iron, copper, lead, manganese, chromium, mercury, molybdenum, nickel, thallium, vanadium, and zinc; and acidic deposition, most importantly the wet and dry deposition of sulfuric and nitric acids. Ozone and sulfuric and nitric acids are termed *secondary air pollutants* because they are synthesized in the atmosphere rather than released directly into the atmosphere. The precursor chemicals, released directly into the atmosphere and causing secondary pollutant formation, include hydrocarbons and nitrogen oxides, in the case of ozone, and sulfur dioxide and nitrogen oxides, in the case of sulfuric and nitric acid. The combustion of fossil fuels for energy production releases hydrocarbons and sulfur dioxide. The heat of all combustion processes causes nitrogen and oxygen to react and form nitrogen oxides. Many combustion activities also generate small particles (approximately 0.1–5 μm diameter). Those activities associated with energy combustion (particularly coal burning) can preferentially contaminate these small particles with trace metals. Because the formation of secondary air pollutants may occur over tens or hundreds of km from the site of precursor release, and because small particles may remain airborne for days or weeks, these pollutants may be transported 100 to more than 1,000 km from their origin. Eventual wet and dry deposition of the pollutants onto lakes, fields, or forests occurs over large rather than small areas.

Evidence is available—satellites, surface deposition of aerosol sulfate and reduced visibility (Chung, 1978; Tong et al., 1976; Wolff et al., 1981)—for long-range transport of acidifying pollutants from numerous sources. During the winter, approximately 20% of the emissions from tall power plant stacks in the northeastern USA may remain elevated and relatively coherent for more than a day and 500 km (USEPA, 1983).

The long-distance transport of regional pollutants means they may have interstate, international, and even intercontinental significance. It means further that the forests subject to their deposition exceed tens of thousands of km$^2$.

## II. Forest Ecosystems: An Overview

Consideration of ecosystem interactions with atmospheric deposition is substantially different than individual organism or species interaction with air pollution stress. Evaluation of ecosystem response to stress requires recognition of all components and processes characteristic of all ecosystems (Smith, 1984).

### A. Ecosystem Perturbation

In order for an ecosystem to be perturbed or stressed, an influence must be imposed on one or more of the 11 elements (Table 6–1) of ecosystems. Because of the interaction among the different elements of an ecosystem, a change in one element can cause a response throughout the ecosystem. A change in energy flow or chemical cycling can influence the abundance of organisms, the metabolic rate at which they live, and the complexity and structure of the ecosystem. Energy and nutrients are transferred from organism to organism through food webs. Removal of or injury to any organism in a food web can, therefore, disrupt energy flow or nutrient cycling. Disruption of these processes in turn may influence production (energy storage by plants), succession, species composition, and regulation of ecosystems.

Disturbances are characteristic of natural ecosystems. Disturbance may be caused by natural or anthropogenic processes. Examples of the former are fire, drought, insect defoliation, volcanic eruption, and destructive wind. Examples of the latter include selective species removal (hunting, thinnings), pesticide application, fertilizer application, controlled fire, irrigation, and atmospheric deposition.

Ecosystem disturbance may be viewed from a variety of levels of interaction: between an individual organism and the environment, between a population and its environment, and between a biological community and its environment (Billings, 1978).

Disturbance may have a positive or negative effect or may produce both positive and negative responses in ecosystems. Perturbation caused by environmental or biotic forces may be necessary to maintain maximum ecosystem diversity or productivity. The natural tendency in temperate forest ecosystems appears to be toward periodic perturbation at some interval that recycles the system and maintains a periodic wave of peak diversity and a corresponding wave of peak primary production (Loucks, 1970).

Responses at the ecosystem level are less well defined and of more extended duration than responses at the population level. Acute stresses over localized areas followed by rapid recovery and return to an unstressed state are presumed to have a different effect on ecosystems than chronic stresses over expansive areas that continue for long time periods.

It is further important to realize that not all ecosystems respond in the same way to disturbance. Each forest ecosystem has distinctive characteristics of inertia and

**Table 6-1.** Ecosystem elements. All units of nature with energy flow and complete chemical cycles, or ecosystems, have common components organized in structural patterns and united by functional processes.

| Components | Structural patterns | Functional processes |
|---|---|---|
| Inorganic substances[a] | Food web[g] | Energy flow[j] |
| Organic substances[b] | Diversity: species varia- | Production = energy |
| Climate[c] | tion in space[h] | storage, biomass |
| Autotrophs (producers)[d] | Succession: species varia- | Biogeochemical cycling[k] |
| Heterotrophs (consumers)[e] | tion in time[i] | |
| Decomposers[f] | | |

[a] Inorganic substances: Elements and compounds required for metabolism and other materials involved in biogeochemical cycles. Examples include carbon, nitrogen, carbon dioxide, and water.

[b] Organic substances: Carbon compounds that join living and nonliving components of ecosystems. Examples include proteins, carbohydrates, lipids, and humic materials.

[c] Climate: The meteorologic context of the troposphere in which the ecosystem is located. Elements involved include temperature, precipitation, solar radiation, and wind.

[d] Autotrophs (producers): Organisms capable of synthesizing their own food materials. Green plants able to make their own food from simple substances are the dominant producers of ecosystems. Trees, of course, are the dominant producers in forests.

[e] Heterotrophs (consumers): Organisms requiring preformed food materials for metabolism. They are extremely variable in size and morphology and range from microorganisms to mammals. Heterotrophs are commonly characterized as herbivores (plant eating) or carnivores (meat eating).

[f] Decomposers: Heterotrophic organisms capable of degrading complex organic debris from decomposing plants and animals (detritus), utilizing some of the decomposition products for their own use, and releasing organic residues and inorganic substances for use by other organisms. Decomposers are primarily microorganisms, including bacteria, fungi, and protozoa.

[g] Food webs: Organisms within ecosystems feed on one another. Energy, elements, and compounds are transferred from organism to organism through food webs consisting of integrated systems of interacting food chains. Organisms are grouped in food webs into trophic levels. All organisms in a food web that are the same number of steps away from the original source of energy are at the same trophic level.

[h] Diversity: The number of species in a given ecosystem is variable. Desert ecosystems have fewer species than do tropical rain forests. Natural ecosystems generally have more species than do agricultural ecosystems.

[i] Succession: The number of species may change as an ecosystem matures and succession proceeds. Succession involves recognizable, repeated, and predictable changes in species composition over time. Primary succession occurs when an ecosystem is initially established, whereas secondary succession involves the reestablishment of an ecosystem following disturbance.

[j] Energy flow: Energy must be continually added to an ecosystem in a usable form. Generally it flows in from outside the ecosystem, mainly through photosynthesis. Within the ecosystem, energy is passed from one organism to another, and some is released as heat. Production is stored energy and is termed *biomass* or *organic matter*.

[k] Biogeochemical cycling: Chemical elements move into and out of ecosystems. This cycling connects biological cycles to geologic cycles. Chemical elements cycle within an ecosystem from organism to organism through water, air, soils, and rocks.

(Source: Adapted from Botkin and Keller, 1982; Odum 1971.)

resilience with regard to particular disturbances. Inertia is the resistance of the ecosystem to disturbance. Resilience is the degree, manner, and pace of restoration of initial structure and function in an ecosystem following disturbance (Westman, 1978). The diversity of forest ecosystems in North America precludes generalization concerning the impact of a specific disturbance on all forests.

## B. Inadequacy of Ecosystem Science

There are significant deficiencies in our understanding of ecosystems that seriously restrict prediction of responses of ecosystems to any stress. The National Research Council, Committee on Nuclear and Alternative Energy Systems, has inventoried these deficiencies (Table 6–2). Uncertainty of ecosystem understanding was characterized as inherent or rectifiable. The former was so labeled as it was difficult to design a research program that could reduce associated deficiencies, and the latter so labeled because it was deemed possible to plan research that could meaningfully reduce uncertainty. It is essential to appreciate that atmospheric deposition effects on ecosystems require a more adequate understanding of fundamental ecosystem structure and function than is presently available.

**Table 6–2.** Major sources of deficiency in predicting the response of ecosystems to stress.

*Inherent*
1. Deficiency of information on effects of long-term, low-level effluents on ecosystems
2. Difficulty of performing controlled, replicable experiments that provide in situ information about ecosystems
3. Lack of models allowing the use of measurable data to predict detailed ecological responses to stress

*Rectifiable (lack of data on and understanding of)*
4. Energy and nutrient needs of organisms (so-called limiting factors)
5. Overconfidence in untested ecological dogma
6. Effects of acute stresses on ecosystem (especially synergistic effects)
7. Critical stability indicators and correlates of stability
8. Population fluctuations
9. Environmental fluctuations
10. Ecological-meteorological interactions
11. Microbial ecology and nutrient chemistry
12. Sources of stress (both gross and effluent levels and pollution levels in the microenvironment of organisms)
13. Genetic parameters governing ecosystem dynamics and response to stress
14. Cumulative effects on populations of successive, small habitat losses

(Source: National Research Council, Committee on Nuclear and Alternative Energy Systems, 1980.)

## III. Forest Ecosystem Interaction with Regional-Scale Atmospheric Deposition

### A. Ecosystem Response Varies with Exposure

Disturbance from air pollutants is exposure related, and dose-response thresholds for a specific pollutant are very different among the various organisms of an ecosystem. Ecosystem response is, therefore, a very complex process. In response to low exposure to air pollution, the vegetation and soils of an ecosystem function as a sink or receptor. When exposed to intermediate loads, individual plant species or individual members of a given species may be subtly and harmfully affected by nutrient stress, impaired metabolism, predisposition to entomological or pathological stress, or direct induction of disease. Exposure to high deposition may induce acute morbidity or mortality of specific plants. At the ecosystem level, the impact of these various interactions would be highly variable. In the first situation, the pollutant would be transferred from the atmosphere to the various elements of the biota and to the soil. With minimal physiological effect, the impact of this transfer on the ecosystem would be undetectable (innocuous effect) or stimulatory (fertilizing effect). If the effect of the pollutant dose on some component of the biota is harmful, then a subtle adverse response may occur. The ecosystem impact in this case could include reduced productivity or biomass, alterations in species composition or community structure, or increased morbidity. Under conditions of high dose, ecosystem impacts may include gross simplification, impaired energy flow and biogeochemical cycling, changes in hydrology and erosion, climate alteration, and major impacts on associated ecosystems (Smith, 1981).

### B. Acidic Deposition and Forest Ecosystem Response

In view of present evidence, the impact of acidic deposition on North American forest ecosystems is highly uncertain. The interaction, however, is presumed to be highly interactive with other stresses, very subtle in effect and probably of long-term nature.

Extraordinary numbers of recent reviews have attempted to provide comprehensive or summary overview of our understanding of acidic deposition and forests (Abrahamsen, 1984; Bell, 1986; Bennett et al.,, 1985; Bruck, 1987; Evans, 1982; Fuhrer and Fuhrer-Fries, 1982; Linthurst, 1984; National Council Paper Industry, 1981; Postel, 1984; Shepard, 1985; Smith, 1985; Society of American Foresters, 1984; Treshow, 1984; U.S. Congress, 1986; Ulrich and Pankrath, 1983; USEPA-USDA Forest Service, 1986; University of Maine, 1984; Woodman and Cowling, 1987). Without exception, all of these reviews emphasize the uncertainty of understanding regarding acidic deposition and forest ecosystem effects. Clearly, contemporary science is at the hypothesis-testing stage with regard to mechanisms of acidic deposition interaction with forests. There is some degree of consensus that the most important hypotheses include those listed in Table 6–3.

**Table 6–3.** Hypotheses for acidic deposition interactions with individual forest trees and populations of forest species that may result in long-term forest ecosystem perturbations in productivity or species composition. Impacts on reproduction, biomass accumulation, competitive ability, or abiotic/biotic stress agent interactions of important tree species will have ecosystem ramifications if a sufficiently large proportion of the population is influenced.

| | Hypotheses | |
| --- | --- | --- |
| Tree population interaction | Forest ecosystem perturbation | Important references |
| 1. Increased available soil aluminum results in fine-root morbidity | Population dynamics, tree competition, species composition | Reuss and Johnson (1986), Cronan and Schofield (1979), Cumming et al. (1985), Foy (1984), Ulrich and Pankrath (1983), Schier (1985), Foy et al. (1978), Steiner et al. (1984), McCormick and Steiner (1978) |
| 2. Increased rate of soil acidification causes altered nutrient availability and root disease | Population dynamics, tree competition species composition | Reuss and Johnson (1986), Binkley and Richter (1986), Krug and Frink (1983), Voigt (1980), McFee (1980), van Breeman et al. (1984), Ulrich and Pankrath (1983) |
| 3. Cation nutrients are leached from foliage to throughfall and stemflow | Biogeochemical cycle rates | Cronan and Reiners (1983), Parker (1983), Reiners and Olson (1984), Reiners et al. (1985), Hoffman et al. (1980), Scherbatskoy and Klein (1983), Raynal et al. (1982a,b), Cronan (1984) |

**Table 6-3.** (*Continued*)

| | Hypotheses | |
|---|---|---|
| Tree population interaction | Forest ecosystem perturbation | Important references |
| 4. Cation nutrients are leached below soil horizons of active root uptake | Biogeochemical cycle rates | Richter et al. (1983), Cronan (1980), Cronan (1979), Cronan et al. (1978), Mollitor and Raynal (1982), Johnson et al. (1983), Johnson et al. (1982), Lewis and Grant (1979), Singh et al. (1980) |
| 5. Increased available heavy metal and hydrogen ion concentrations in soil result in enhanced root uptake or impact on soil microbiota | Decomposer impact, biogeochemical cycle rates, species composition | Crist et al. (1985), Jackson and Watson (1977), Jackson et al. (1978), Doelman (1978), Smith (1981), Cronan (1985), Nordgren et al. (1985), Firestone et al. (1984) |
| 6. Deposition causes alteration of carbon allocation to maintenance respiration/repair or to aboveground as opposed to belowground tissues | Productivity, energy storage | McLaughlin et al. (1982), Amthor (1986), McCool and Menge (1983) |
| 7. Deposition increases/decreases phytophagous arthropod activity | Consumer impact, insect population dynamics | Alstad et al. (1982), Montgomery and Meyer (1985), Smith et al. (1984) |

| | | |
|---|---|---|
| 8. | Deposition increases/decreases microbial pathogen activity | Consumer impact, pathogen population dynamics | Campbell et al. (1985), Martin et al. (1985), Bruck and Shafer (1984), Smith et al. (1984), Huttunen (1984) |
| 9. | Deposition increases/decreases abiotic stress influence (temperature, moisture, wind, nutrient stresses) | Population dynamics, tree competition, species competition | Johnson (1983), McLaughlin et al. (1983) |
| 10. | Increased soil weathering alters soil cation availability | Biogeochemical cycle rates | Cronan (1985), Johnson et al. (1972) |
| 11. | Increased nitrogen/sulfur deposition alters nitrogen/sulfur cycle dynamics | Biogeochemical cycle rates | Bewley and Stotzky (1983a, 1983b), Bryant et al. (1979) |
| 12. | Deposition increases/decreases microbial symbioses | Productivity, energy storage | Shriner and Johnston (1981), Reich et al. (1986), Reich et al. (1985), Shafer et al. (1985), Stroo and Alexander (1985), Graw (1979), Killham and Firestone (1983) |
| 13. | Deposition impacts one or more processes of reproductive or seedling metabolism | Population dynamics, tree composition, species composition | Lee and Weber (1979), Raynal et al. (1982a, 1982b), Smith (1981), Schier (1985) |
| 14. | Deposition impacts a critical metabolic process, e.g., photosynthesis, respiration, water uptake, translocation, evapotranspiration | Population dynamics, tree competition, species composition | Chevone et al. (1986), Evans and Curry (1979), Evans et al. (1978), Dochinger and Jensen (1985), Klein (1984), Evans (1982), Hutterman (1983) |

The evidence presented in the various references cited in Table 6–3 do not allow us to accept any of the hypotheses as the most important or dominant interaction of acidic deposition with forest systems. At the same time the evidence does not allow total rejection of any of these hypotheses. Our present level of forest ecosystem understanding and our present level of forest ecosystem monitoring do not allow us to assess the impact of acidic deposition on most forest systems. The hypotheses identified are all highly interactive with other stresses, subtle in manifestation, and of chronic, long-term nature.

## IV. North American Forest Ecosystem Decline

### A. Definition and Examples of Forest Decline

One of the very most important phenomena of temperate zone forests are extended-term stresses termed *dieback-decline diseases* or more simply *declines*. These occur when a large proportion of a tree population exhibits visible symptoms of stress or consistent growth decreases, typically over a widespread area. The importance of dieback-decline stress in long-term forest ecosystem development may be very great. The shorter-term influence on forest management objectives is appreciated to be very severe in cases where the tree population exhibiting decline phenomena is the focus of management efforts.

Presently decline stress is especially important in portions of the northern forest on sugar maple (*Acer saccharum*) (McLaughlin et al., 1985; McIlveen et al., 1986; Parker and Houston, 1971; Wargo and Houston, 1974) and on American beech (*Fagus grandifolia*) (Houston, 1975; Houston et al., 1979; Lacki, 1985). Very widespread decline and mortality of yellow birch (*Betula alleghaniensis*) occurred throughout New England, New York, and southeastern Canada from 1940 to 1965 (Greenidge, 1953; Hansbrough et al., 1950; Hawboldt, 1952). An important decline of white ash was most severe in the Northeast during the 1950s and 1960s (Brandt, 1961; Hibben and Silverborg, 1978; Sinclair, 1965) and may again be significant (Castello et al., 1985).

The most widespread and serious declines over the past several decades in the central hardwood forest have been associated with members of the red oak group (scarlet [*Quercus coccinea*], pin [*Quercus palustris*], red [*Quercus rubra*], and black oak [*Quercus velutina*]) (Nichols, 1968; Skelly, 1974; Staley, 1965). The present distribution of oak decline is highly variable, but mortality can vary from 20% to 50% in severely impacted areas. Significant mortality has occurred in the George Washington and Jefferson National Forests in Virginia (Shriner, 1986). There is evidence that similar declines occurred during earlier periods in this century in the southern Appalachians (Balch, 1927). Declines of lesser importance that have occurred in the central and southern hardwood forests include sweetgum (*Liquidambar styraciflua*) decline (Miller and Gravatt, 1952), yellow poplar (*Liriodendron tulipfera*) decline (Toole and Huckenpahler, 1954), and magnolia (*Magnolia grandiflora*) decline (McCracken, 1985).

Conifers also exhibit wide-area decline stress. Important North American examples of the recent past include: Alaska cedar (*Chamaecyparis nootkatensis*) (Shaw et al., 1985), western white pine (*Pinus monticola*) (Leaphart and Copeland, 1957), and white spruce (*Picea glauca*) (Molnar and Silver, 1959) in the west; and shortleaf pine (*Pinus echinata*) (Copeland and McAlpine, 1955), other southern pines (Sheffield et al., 1985), and high elevation red spruce (*Picea rubens*) (Weiss et al., 1985) in the east.

The decline stress is highly variable in distribution, discontinuous and recurrent in time, and the result of multiple interactive individual stress agents. Nevertheless, these phenomena have sufficient common characteristics in various forest tree species to allow some generalization and synthesis of concept. Decline stress is more characteristic of mature forests than of immature forests. Generally declines result from sequential influence of multiple stress factors. Stress factors are both abiotic and biotic in nature. Abiotic stress factors common to numerous declines include drought and low- and high-temperature stress. Biotic agents of particular importance include defoliating insects, borers, and bark beetles, along with root-infecting and canker-inducing fungi. Typically declines are initiated by an abiotic stress with mortality ultimately caused by a biotic stress agent. Quite commonly the abiotic stress responsible for decline initiation is the direct or indirect influence of change in some climatic parameter, for example, less than normal precipitation. Decline stress is characterized by a common progression of symptoms (Table 6–4). Excellent reviews of the decline phenomenon have been provided by Houston (1974, 1981) and Manion (1981). Examples of decline stresses of significance over the past century in the United States, with a suggestion of primary stress factors involved, are presented in Table 6–5.

## B. Atmospheric Deposition and Forest Decline

In view of recent evidence documenting forest tree growth decline (Hari et al., 1986; Lucier, 1986; McClenahen and Dochinger, 1985; Sheffield et al., 1985) and visibly symptomatic tree dieback and decline (Friedland et al., 1984; Johnson and

**Table 6-4.** Symptoms characteristic of decline stress of forest trees.
1. Reduced growth—Reduced shoot elongation and diameter increment
2. Shorter internodes—Tufted foliage
3. Fine root and mycorrhizal root destruction
4. Reduced food storage in roots
5. Premature development of foliar color in late summer/early fall
6. Reduced size of foliage and yellowing
7. Twig and branch dieback typically caused by weakly parasitic fungi
8. Outer crown dieback
9. Stimulation of stem and branch adventitious buds
10. Root infection by decay fungi, notably *Armillaria mellea*

(Source: Manion, 1981.)

**Table 6-5.** Important forest tree declines of the past century in North America.

| Species | Initial stress | Contributing stress | Final stress |
|---|---|---|---|
| Yellow and white birch | Warm summer temperatures, drought | Foliage feeding insects | Bronze birch borer, *Armillaria mellea* (root disease) |
| White ash | Drought | | *Cytophoma pruinosa* and *Fusicoccum* spp. (canker disease) |
| Sugar maple | Drought | Insect defoliation, harvesting, nutrient deficiency | *Armillaria mellea* (root disease); *Verticillium dahliae* (vascular disease) |
| American beech | Beech scale, *Cryptococcus fagisuga* | ? | *Nectria* spp., canker fungi |
| Oak | Drought, low temperatures in spring | Gypsy moth or oak leaf roller defoliation (insect) | *Armillaria mellea* (root disease), *Agrilus bilineatus* (insect borer) |
| Sweetgum | Drought | Nutrient stress | ? |
| Western white pine | Drought | | *Armillaria mellea, Leptographium* spp. (root disease), *Europhium trinacriforme* (canker disease) |
| Shortleaf pine | Drought and excessive soil moisture | Nutrient stress, *Phytophthora cinnamomi* (root disease) | *Armillaria mellea, Pythim* spp (root disease)?? |
| Balsam fir | Drought | Eastern spruce budworm | *Armillaria mellea* (root disease) |
| White spruce | Drought (?) | Western spruce budworm (bud mortality) | *Aureobasidum pullulans* (twig, branch disease) |
| Red spruce (high elevation) | Drought, climate warming | Abiotic winter stresses | Atmospheric deposition, biotic agents |

Siccama, 1983; Siccama et al., 1982; Weiss et al., 1985), sometimes involving more than one tree population, it is appropriate to examine critically the possible role that regional-scale atmospheric contaminants may play in these wide-area forest stresses. As indicated in section III, a variety of mechanisms have been identified (Table 6–3) that would potentially allow acidic deposition to function as an initial, contributing, or final stress in decline phenomena. Fine root dysfunction resulting from an extended-term, gradual increase in available soil-solution aluminum or heavy metals such as lead, manganese, zinc, or others could function as an initial or contributing stress in wide-area forest tree decline. Nutrient stress resulting from long-term leaching loss via cation migration out of the available nutrient component, in the absence of adequate replacement, could also function as an initial or contributing stress in forest decline. Predisposition to biotic stress, for example, fungal root infection, insect defoliation, or bark beetle infestation, caused by atmospheric deposition stress could allow regional pollutants to function as a contributing stress. Acute episodes of regional pollutant deposition, especially multiple pollutant episodes, for example, severe oxidant dose followed by an extreme acidic deposition event, could allow regional pollutants to function as a final stress. At this time, however, no North American forest tree decline has been conclusively shown to involve acidic deposition. Evidence is available to support the importance of oxidants (ozone) in wide-area forest decline, for instance, ponderosa and Jeffrey pine decline in southern California (Miller et al., 1982; Smith, 1981; USEPA, 1986).

## V. Global Scale Air Pollutant Stress on Forest Ecosystems

In the past 25 years, we have become concerned with a new scale of air pollution—global. A class of global air pollutants of special importance are trace gases released into the troposphere have the potential to influence the influx or efflux of radiation to or from the earth. These trace gas pollutants include nitrous oxide, methane, ozone, carbon monoxide, carbon dioxide, and halocarbons. The potential of atmospheric accumulation of carbon dioxide and halocarbons to influence forest ecosystem health will be discussed in order to illustrate the potential significance of these contaminants to ecosystem integrity.

Careful monitoring of carbon dioxide during the past three decades in Hawaii, Alaska, New York, Sweden, Austria, and the South Pole has firmly established that carbon dioxide is steadily increasing in the global atmosphere. This increase is due to anthropogenic activities including fossil fuel combustion. It may also be caused by altered land-use management, such as forest destruction in the tropics. The atmospheric carbon dioxide concentration has been estimated to have been approximately 290 ppm ($5.2 \times 10^4$ $\mu$g m$^{-3}$) in the middle of the nineteenth century. Today, the carbon dioxide concentration approximates 340 ppm ($6.1 \times 10^5$ $\mu$g m$^{-3}$) and is increasing very approximately two ppm ($3.6 \times 10^3$ $\mu$g m$^{-3}$) per year. In the year 2030, if the increasing rate continues, the carbon dioxide amount in the global atmosphere may be nearly twice the present value (Hansen, 1987; Holdgate et al., 1982).

Naturally occurring stratospheric ozone is important because it screens the earth from biologically damaging ultraviolet radiation—light with wavelengths between 290 and 320 nanometers—released by the sun. Halocarbons released by humans may be capable of reducing the natural ozone layer surrounding the earth. In summary, halocarbon molecules, for example, chlorofluoromethanes, released by various human activities, are slowly transported through the troposphere. They pass through the tropopause and lower stratosphere and are decomposed in the mid to upper stratosphere. Free chlorine, resulting from decomposition, causes a rapid, catalytic destruction of ozone. In 1979, the National Academy of Sciences estimated that release of halocarbons to the atmosphere, at rates inferred for 1977, would eventually deplete stratospheric ozone by 5% to 28%, most probably 17% (NAS, 1979). In 1982, the National Academy revised its previous estimate and suggested a depletion of from 5% to 9% (NAS 1982c).

Increasing carbon dioxide and halocarbon concentrations and decreasing stratospheric ozone concentration of the atmosphere may alter global radiation fluxes. Presumably a primary result of more carbon dioxide, halocarbon, and other trace gases in the atmosphere will be warming. Although incoming solar radiation is not absorbed by carbon dioxide and certain other trace gases, portions of infrared radiation from earth to space are. Over time, the earth could become warmer. Although the forces controlling global temperature are varied and complex, the increase of 0.5°C since the mid-1800s is generally agreed to be at least partially caused by increased carbon dioxide. By 2000, it may increase an additional 0.5°C. Numerous models advanced to estimate the average global warming per doubling of carbon dioxide and related climatic feedbacks predict an increase of 1.0 to 5.0°C (Hansen, 1987). Natural impacts on climate, such as solar variability, remain important and of unclear relationship to anthropogenic causes. A mean global average surface warming, however, of 3 ± 1.5°C in the next century appears possible (National Academy of Sciences 1982a, 1982b).

Actually, in geologic time, plant species in disequilibrium with their climate have been extremely common. During the 2 million years of Quaternary time, during which the modern flora evolved, there were at least 16 glaciations lasting 50,000 to 100,000 years, with interglacial periods lasting 10,000 to 20,000 years. These interglacial periods were characterized by dynamic changing climates and dynamic changing, constantly adjusting forests. The present interglacial period (the Holocene epoch) has been characterized by numerous examples of forest tree distributions not in equilibrium with climate (Davis, 1981). The impressive average rates of Holocene range extensions in eastern North America tree species emphasize these disequilibria (Table 6–6).

A serious consequence of anthropogenic release of halocarbons to the atmosphere is the depletion of naturally occurring stratospheric ozone. Some reduction in halocarbon release has been achieved in the USA and a few other countries. Immediate termination of all release worldwide, however, may still leave the world with important stratospheric ozone reductions during the next decade. Reduced upper-air ozone would increase ultraviolet radiation reaching the surface of the earth. Current understanding does not allow an inventory of the impacts of increased ultraviolet radiation on forests. Vegetation can acclimatize to changes in ultraviolet radiation. Higher plants also vary substantially in resistance to ultraviolet radiation. Because some plants are sensitive, however, reductions in biomass production and/or competitive strengths may be altered by changes in shortwave radiation flux at the surface of the earth (Caldwell, 1981). Studies of more than 100 agricultural species showed that increased ultraviolet exposure reduces plant dry

**Table 6-6.** Average rates of Holocene tree range extensions in eastern North America.

| Species | Rate (m yr$^{-1}$) |
|---|---|
| Jack/red pine | 400 |
| White pine | 300–350 |
| Oak | 350 |
| Spruce | 250 |
| Larch | 250 |
| Elm | 250 |
| Hemlock | 200–250 |
| Hickory | 200–250 |
| Balsam fir | 200 |
| Maple | 200 |
| Beech | 200 |
| Chestnut | 100 |

(Source: Davis, 1981.)

weight and changes the proportion of root, shoot, and leaf tissue. Studies of more than 60 aquatic organisms showed that many were quite sensitive to current levels of ultraviolet radiation at the water surface (Maugh, 1980). Chlorofluorocarbons can also contribute to global warming in a manner similar to carbon dioxide, as previously indicated.

# VI. Future Assessment of Forest Ecosystem Stress Related to Atmospheric Contamination

Assessment of the subtle influences of regional and global air pollutants cannot be based on visible symptom expression. It must be based on systematic monitoring of forest health and growth indices. Waring and Schlesinger (1985) have provided an excellent list of potentially useful forest ecosystem stress indicators (Table 6–7). Canopy leaf area and its duration of display are a very appropriate general index of forest ecosystem stress. Canopy quantity and quality are an indicator of productivity. Inventory techniques from the air (multispectral scanning, micro-wave transmission, radar, laser) and ground (correlations with stem diameter, sapwood cross-sectional area) for canopy leaf area are available. At a given site, detection of an increase in leaf area would suggest an improving environment, whereas a decrease in leaf area would suggest the system is under stress (Waring, 1983, 1985). Baes and McLaughlin (1984) have proposed that trace metal analyses of tree rings can provide information on temporal changes in air pollutant deposition and tree health.

In addition to forest health assessment, systematic determinations must be made of the exposure to air pollutants experienced by forests. This requires a rural air-quality monitoring network capable of measuring all important phytotoxic pollutants deposited by both wet and dry mechanisms. This could be accomplished

**Table 6-7.** Forest ecosystem stress indicators.

| Indication | Examples |
|---|---|
| Canopy limitation | Maximum leaf area index<br>Duration of leaf display |
| Production limitation | Diameter growth and cell division<br>Growth efficiency per unit leaf area |
| Susceptibility to insect<br>and disease attack | Starch content of twigs and large roots<br>Tannin and terpene content of tissues<br>Growth efficiency per unit leaf area |
| Moisture limitations | Predawn water stress<br>Noon leaf turgor<br>Sapwood relative water content<br>Stomatal closure |
| Nutrient limitations | Foliar nutrient content<br>Foliar nutrient ratios<br>Nutrient retranslocations<br>Mineralization indices |
| Physical stress | Sapwood/diameter index<br>Bole taper<br>Symmetry of wood growth<br>Leaf-edge tatter |

(Source: Waring and Schlesinger, 1985.)

by expanding selected National Acid Precipitation Assessment Program wet-deposition monitoring stations to include dry-deposited pollutants.

Assessment of both forest health and atmospheric deposition would be most appropriately implemented by establishing a permanent plot network not unlike the Continuous Forest Inventory plots of the USDA Forest Service. Trends over time in both health and deposition would be regularly and continuously recorded.

A final component of the assessment program is research directed to understanding the mechanisms of low-exposure influence on vegetative health imposed by air contaminants, the influence of enhanced carbon dioxide and warming, and the influence of increased shortwave radiation on trees. This research program must be simultaneously conducted in managed environments (e.g., laboratory and greenhouse chambers, open-top chambers, and other regulated environment facilities) and in natural environments (e.g., field studies conducted in association with long-term ecosystem research areas, experimental forests of the USDA, national laboratory facilities, or other sites presenting unique characteristics, ancillary data, security, and long-term perspective). Permanent assessment plots could be used to validate models developed in seedling and sapling studies. Stand-level models could be verified with permanent plot data. Much of the active research being conducted under the federally supported Forest Response Program,

National Acid Precipitation Assessment Program, is addressing the questions associated with regional-scale pollutants.

Forests are variable in species, topography, elevation, soils, and management. Air pollution deposition and influences are also variable and poorly documented in the field. Monitoring of species dynamics and productivity, necessary to detect effects of regional and global air pollutants or any other environmental stress, are presently rarely available. Dendrochronological or other tree-ring analytical techniques alone are subject to enormous difficulty when they are used in the attempt to partition the relative importance of forces that may influence tree growth. Growth is regulated by precipitation, temperature, length of growing season, frost, drought, developmental processes such as succession and competition, and stochastic events such as insect outbreaks, disease epidemics, fire, windstorms, and anthropogenic activities such as thinning, fertilization, harvesting, *and* finally air quality.

For a long time, dieback and decline of specific forest species, somewhere in the temperature zone, have been common. Age, climate, or biotic stress factors have frequently been judged to be the principal causes for declines. Again, however, it is difficult to assign responsibility for specific cause and effect. Trees are large and long-lived, and their health integrates all the stresses to which they are exposed over time.

The risks associated with regional and global air pollution stress and temperate-zone forest ecosystem health are high. The evidence available to describe the total boundaries of the problem for all pollutants is incomplete. There is enormous uncertainty about specific effects on forests from regional and global air pollutants.

Individual forest tree processes at greatest risk from wide-area air pollutant stress include foliar metabolism (especially photosynthesis and respiration), root metabolism, reproductive physiology, pest (arthropod and microbial) interactions, and growth. At the ecosystem level, processes at greatest risk from wide-area air pollutant deposition include nutrient cycling (especially mineralization and nutrient-uptake processes), biomass production, and population dynamics of insect and microbial pests. Major structural changes may be caused in impacted forests over the long term due to alteration of species composition and patterns of succession.

The primary needs to enable a more complete understanding and assessment of air quality and forest ecosystem decline include the following:

1. Complete monitoring of gas and particulate, wet and dry pollutant deposition in forest ecosystems.
2. Systematic monitoring of forest growth and health parameters on permanent plots over extended time.
3. Coordinated field, controlled environment, and laboratory studies directed toward tree and ecosystem processes identified at risk, using single pollutant and pollutant mixtures in realistic (ambient and above and below ambient) exposure patterns and in association with varying climatic scenarios.

4. Modeling and systems analysis studies to assist with experimental designs and to extend experimental conclusions to larger forest areas.

Implementation of wide-area forest monitoring of any nature involves two challenges. First, detection of stress does not suggest cause. We are keenly aware that tree and forest health are controlled by many factors in addition to air quality. We desperately need procedures to partition the relative importance of influencing variables for a given site. Fortunately we are making research progress toward this resolution (Fritts and Stokes, 1975; Waring, 1985). The second challenge is to convince natural resource managers that the time and cost of systematic forest health monitoring is justified. I feel it is not only justified but also essential for intelligent decisions regarding regulation of regional and global air pollutants, for understanding the health and productivity of our forests over the next century, and for deliberations on national energy policy.

## VII. Conclusions

1. Acidic deposition appropriately includes all forms of acidic or acidifying materials transferred from the atmosphere to the earth in the dry (gas, particulate) or wet (precipitation) form.
2. Acidic deposition is appropriately designated a regional-scale air contaminant, along with oxidants and heavy metals, as the formation of these secondary air pollutants occurs over tens or hundreds of km from the site of precursor release and because these pollutants may be deposited 100 to more than 1,000 km from precursor origin.
3. Various North American forest ecosystems are subject to the deposition of regional-scale air contaminants, singly or in combination, at certain times during the year.
4. Our understanding of fundamental forest ecosystem biology is incomplete. This makes assessment of subtle, long-term stress imposed by regional-scale air pollutant deposition very difficult. Some important ecosystem generalizations, central to air pollution stress understanding, are as follows:
   a. Ecosystems are dynamic, not static. They are characterized by variability rather than constancy and tend to change with time and space. Adaptation, adjustment, and evolution take place with time as the biotic and abiotic components interact. The dominants of a community are replaced by new species as energy flow and nutrient cycling are altered. As the sequential changes in species composition and community structure known as *succession* occur, ecosystems originate, develop, mature, and ultimately decline and disaggregate.
   b. Disturbances are characteristic of natural ecosystems. Disturbance may be caused by natural or anthropogenic processes. Examples of the former are fire, drought, microbial infection, insect infestation, volcanic eruption, and destructive wind. Examples of the latter include selective species removal (hunting, thinnings), pesticide application, fertilizer application, controlled

fire, construction (e.g., alteration of moisture availability), air pollution, and urbanization.

c. Increasing evidence suggests that numerous terrestrial ecosystems require major disturbance, such as fire, drought, and windstorms, to retain their characteristics. In the absence of disturbance, some ecosystems appear to degrade. With moderate rates of disturbance, ecosystems may be most productive and have the largest number of species and biomass.

d. In order for an ecosystem to be perturbed or stressed, an influence *must* be imposed on one or more of the 11 elements of ecosystems. Because of the interaction among the different elements of an ecosystem, a change in one element can cause a response throughout the ecosystem.

e. Not all ecosystems respond in the same way to stress. Some ecosystems may be more sensitive to a given perturbation at one stage of development than at another. Each ecosystem has distinctive characteristics of inertia and resilience.

f. Two groups of organisms particularly critical to the maintenance of an ecosystem are the producers, through which solar energy, carbon, and other nutrients enter living systems, and the decomposer trophic level, through which nutrients bound up in other organisms are released for reuse. Loss of either of these groups results in the collapse of the entire system.

5. Hypotheses advanced to describe the interaction between regional air pollutants and forest ecosystems emphasize the importance of long-term impacts on nutrient cycling, tree species composition, and productivity (biomass accumulation). Gradual potential for alteration of species composition, energy storage, and rates of biogeochemical cycling are emphasized over severe tree morbidity or mortality with dramatic symptoms.

6. Regional air contaminants may influence reproductive processes, nutrient uptake or retention, metabolic rates (especially photosynthesis and respiration), and insect pest and pathogen interactions of individual trees. At the ecosystem level, regional air pollutants may infleunce nutrient cycling, population dynamics of arthropod or microbial species, succession, species composition, and biomass production. In the case of low-dose regional-scale pollution, symptoms are typically not visible (at least initially), undramatic, and not easily measured. The integration of regional pollutant stresses can be slower growth, altered competitive abilities, and changed susceptibility to pests. Ecosystem symptoms may include altered rates of succession, changed species composition, and biomass production. Visible symptom expression is not a useful assessment strategy.

7. For a long time, dieback and decline of specific forest species, somewhere in the temperate zone, have been common. Age, climate, or biotic stress factors have frequently been judged to be the principal causes for declines. Again, however, it is difficult to assign responsibility for specific cause and effect. Trees are large and long-lived, and their health integrates all the stresses to which they are exposed over time.

8. Forests are variable in species, topography, elevation, soils, and manage-

ment. Air pollution deposition and influences are also variable and poorly documented in the field. Monitoring of species dynamics and productivity, necessary to detect effects of regional and global air pollutants, or any other environmental stress, are presently rarely available.

9. The potential stresses imposed by regional-scale air pollutants must be viewed against the background of uncertainty associated with the changes in radiation balances associated with global-scale air contaminants. If global warming does accelerate in the near term, and especially if expansive forest regions are subjected to reduced moisture availability, the expansion of forest tree decline could be very great. Such an occurrence would pale the impact of regional-scale pollutant deposition.

10. The primary needs to enable a more complete understanding and assessment of air quality and forest ecosystem health include the following:

   a. Complete monitoring of gas and particulate, wet and dry pollutant deposition in forest ecosystems.

   b. Systematic monitoring of forest growth and health parameters on permanent plots over extended time.

   c. Coordinated field, controlled environment, and laboratory studies directed toward tree and ecosystem processes identified at risk using single pollutant and pollutant mixtures in realistic ambient, and above and below ambient, dose patterns, and in association with varying climatic scenarios.

   d. Modeling and systems analysis studies to assist with experimental designs and to extend experimental conclusions to larger forest areas.

# References

Abrahamsen, G. 1984. Phil Trans R Soc London 305:369–382.

Alstad, D.N., G.F. Edmunds, and L.H. Weinstein. 1982. Ann Rev Entomol 27:369–384.

Amthor, J.S. 1986. New Phytol 102:359–364.

Baes, M., and S.B. McLaughlin. 1984. Science 224:494–497.

Balch, R.E. 1927. For Worker 3:13.

Bell, J.N.B. 1986. Experientia 42:363–371.

Bennett, D.A., R.L. Goble, and R.A. Linthurst. 1985. *The acidic deposition phenomenon and its effects.* USEPA 600-8-85-001, Washington, D.C. 159 p.

Bewley, R.J.F., and G. Stotzky. 1983a. Soil Biol Biochem 15:431–437.

Bewley, R.J.F., and G. Stotzky. 1983b. Soil Biol Biochem 15:425–429.

Billings, W.D. 1978. Plants and the ecosystem, Wadworth Publish. Co., Belmont, Ca, 62 p.

Binkley, D., and D. Richter. 1986. Advan Ecol Res (in press).

Botkin, D.B., and E.A. Keller. 1982. *Environmental studies.* Merrill Publishing Co., Columbus, Ohio. 506 p.

Brandt, R.W. 1961. USDA Forest Service, Northeastern For Exp Sta Paper 163, Upper Darby, Pa. 8 p.

Bruck, R.I. 1987. *Forests and acid deposition: Status of knowledge* (in press).

Bruck, R.I., and S.R. Shafer. 1984. *In* J.I. Teasley and R.A. Linthurst, eds. *Direct and indirect effects of acidic deposition on vegetation*, 19–32. Butterworth, Boston.

Bryant, R.D., E.A. Gordy, and E.J. Laishley. 1979. Water Air Soil Pollut 11:437–445.

Caldwell, M.M. 1981. Encyclopedia Plant Physiol NS 12A:169–197.

Campbell, C.L., S.B. Martin, J.F. Sinn, and R.I. Bruck. 1985. Phytopathology (in press).

Castello, J.D., S.B. Silverborg, and P.D. Manion. 1985. Pl Dis 69:243–246.

Chevone, B.I., D.E. Herzfeld, S.V. Krupa, and A.H. Chappelka. 1986. J Air Pollu Cont Assoc 36:813–815.

Chung, Y.S. 1978. Atmos Environ 12:1471–1480.

Copeland, O.L., and R.G. McAlpine. 1955. Ecology 36:635–641.

Crist, T.O., N.R. Williams, J.S. Amthor, and T.G. Siccama. 1985. Environ Poll 14: 38–40.

Cronan, C.S. 1979. Oikos 34:272–281.

Cronan, C.S. 1980. Plant Soil 56:301–322.

Cronan, C.S. 1984. *In* J.J. Teasley and R.A. Linthurst, eds. *Direct and indirect effects of acidic deposition on vegetation*, 65–79. Butterworth, Boston.

Cronan, C.S. 1985. *In* J.I. Drever, ed. *The chemistry of weathering*, 175–195. D. Reidel, New York.

Cronan, C.S., and W.A. Reiners. 1983. Oecologia 59:216–223.

Cronan, C.S., W.A. Reiners, R.C. Reynolds, and G.E. Lang. 1978. Science 200: 309–311.

Cronan, C.S., and C.L. Schofield. 1979. Science 204:304–305.

Cumming, J.R., R.T. Eckert, and L.S. Evans. 1985. Can J Bot 63:1099–1103.

Davis, M.B. 1981. *In* D.C. West, H.H. Shugart, and D.B. Botkin, eds. *Forest succession concepts and application*, 132–153. Springer-Verlag, New York.

Dochinger, L.S., and K.F. Jensen 1985. USDA For Ser Res Paper NE-572, Broomall, Pa. 104 p.

Doelman, P. 1978. *In* J.O. Nriagu, ed. *The biogeochemistry of lead in the environment*, Part B, 343–353. Elsevier, New York.

Evans, L.S. 1982. Environ Exp Bot 22:155–169.

Evans, L.S., and T.M. Curry. 1979. Amer J Bot 66:953–962.

Evans, L.S., N.F. Gmur, and F. DaCosta. 1978. Phytopathology 68:847–856.

Firestone, M.K., J.G. McColl, K.S. Killham, and P.D. Brooks. 1984. *In* J.I. Teasley and R.A. Linthurst, eds. *Direct and indirect effects of acidic deposition on vegetation*, 51–63. Butterworth, Boston.

Foy, C.D. 1984. *In* F. Adams, ed. *Soil acidity and liming*, 57–97. Soil Sci Soc Amer, Madison, Wisc.

Foy, C.D., R.L. Chaney, and M.C. White. 1978. Ann Rev Pl Physiol 29:511–566.

Friedland, A.J., R.A. Gregory, L. Karenlampi, and A.H. Johnson. 1984. Canad J For Res 14:963–965.

Fritts, H.C., and M.A. Stokes. 1975. Tree Ring Bull 35:15–24.

Fuhrer, J., and C. Fuhrer-Fries. 1982. Eur Jour For Pathol 12:377–390.

Graw, D. 1979. New Phytol 82:687–695.

Greenidge KNH (1953) Canad J Bot 31:548–559.

Hansbrough, J.R., V.S. Jensen, H.J. Macaloney, and R.N. Nash. 1950. Forest Pest Leaflet 52, Society of American Foresters, Hillsboro, N.H., 4 p.

Hansen, J. 1987. Personal communication, Goddard Instit. Space Studies, NASA, Columbia Univ., N.Y., Feb. 6, 1987.

Hari, P., T. Raunemaa, and A. Hautojarvi. 1986. Atmos Environ 20:129–137.

Hawboldt, L.S. 1952. Dept Lands and Forests Bull No 6, Province of Nova Scotia, Canada. 37 p.

Hibben, C.R., and S.B. Silverborg. 1978. Jour Arboricul 4:274–279.

Hoffman, W.A., S.E. Lindberg, and R.R. Turner. 1980. J Environ Qual 9:95–100.

Holdgate, M.W., M. Kassas, and G.F. White. 1982. Environ Conserv 9:11–29.

Houston, D.R. 1974. Arborists News 49:73–76.

Houston, D.R. 1975. Jour For 73:660–663.

Houston, D.R. 1981. USDA Forest Service Pub NE-INF-41-81, Washington, D.C. 36 p.

Houston, D.R., E.J. Parker, R. Perrin, and K.J. Lang. 1979. Eur Jour For Pathol 9:199–211.

Hutterman, A. 1983. Allgem Forstz Ztg 27:663–664.

Huttunen, S. 1984. *In* M. Treshow, ed. *Air pollution and plant life*, 321–356. John Wiley, New York.

Jackson, D.R., W.J. Selvidge, and B.S. Ausmus. 1978. Water Air Soil Pollut 10:13–18.

Jackson, D.R., and A.P. Watson. 1977. J Environ Qual 6:331–338.

Johnson, A.H. 1983. J Air Pollu Cont Assoc 33:1049–1054.

Johnson, A.H., and T.G. Siccama. 1983. Environ Sci Technol 17:294–305.

Johnson, D.W., D.D. Richter, H. Van Miegroet, and D.W. Cole. 1983. J Air Pollu Cont Assoc 33:1036–1041.

Johnson, D.W., J. Turner, and J.M. Kelly. 1982. Water Res Res 18:449–461.

Johnson, N.M., R.C. Reynolds, and G.E. Likens. 1972. Science 177:514–516.

Killham, K., and M.K. Firestone. 1983. Plant Soil 72:39–48.

Klein, R.M. 1984. *In* J.I. Teasley, and R.A. Linthurst, eds. *Direct and indirect effects of acidic deposition on vegetation*, 1–11. Butterworth, Boston.

Krug, E.C., and C.R. Frink. 1983. Science 221:520–525.

Lacki, M.J. 1985. Bull Torr Bot Club 112:398–402.

Leaphart, C.D., and O.L. Copeland 1957. Soil Sci Soc Amer 21:551–554.

Lee, J.J., and D.E. Weber. 1979. For Sci 25:393–398.

Lewis, W.M., and M.C. Grant. 1979. Ecology 60:1093–1097.

Linthurst, R.A., ed. 1984. *Direct and indirect effects of acidic deposition on vegetation*. Acid Precipitation Series, Vol. 5. Butterworth, Boston. 117 p.

Loucks, O.L. 1970. Amer Zool 10:17–25.

Lucier, A.A. 1986. National Council of the Paper Industry for Air and Stream Improvement Tech Bull 508, New York. 15 p.

McClenahen, J.R., and L.S. Dochinger. 1985. J Environ Qual 14:274–280.

McCool, P.M., and J.A. Menge. 1983. New Phytol 94:241–247.

McCormick, L.H., and K.C. Stiner. 1978. For Sci 24:565–568.

McCracken, F.I. 1985. J Arboricul 11:253–256.

McFee, W.W. 1980. *Sensitivity of soil regions to acid precipitation*. USEPA Pub 600-3-80-013, Corvallis, Oreg. 178 p.

McIlveen, W.D., S.T. Rutherford, and S.N. Linzon. 1986. Ministry of the Environment Report ARB-141-86-Phyto, Ontario, Canada. 40 p.

McLaughlin, D.L., S.N. Linzon, D.E. Dimma, and W.D. McIlveen. 1985. Ministry of the Environment Report ARB-144-85-Phyto, Ontario, Canada. 18 p.

McLaughlin, S.B., T.J. Blasing, L.K. Mann, and D.N. Duvick. 1983. J Air Pollu Cont Assoc 33:1042–1049.

McLaughlin, S.B., R.K. McConathy, D. Duvick, and L.K. Mann. 1982. For Sci 28:60–70.

Manion, P.D. 1981. *Tree disease concepts*. Prentice-Hall, Englewood Cliffs, N.J. 399 p.

Martin, S.B., C.L. Campbell, and R.I. Bruck. 1985. Phytopathology (in press).

Maugh, T.H. 1980. Science 207:394–395.

Miller, P.R., and G.F. Gravatt. 1952. Pl Dis Rptr 36:247–252.

Miller, P.R., O.C. Taylor, and R.G. Wilhour. 1982. U.S. Environmental Protection Agency Pub EPA-600/D-82-276, Corvallis, Oreg. 10 p.

Mollitor, A.V., and D.J. Raynal. 1982. Soil Sci Soc Amer 46:137–141.

Molnar, A.C., and G.T. Silver. 1959. For Chron 35:227–231.

Montgomery, M.E., and G.A. Meyer. 1985. Proc. 4th IUFRO Conf., Curitiba, Brazil.

National Academy of Sciences. 1979. *Stratospheric ozone depletion by halocarbons.* NAS, Washington D.C. 249 p.

National Academy of Sciences. 1982a. *Solar variability, weather, and climate.* NAS, Washington, D.C. 120 p.

National Academy of Sciences. 1982b. *Carbon dioxide and climate: A second assessment.* NAS, Washington, D.C. 92 p.

National Academy of Sciences. 1982c. *Stratospheric ozone depletion by halocarbons.* NAS, Washington, D.C. 55 p.

National Council of the Paper Industry for Air and Stream Improvement Inc. 1981. *Acidic deposition and its effects on forest productivity—A review of the present state of knowledge, research activities, and information needs.* Tech Bull 110. Author, New York. 66 p.

National Research Council, Committee on Nuclear and Alternative Energy Systems. 1980. Supporting Paper No. 8, National Academy Press, Washington, D.C. 399 p.

Nichols, J. 1968. J For 66:681–694.

Nordgren, A., E. Baath, and B. Soderstrom. 1985. Can J Bot 63:448–455.

Odum, E.P. 1971. *In* J.A. Wiens, ed. *Proceedings of the 31st Annual Biology Colloquium,* 11–24. Oregon State Univ Press, Corvallis, Oreg.

Parker, G.G. 1983. Ad Ecol Res 13:57–133.

Parker, J., and D.R. Houston. 1971. For Sci 17:91–95.

Postel, S. 1984. *Air pollution, acid rain and the future of forests.* Worldwatch Paper No. 58, Washington, D.C.

Raynal, D.J., J.R. Roman, and W.M. Eichenlaub. 1982a. Environ Exp Bot 22:377–383.

Raynal, D.J., J.R. Roman, and W.M. Eichenlaub. 1982b. Environ Exp Bot 22:385–392.

Reich, P.B., A.W. Schoettle, H.F. Stroo, J. Troiano, and R.G. Amundson. 1985. Canad J Bot 63:2049–2055.

Reich, P.B., H.F. Stroo, A.W. Schoettle, and R.G. Amundson. 1986. Air Pollu Cont Assoc 36:724–726.

Reiners, W.A., G.M. Lovett, and R.K. Olson. 1985. *In Proc Forest-Atmos Interaction Workshop.* Lake Placid, N.Y. 27 p.

Reiners, W.A., and R.K. Olson. 1984. Oecologia 63:320–330.

Reuss, J.O., and D.W. Johnson. 1986. *Acid deposition and the acidification of soils and waters.* Springer-Verlag, New York. 119 p.

Richter, D.D., D.W. Johnson, and D.E. Todd. 1983. J Environ Qual 12:263–270.

Scherbatskoy, T., and R.M. Klein. 1983. J Environ Qual 12:189–194.

Schier, G.A. 1985. Can J for Res 15:29–33.

Shafer, S.R., L.F. Grand, R.I. Bruck, and A.S. Heagle. 1985. Canad J For Res 15:66–71.

Shaw, C.G., A. Aglitis, T.H. Laurent, and P.E. Hennon. 1985. Pl Dis 69:13–17.

Sheffield, R.M., N.D. Cost, W.A. Bechtold, and J.P. McClure. 1985. USDA Forest Service, Southeastern For Exp Sta Res Bull SE-83, Asheville, N.C. 112 p.

Shepard, M. 1985. EPRI Jour (September):17–25.

Shriner, D.S. 1986. USDA Forest Service, US Environmental Protection Agency, Cooperative Research Plan, fiscal 1986. Broomall, Pa. 44 p.

Shriner, D.S., and J.W. Johnston. 1981. Environ Exp Bot 21:199–209.

Siccama, T.G., M. Bliss, and H. Vogelmann. 1982. Bull Torrey Bot Club 109:162–168.

Sinclair, W.A. 1965. Cornell Plantations 20:62–67.

Singh, B.R., G. Abrahamsen, and A. Stuanes. 1980. Soil Sci Soc Amer J 44:75–80.

Skelly, J.M. 1974. Pl Dis Rptr 58:396–399.

Smith, W.H. 1981. *Air pollution and forests*. Springer-Verlag, New York. 379 p.

Smith, W.H. 1984. For Ecol Manag 9:193–219.

Smith, W.H. 1985. Jour For 83:82–92.

Smith, W.H., G. Geballe, and J Fuhrer. 1984. *In* J.I. Teasley and R.A. Linthurst, eds. *Direct and indirect effects of acidic deposition on vegetation*, 33–50. Butterworth, Boston.

Society of American Foresters. 1984. *Acidic deposition and forests*. Washington, D.C. 48 p.

Staley, J.M. 1965. For Sci 11:2–17.

Steiner, K.C., J.R. Barbour, and L.M. McCormick. 1984. For Sci 30:404–410.

Stroo, H.F., and M. Alexander. 1985. Water Air Soil Pollut 25:107–114.

Thompson, G.W., and R.J Medve. 1984. Appl Environ Microbiol 48:556–560.

Tong, E.Y., G.M. Hidy, T.F. Lavery, and F. Berlandi. 1976. *Third symposium on turbulence, diffusion and air quality*. Amer Meteor Soc, Boston.

Toole, E.R., and B.J. Huckenpahler. 1954. Pl Dis Rptr 38:786–788.

Treshow, M. 1984. *Air pollution and plant life*. John Wiley, New York. 486 p.

Ulrich, B., and J. Pankrath. 1983. *Effects of accumulation of air pollutants in forest ecosystems*. D. Reidel, Boston. 389 p.

University of Maine. 1984. *Forest responses to acidic deposition*. Univ. of Maine, Orono. 117 p.

U.S. Congress, House of Representatives. Subcommittee on Forests, Family, Farms and Energy. 1986. *Review the effects of acid deposition and other air pollutants on forest productivity; forest ecosystems and atmospheric pollution research act of 1985; and the endangered forest research act of 1985*. U.S. Government Printing Office, Washington, D.C. 381 p.

U.S. Environmental Protection Agency. 1983. *The acidic deposition phenomenon and its effects*, vol I, 3–92. Pub 600-80-83-016A. Washington, D.C.

U.S. Environmental Protection Agency. 1986. USEPA Pub EPA-600-8-84-020. Research Triangle Park, N.C.

USEPA and USDA Forest Service. 1986. *Responses of forests to atmospheric deposition: National research plan for the forest response program*. Washington, D.C. 67 p.

van Breeman, N., C.T. Driscoll, and J. Mulder. 1984. Nature 307:599–604.

Voigt, G. 1980. *In* Proc Int Conf Impact acid precip, 53–57. SNSF Project, Acid Precipitation Effects on Forest and Fish, Oslo, Norway.

Wargo, P.M., and D.R. Houston. 1974. Phytopathology 64:817–822.

Waring, R.H. 1983. Adv Ecol Res 13:327–354.

Waring, R.H. 1985. For Ecol Manage 12:93–112.

Waring, R.H., and W.H. Schlesinger 1985 *Forest ecosystems concepts and management*. Academic Press, New York. 340 p.

Weiss, M.J., L.R. McCreery, I. Millers, J.T. O'Brien, and M. Miller-Weeks. 1985. *Cooperative survey of red spruce and balsam fir decline and mortality in New Hampshire, New York, and Vermont—1984*. Interim Report. USDA Forest Service, Durham, N.H. 130 p.

Westman, W.E. 1978. BioScience 28:705–710.

Wolff, G.T., N.A. Kelly, and M.A. Furman. 1981. Science 211:703–705.

Woodman, J.N., and E.B. Cowling. 1987. Environ Sci Technol 21:120–126.

# Effects of Acidic Precipitation on Forest Ecosystems in Europe

## B. Ulrich*

## Abstract

The system approach that underlies the research is presented. The input–output relations of forest ecosystems in regard to acidity, heavy metals, and nutrients are discussed in relation to the buffering ability and nutrient demand of the systems. During the last decades, acidity of forest soils has increased with a concomitant decrease in nutrient cation storage. There are indications that litter decomposition has been retarded. The direct effects of acidic deposition, $SO_2$, $NO_x$, and ozone on the leaves of trees well supplied with nutrients and water cannot explain the defoliation symptoms observed in German forests. A typical symptom seems to be damage in the vascular bundle caused by mineral deficiencies. Soil solution parameters characterizing proton and aluminum stress have been defined and tested in solution culture studies. Extensive field studies on the depth gradient of fine and coarse roots, on the dynamics of fine-root biomass and necromass, on damage symptoms of roots, on the element content of roots, and on the acidic stress parameters in the soil solution have been carried out. The results suggest that during the last decades the fine-root system has shifted to the topsoil as a consequence of the increased acidity in the subsoil. In the case of Norway spruce, this shift causes chronic water stress, which in turn causes crown thinning. Magnesium deficiency in Norway spruce, manifested as needle yellowing, can be attributed to the blockage of magnesium ion uptake by aluminum ions.

## I. Introduction

In respect to the mathematical description of systems, ecosystems can be characterized as dynamic. Dynamic systems are determined by causal relationships that have a memory, that is, a storage capacity.

---

*Institute of Soil Science and Forest Nutrition, University of Göttingen, D-3401 Göttingen, Büsgenweg 2, FRG.

Dynamic compartmental systems $E$ are defined by the quintuple (Ludyck, 1977)

$$E = (u, y, x, f, g) \tag{1}$$

where $u$ represents the input vector, $y$ the output vector, $x$ the state vector, f the transfer function, and g the output function. In addition, the state of rest of the system, comparable to an equilibrium state, has to be defined.

*Input* and *output* have their usual meanings. In respect to effects of acidic deposition, the input vector consists of all independent variables that have an influence on the acid-base status of the ecosystem. The output vector consists of all dependent variables that connect the system with its environment and that in turn are influenced by the input. The state vector represents all state variables of the system that are influenced by the input and that allow, in connection with the input vector, calculation of the output vector. The transfer functions allow calculation of the change in state variables due to an input. The output function allows the calculation of the output as a function of changes in state variables.

The state variables of forest ecosystems can be listed only at a high level of aggregation. They are (Ulrich, 1987):

The primary producers, which carry out photosynthesis. In forest ecosystems, trees and the ground vegetation are the most important components.

The secondary producers, which utilize organic matter as an energy source. In forest ecosystems, decomposers (soil animals, microorganisms, etc.) and herbivores, as well as plant pathogens, are of major importance.

The soil solution with the dissolved ion pool.

The mobilizable ion pool bound on surface of the solid soil phase and in soil organic matter.

The environment of a forest ecosystem consists of:

The atmosphere.

The ion pool bound in the interior of soil minerals and being mobilized by weathering. With this definition, the release of ions by weathering is part of the input vector.

The seepage water leaving the rooted soil and being the main carrier of the output of dissolved ions.

Neighboring ecosystems. Another ecosystem begins if there is a change in state variables exceeding the spatial and temporal variation that is a characteristic feature for a given ecosystem.

Anthropogenic air pollutants may mean a change in the material (ion) input that reaches the forest ecosystem. A change in material input creates an automatic change in the material (mass) balance of the ecosystem.

The flow of materials in the ecosystem can be expressed by the following equation:

$$\text{Photosynthesis and formation of organic substances} \longrightarrow \tag{2}$$
$$\text{Respiration and mineralization of organic substances} \longleftarrow$$

$$a CO_2 + x M^+ + y A^- + (y-x)H^+ + z H_2O + \text{Energy} \xrightleftharpoons$$

$$\xrightleftharpoons (C_a H_{2z} O_z M_x A_y)_{\text{organic matter}} + (a + \ldots) O_2$$

where $a$, $x$, $y$, $z$ represent stoichiometric coefficients, $M^+$ is cations, and $A^-$ is anions of unit charge. In this equation, *organic matter* represents the primary and secondary producers, that is, the total amount of organisms, as well as undecomposed forms of organisms (humus). The cations $M^+$, the anions $A^-$, and the protons $H^+$ are present in the soil within the dissolved and mobilizable ion pool.

The *state of rest* can be defined by equal rates of the forward and backward reaction. This indicates that the system is in steady state. Many scientists maintain that terrestrial ecosystems can approach a steady state in which the species composition and the organic matter storage in biomass and in soil show no trend. In terms of Equation 2, *steady state* means that the material fluxes caused by the activity of primary producers are balanced by the material fluxes caused by the activity of secondary producers. The steady state represents an ideal state in which ecosystems could exist. Terrestrial ecosystems have thus, in the ideal state, the property of keeping the internal ion cycle closed. This property is important for maintaining the chemical state within the ecosystem and in its environment. As shown in Equation 2, a net flow of ions may be connected with the production or consumption of protons and result in the acidification or alkalinization of the soil, respectively. Organisms therefore have the property of changing the chemical state of their environment. In a metaphorical sense, man-made pollution is a consequence of this property. If terrestrial ecosystems approach the ideal (steady) state, however, the sink and source effects of primary and secondary producers in respect to ions tend to balance each other and result in a net effect of proton turnover close to zero. Acidic deposition unquestionably influences the proton balance of the ecosystem. The long-term effect of acidic deposition on forest ecosystems cannot be assessed without identifying and quantifying its influence on the proton balance.

Equation 2 shows that all ions involved in the turnover have an effect on the proton balance. From this it follows that the input and output vectors involve all substances existing as ions or being transferred into ions (like $SO_2$, $NO_x$, $NH_3$, organic substances). The transfer of ions across phase boundaries occurs almost exclusively in their dissolved form. This component of the input and output vector can be assessed by measuring the water flows (precipitation, seepage) and analyzing the major cations and anions in the solutions H, Na, K, Mg, Ca, Mn, Al, Fe, $NH_4$, $NO_3$, Cl, $SO_4$, $N_{org}$, TOC, and organic anions (that are usually calculated from the charge balance of the solution).

Then the actual state of a forest ecosystem can be described by

Its species composition regarding primary producers (vegetation form).
Its species composition regarding secondary producers (the humus form can serve as a characteristic feature).
The chemical soil state (spatial and temporal variability of the composition of the soil solution and the mobilizable ion pool).

The deviation of the material balance of the ecosystem from the steady state. The steady state is characterized by quantitative equality of the input and output vector. With regard to acidification through acidic deposition, the input and output of ions has to be compared. This can be done by measuring deposition (input) and the flow of ions with seepage water (output).

The measurement of ion output with seepage water requires knowledge of the seepage rate. The seepage rate can be calculated with aid of mathematical models of the transport of water through ecosystems. For the data presented later, the model described by Hauhs (1986) was adopted. For chemical analysis, the seepage water is collected with suction lysimeter candles or plates located below the root zone in the undisturbed soil. This is now a standard technique.

Until data on the deviation between input and output of ions are available, the long-term effect of acidic deposition on forest ecosystems cannot be assessed.

The approach outlined so far is the basis for the following review on the effects of acidic deposition on forest ecosystems in Europe. First of all, however, the approach used to assess the input vector has to be discussed.

## II. Assessment of Deposition

Air pollutants may be natural or anthropogenic in origin. They can be classified according to their ecochemical function as follows:

Neutral substances: $Na^+$, $Cl^-$ (sea spray)
Nutrients and essential elements: $SO_4^{2-}$, $Mg^{2+}$, and so forth (sea spray); $NH_4^+$, $NO_3^-$, $Ca^{2+}$, trace elements
Acidifiers: $SO_2$, $NO_x$
Potential toxins; $SO_2$, HF, heavy metals, metal oxides, hydrocarbons
Oxidants: $NO_x$, ozone, hydrocarbons

Many anthropogenic air pollutants also occur naturally and are in some cases vital for the development of ecosystems. In central Europe, anthropogenic emissions do, however, exceed natural emissions by a factor of around 10 in the case of many air pollutants.

The following deposition processes can be distinguished:

1. Wet deposition in the form of rain or snow
2. Sedimentation (gravimetric fall) of particles (dry deposition)
3. Impaction of aerosols (dry deposition)
4. Impaction of mist, fog, cloud droplets, (occult deposition)
5. Absorption of gases, for example, $SO_2$ (dry deposition)
    On wet surfaces like foliage, bark, wet snow
    Inside stomates
      On cell walls
      On mesophyll and palisade cell surfaces
6. Reemission

Wet deposition of rain or snow is not influenced quantitatively by the receiving surface. Deposition of particles (sedimentation, impaction) is strongly dependent on particle size. Fog, mist, and cloud droplets could be regarded as large particles and are very efficiently captured by a canopy, making the rate of deposition dependent mostly on the wind speed and thus on the exposure of the receiving surface to wind action. Deposition of gases varies with the type and state of the receiving surface.

Deposition can be viewed in two different ways: either from the deposited compound or from the receiving surface. As to deposition compounds, there are wet and dry forms. In the case of the latter it is prudent to distinguish between precipitation deposition $(1 + 2)$ and interception deposition (3 to 5) (Ulrich et al., 1979; Ulrich, 1983b).

It is evident that the processes summarized under the term *interception deposition* depend upon size, type and (chemical) state of the receiving surface. This has important consequences regarding the measuring techniques.

After deposition, chemical reactions can occur. The following reaction types can be expected:

Acid production (e.g., $SO_2$ reacts to $H_2SO_4$)
proton buffering (e.g., cation exchange on cell walls)
precipitation or dissolution, depending on pH (e.g., heavy metals)
reactions with living plant cells (reactions with cell membranes, uptake into cells,
    assimilation in the cell metabolism)
formation of gaseous compounds (e.g., sulfides)

After deposition, translocation within the canopy may take place. Due to leaching by rain, removal of elements from the surface and interior of leaves and needles may also take place. Much of the deposited material could be transferred from the canopy to the ground by this process.

Tree leaves can buffer acidity deposited on the leaf via exchange with Ca (Ulrich, 1983b; Leonardi and Flückinger, 1987). It is assumed that Ca bound on acidic groups in the cell wall is exchanged by the protons resulting from the formation of $H_2SO_4$. During stomatal closure, this exchange may be reversed due to an input of $Ca(HCO_3)_2$ (or other Ca salts of weak acids). Because the Ca is finally supplied from the exchangeable pool in the soil, the amount of basicity fed in the transpiration stream is balanced by an equivalent amount of acidity formed in the soil. Proton buffering occurring in leaves is thus transferred to the roots and to the soil, as demonstrated by Arndt and others (1985), Jurat and others (1986), Leonardi and Flückinger (1988), and Seufert and Arndt (1986).

Acidity can be deposited as $H^+$ (including $HNO_3$), $SO_2$, $NO_x$, $NH_4^+$, cation acids like $Mn^{2+}$ and $Al^{3+}$, and organic acids. The contribution of cation acids and organic acids to the total acidity in deposition is usually negligible. No data are available on deposition rates of $NO_x$ and $HNO_3$.

In solutions of pH $<5.0$, typical for acidic deposition, the alkalinity (concentration of $OH^-$, $HCO_3^-$) approaches zero. Such solutions are mixtures of acids, acidic salts, and neutral salts. Their acidity can be characterized by the percentage of acids and acidic salts ($H^+$ + cation acids) of the total cation equivalent sum (acidity degree, AD in percent).

Basicity can be deposited as dust (restricted to small forests located in rural and urban areas), as $NH_3$ (restricted to forests close to large emission sources like feedlots), and as $NO_3^-$. Nitrate behaves as a base if it is taken up by organisms and, together with a proton, transformed to organic bound nitrogen.

Therefore, the assessment of the rate of acidic deposition requires:

Using the forest canopy under the natural climatic conditions as a collecting surface. Artificial collectors cannot imitate the indigenous properties of a living forest canopy. In case of the data presented later, the ion fluxes with throughfall and stemflow have been used to calculate deposition. This approach requires, however, that the fluxes of ions leached from or retained in the canopy are assessed separately. Several approaches have been developed to calculate, on the basis of the flux balance of the canopy, the leaching rate from the canopy separately (Mayer and Ulrich, 1974; Lakhani and Miller, 1980; Ulrich et al., 1979; Ulrich, 1983b). These approaches give comparable results. For the data presented later, the approach of Ulrich (1983b; see also Bredemeier, 1988) was adopted, using Na as a reference basis and making following assumptions: (1) that Na from sea spray is a main component of fog and cloud water interception, (2) that the fog and cloud water droplets are in a steady state with the ambient gaseous and particulate air pollution, and (3) that Na is not leached from leaves. In areas close to the seacoast, Na deposition is usually high but the second condition may not be fullfilled. In southern Germany, Na deposition is very low and the first condition may not be fulfilled. Thus, the use of Na as a reference basis may be restricted to the northern and Central parts of Germany (Hauhs et al., 1989).

Assessing the rate of proton buffering in the leaves. This process diminishes the acidity in throughfall; therefore, the acidity in throughfall cannot be taken as a measure of acidic deposition. In addition, this process represents a direct acidic load of the tree that has to be assessed separately. The leaching of nutrient cations like Ca and Mg from the leaves is a direct consequence of this process. No method exists to measure the rate of proton buffering in the leaves directly. The data presented later have been calculated using an indirect calculation based on the ion flux balance of the canopy (see Ulrich, 1983b). The proton buffering by the canopy may be overestimated if the forest receives dry deposition of basic particles such as soil dust (Richter and Lindberg, 1988).

Assessing the reactions of $NH_4^+$ and $NO_3^-$ deposited within the ecosystem. A first approach is based on the assumption that ammonium and nitrate are either transformed into uncharged compounds (organic N, gaseous N compounds) or leached with the seepage water. Thus the difference between input and output can be used to calculate the proton production due to $NH_4^+$ deposition and turnover, and the proton consumption due to $NO_3^-$ deposition and turnover. This approach requires that besides the measurement of deposition (input), also the flux of $NH_4^+$ and $NO_3^-$ with seepage water is measured. The acidic load of the ecosystem due to acidic deposition can therefore only be assessed if the output vector is at least partially ($NH_4^+$, $NO_3^-$) known.

The presence of acidic aerosols in the air has the consequence that nutrient cations like Ca, Mg, and K present in soil dust or other particles pass over into a dissolved form and are deposited together with acidic deposition. Thus the input of dissolved nutrient salts into the ecosystem is increased not only directly ($NH_4^+$, $SO_4^{2-}$, $NO_3^-$, heavy metals as trace nutrients) but also indirectly ($Ca^{2+}$, $K^+$). The fact that these cations are deposited concurrently with sulfate and nitrate as anions proves that their existence in a dissolved form is due to the emission of acidic precursors. In central Europe, around 20% of the acidity due to $SO_2$ and $NO_x$ emission seems to be neutralized by the dissolution and/or exchange from basic Ca, Mg, and K compounds before their deposition (Ulrich, 1984). The effect of dissolved nutrients has to be considered as a part of acidic deposition. This means that the deposition not only of acidity but also of nutrients and heavy metals must be assessed.

Most data available are restricted to precipitation deposition (wet deposition collected with bulk samplers) and throughfall (including stemflow in the case of beech (*Fagus sylvatica*) forests). Throughfall data do not represent deposition values, however, because they contain canopy leaching and proton buffering in the canopy.

## III. Rates of Deposition

In a dose-effect relationship of acidic deposition, the dose has to be specified as the flux of definite chemical compounds reaching a specified surface within the forest ecosystem. For those chemical compounds that reach the ecosystem by deposition, the flux corresponds to the rate of deposition. This chapter is subdivided according to the type of chemical compounds: acidity (section III,A) and nutrients (section III,B). Deposition rates of heavy metals, which represent trace cation acids, are dealt with in section X. The surface reached by the flux can be the cuticula of leaves, the apoplast of the stomate, cell membranes at the leaf surface and in the stomate, the bark, the soil surface exposed to the atmosphere, and the soil surface exposed to the roots. In case of protons, the effective flux to the apoplast and also to the soil surface exposed to the roots is approximated by the rate of buffering of protons in the canopy. A flux of dissolved $SO_2$ to cell membranes should be bound to the condition that the buffering of protons by exchange with cations in the apoplast of the stomate is limited. The flux to the cuticula and the soil surface exposed to the atmosphere is approximated by the deposition rate in throughfall. The flux to the bark is approximated by the deposition rate in stemflow.

### A. Rate of Deposition of Acidity

Table 7–1 shows the deposition of acidity from 24 sites in northwestern Germany. Data from other authors are recalculated according to Ulrich (1983b). The wet

**Table 7-1.** Deposition of acidity (multiannual mean values in kmol acid equivalents · ha$^{-1}$ · yr$^{-1}$) of 24 forests in northern Germany.

| Wet deposition | Rural areas vicinity of cities 0.2-0.8 | | Large closed forest areas 0.6-1.1 | |
|---|---|---|---|---|
| | Deciduous forests | | Coniferous forests | |
| | Neutral soil | Acidic soil | Acidic soil | |
| Dry deposition of SO$_2$ | 0.6-1.3 | 0.3-1.0 | 0.1-3.0 | |
| | Deciduous forests | | Coniferous forests | |
| | Rural areas vicinity of cities sheltered areas | Exposed mountains | Rural areas vicinity of cities sheltered areas | Exposed mountains |
| Particle deposition | 0-0.3 | 0.5-1.8 | 0-0.5 | 0.5-3 |
| Total H$^+$ deposition | 0.8-1.0 | 1.1-2.4 | 1.2-2.0 | 3.9-5.5 |
| H$^+$ Buffer quota | 10-87% | 0-70% | 30-60% | 0-42% |
| NH$_4^+$ Deposition | 0.4-1.0 | 0.9-1.3 | 0.8-1.0 | 1.0-1.6 |
| Total deposition of acidity | 1.2-2.6 | 2.0-3.5 | 2.0-2.7 | 2.9-6.4 |

Adopted from Ulrich, 1985.

deposition can be quite low in dry or warm years, especially in rural areas and in the vicinity of cities supplying neutralizing dust. In forested areas where settlements and agriculture are of minor importance, the wet deposition amounts to around 0.8 kmol $H^+ \cdot ha^{-1} \cdot yr^{-1}$. This corresponds to a rain pH of around 4.1. According to Winkler (1982), there has been no trend in rain pH during the last 5 decades. Dry deposition of $SO_2$ can substantially contribute to the acidic input, thereby substantially decreasing the $SO_2$ concentration in the air above and below a forest canopy (Baumbach and Käss, 1985). The dry deposition of gaseous $SO_2$ depends upon tree species (for example, coniferous trees have a greater absorbing surface in winter), the roughness of the canopy, and the vitality of the tree (the extent to which the acidity formed after absorption of $SO_2$ can be buffered).

Asche and Beese (1987) found a positive correlation between $SO_2$ concentration in air and sulfate concentration in stemflow. Low rates of stemflow at higher $SO_2$ concentrations in air (100–200 mg/m$^3$) resulted in pH values as low as 2.1 in stemflow. On neutral soils, where the uptake of basicity may not be limiting acidic buffering in the canopy, dry deposition of $SO_2$ is always of significance.

The deposition of particles, where fog and cloud waters make their greatest contribution (Unsworth and Crossley, 1987), depends upon the exposure and inclination of the site, and tree species, the tree height, and the compactness of the canopy (stand closure, leaf area index). On exposed mountains, and especially in the case of coniferous forests, the deposition of particles (mainly cloud water) may be the most important pathway of deposition. The increased level of deposition in hillside regions can be explained in terms of increased turbulence levels found there (Ruck and Adams, 1988). In sheltered positions, including young stands of low height sheltered by higher stands, particle deposition approaches zero. Verhoeven and others (1987) report for central Europe (Fichtelgebirge) that acidity in fog is much higher than in rain, whereas no difference could be observed in New Zealand due to the lack of strong mineral acids. On a mountain station near a large city, Schmitt (1987) found pH <3.8 in 50% of all fog samples collected (equivalent composition of anions: 23% sulfate, 19% nitrate; of cations; 10% protons, 31% ammonium).

The buffer quota of protons may vary with the vitality of the tree and show considerable year-to-year variance (Ulrich, 1983b). This becomes apparent in the zero values of exposed mountain areas. Even in these regions, however, the buffer quota can reach 40% (coniferous stands) to 70% (deciduous stands). The buffering of protons in the canopy is balanced by uptake of basicity from soil. Thus, these data show that up to 87% (the mean is around 50%) of the protons deposited reach the soil at the root surface. This fraction of acidity has had a direct influence on the tree before reaching the soil.

Of total acidic deposition, $NH_4^+$ contributes 25 to 50%. Extreme values (>3 kmol $NH_4$ deposition in spruce stands) have been reported from areas with intensive animal husbandry, that is, with high local $NH_3$ emission (Van Breemen and Jordens, 1983; Schuurkes et al., 1988; Irens et al., 1988; for a case study, see section IV). This is clearly reflected in the distribution of $NH_4$ deposition in throughfall in Europe (Hauhs et al., 1989). The proton is released during the uptake of $NH_4^+$ by microorganisms or plant roots or during nitrification.

To assess long-term effects of acidic deposition, the cumulative dose since the beginning of industrialization needs to be known. It can be calculated from the time course of the emission of $SO_2$ and $NO_x$. For the area of the FRG, the values are presented in Figure 7–1. The broken line represents the estimated decline in emission of $SO_2$ due to emission control. In addition to the total emission per year in millions of tons, the emission density per ha and year is given. Because $SO_2$ and $NO_x$ are the precursors of the the the acids $H_2SO_4$ and $HNO_3$, respectively, the emission densities can also be expressed as acidic equivalents $H^+$. An emission density of 69 kg $S \cdot ha^{-1} \cdot yr^{-1}$ corresponds to 4.3 kmol $H^+ \cdot ha^- \cdot yr^{-1}$; 38 kg $N \cdot ha^{-1} \cdot yr^{-1}$ correspond to 2.7 kmol $H^+$. The total emission density of acidity around 1980 amounted to 7 kmol $H^+ \cdot ha^{-1} \cdot yr^{-1}$. The time course of accumulated acidic emission is given in the uppermost curve in kmol $H^+$/ha. From 1850 to 1987, the accumulated acidic emission in the area of West Germany amounts to 370 kmol $\cdot ha^{-1}$. If central Europe would be considered (including the German

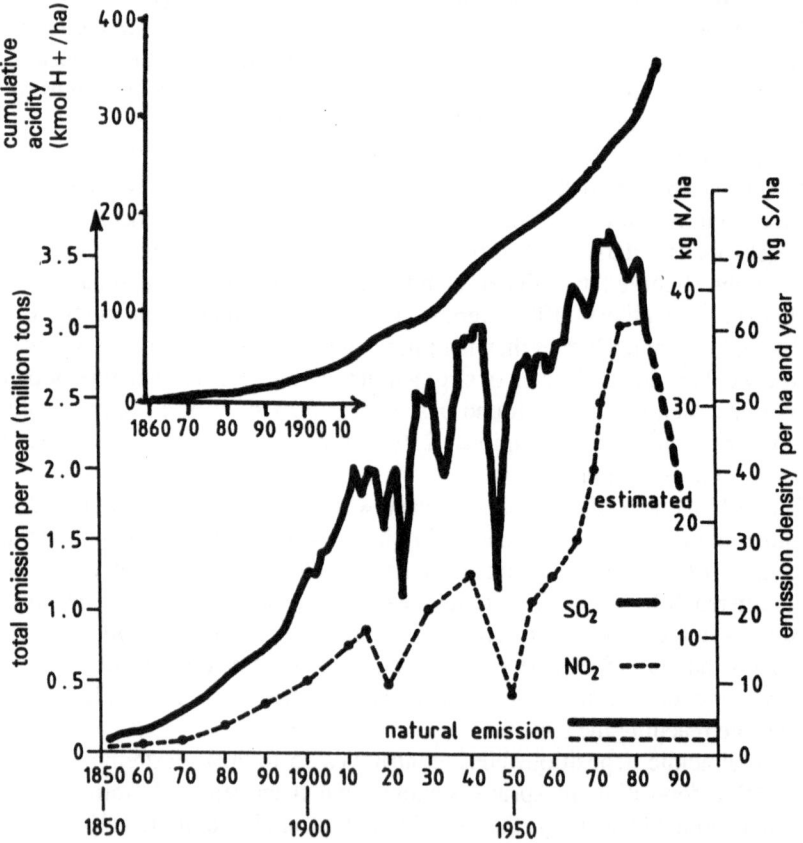

**Figure 7–1.** Development of the annual rates of emission of $SO_2$ and $NO_x$ (in million tons per year and in kg per ha and year) and of the cumulative emission of acidity due to $SO_2$ and $NO_x$ (in kmol $H^+ \cdot ha^{-1}$) in the area of the Federal Republic of Germany since 1850.

Democratic Republic, the western part of Poland, and Czechoslovakia), the figure would increase considerably (probably $>450-500$ kmol $\cdot$ ha$^{-1}$) due to higher emission densities in these areas. One-third of this amount has been emitted since the beginning of the deposition measurements in the Solling in 1969, and 60% since 1945.

For a Norway spruce and a European beech stand in the Solling Mountains (ecosystem 6 and 7 of Table 7–2), the time course of deposition of H$^+$ and SO$_4$-S has been followed since 1969 (see Ulrich, 1985; the data for nitrogen are given by Matzner, 1989). Both forests are in an exposed position with deposition rates at the upper limit of the range given in Table 7–1. The rates of precipitation deposition show only a small annual variation; the deposition into the forests, however, can vary greatly from year to year. With values around 50 (beech) to 80 kg S $\cdot$ ha$^{-1}$ $\cdot$ yr$^{-1}$ (spruce), the deposition rates are within the range of emission density (Figure 7–1: 60–75 kg S $\cdot$ ha$^{-1}$ $\cdot$ yr$^{-1}$ within this period). The data of S and H$^+$ show no clear trend. It may be, however, that a low sulfur deposition in the spruce ecosystem in 1985 is a first sign of a decrease in deposition. With a declining emission of SO$_2$, the frequency with which cloud droplets with high sulfate content and low pH occur should decrease, and thus the rate of interception deposition of particulate sulfate and H$^+$ should also decrease. This would be visible most strongly in coniferous forests of exposed mountains.

With mean values of 2.0 (beech) to 3.8 (spruce) kmol H$^+$ $\cdot$ ha$^{-1}$ $\cdot$ yr$^{-1}$, the deposition rate of H$^+$ is considerably smaller than the emission density (7 kmol H$^+$ $\cdot$ ha$^{-1}$ $\cdot$ yr$^{-1}$). This is a consequence of buffer reactions in the atmosphere. Part of the buffering is due to reactions with NH$_3$ (cf. Table 7–1: 0.9 to 1.6 kmol NH$_4^+$ $\cdot$ ha$^{-1}$ $\cdot$ yr$^{-1}$). In case of the spruce stand in the Solling, the atmospheric buffering by NH$_3$ accounts for 1.1, and the reaction with basic dust particles in the atmosphere accounts for 1.6 kmol H$^+$ $\cdot$ ha$^{-1}$ $\cdot$ yr$^{-1}$. The sum (3.8 + 1.1 + 1.6 = 6.5) approaches at this site the emission density, comparable to sulfur.

Between 1969 and 1985, 54% of the acidic emission per ha within this period has been deposited in the spruce forest in the Solling as H$^+$, and 70% as a sum of H$^+$ + NH$_4^+$. If 70% is assumed to be valid for the whole period from 1852 to 1987, a minimum estimate of accumulated acidic deposition in the spruce forest in Solling yields 260 kmol H$^+$ $\cdot$ ha$^{-1}$ (maximum estimate 400). Using the data on total deposition of acidity in Table 7–1, the range of accumulated acidic deposition into central European forests varies between 60 and 340 kmol H$^+$ $\cdot$ ha$^{-1}$ (minimum values). The low value corresponds to deciduous forests in sheltered areas.

## B. Rate of Deposition of Nutrients

Again, only data sets could be used where the flux of nutrients in throughfall has been subdivided into deposition and leaching. Due to missing data sets, bulk precipitation data have to be used for assessing the regional variability in the case of Mg, Ca, and K. According to the existing data, deposition rates into forests may be around twice as high due to the high degree of roughness and exposure of the forest canopy.

**Table 7-2.** Characterization of the forest ecosystems used as case studies for ion budgets.

| Ecosystem | 1<br>GW/Bu | 2<br>He/Bu | 3<br>LH/Ei | 4<br>SP/Fi |
|---|---|---|---|---|
| Location | Göttinger Wald | Harste (Göttingen) | Lüneburger Heide | Spanbeck (Göttingen) |
| Tree species | Beech | Beech | Oak | Spruce |
| Age (years) | 110 | 95 | 103 | 85 |
| Soil parent material | Lime stone | Loess + limestone | Glacial loamy sands | Loess + sandstone |
| Soil type | Rendzina-Terrafusca | Orthic luvisol | Spodo-dystric podzoliza-tion | Cambisol |
| A Hor. | Cation | | A1 | Fe/A1 |
| Buffer range | exchange | Cation | exchange/ | |
| B Hor. | Carbonate | exchange | A1 | A1 |
| Exchangeable Ca+Mg in 0–50 cm [kmol (+)/ha] | 1100 | 200 | 9 | 18 |
| Deposition rate [kmol (+) ha$^{-1}$ yr$^{-1}$] | | | | |
| pH | 4.1 | 4.02 | 4.26 | 3.42 |
| H$^+$ + NH$_4$ | 2.3 | 2.2 | 1.6 | 3.7 |
| Ca+Mg | 1.1 | 0.7 | 0.5 | 1.0 |
| Accumulated acidic deposition [kmol H$^+$/ha] | 120 | 120 | 80 | 200 |
| Depth of output measurement | 1 m | 0.4 m | 1.5 m | 0.4 m |
| Measurement period | 1983/84 | 1982/85 | 1980/85 | 1982/85 |
| pH of seepage | 7.8 | 4.45 | 4.55 | 3.96 |
| Acidity degree of seepage (ASS) (see text) | 0% | 14% | 14% | 66% |
| SO$_4$-S/Cl [g/g] | | | | |
| Deposition | 1.6 | 1.7 | 0.8 | 2.3 |
| Output | 1.4 | 1.5 | 0.9 | 2.6 |
| NO$_3$-N/Cl [g/g] | | | | |
| Deposition | 0.48 | 0.52 | 0.14 | 0.50 |
| Output | 0.86 | 0.79 | 0.004 | 0.56 |
| Source | Meiwes 1983, 1985 | Cassens-Sasse 1987 | Bredemeier 1987a | Cassens-Sasse 1987 |

**Table 7-2.** (*Continued*)

| Ecosystem | 5 LH/Ki | 6 SO/Bu | 7 SO/Fi | 8 HI/Fi | 9 Wi/Fi |
|---|---|---|---|---|---|
| Location | Lüneburger-Heide | Solling B1 | Solling F1 | Hilskamm | Wingst Westerberg |
| Tree species | Pine | Beech | Spruce | Spruce | Spruce |
| Age (years) | 98 | 135 | 100 | 14 | |
| Soil parent material | Glacial sands | Loess + weathered sandstone | | Poor sandstone | Glacial sands |
| Soil type | Podzolization | Spodo-dystric cambisol | | Orthic podzol | |
| A Hor. Buffer range | Fe/Al | Fe/Al | Fe/Al | Fe/Al | Fe/Al |
| B Hor. | Al | Al | Al | Fe/Al | Al |
| Exchangeable Ca+Mg in 0–50 cm [kmol (+)/ha] | 6 | 9 | 8 | 6 | 8 |
| Deposition rate [kmol (+) ha$^{-1}$ yr$^{-1}$] | | | | | |
| pH | 3.81 | 3.81 | 3.88 | 4.0 | 4.3 |
| H$^+$ + NH$_4$ | 2.0 | 2.8 | 4.9 | 1.5 | 4.5 |
| Ca+Mg | 0.7 | 1.1 | 1.4 | 0.8 | 1.1 |
| Accumulated acidic deposition [kmol H$^+$/ha] | 110 | 150 | 260 | >200 | 150 |
| Depth of output measurement | 1.5m | 1m | 1m | 1m | 1m |
| Measurement period | 1980/85 | 1979/83 | 1979/83 | 1984/86 | 1983/86 |
| pH of seepage | 4.18 | 4.18 | 4.04 | 3.6 | 4.1 |
| Acidity degree of seepage (ASS) (see text) | 52% | 72% | 68% | 67% | 60% |
| SO$_4$-S/Cl [g/g] | | | | | |
| Deposition | 0.8 | 1.6 | 2.0 | 1.7 | 0.6 |
| Output | 1.2 | 2.0 | 3.0 | 2.7 | 0.8 |
| NO$_3$-N/Cl [g/g] | | | | | |
| Deposition | 0.19 | 0.33 | 0.40 | 0.50 | 0.10 |
| Output | 0.005 | 0.05 | 0.35 | 0.60 | 0.22 |
| Source | Bredemeier 1987a | Matzner 1988 Ellenberg et al. 1986 | | Wiedey and Gerriets 1986 | Büttner et al. 1986 |

The rate of deposition of nutrients varies according to the distance from emission sources as well as according to the precipitation. The emission source for Mg is mainly sea spray. Deposition rates of Mg therefore decrease proportionally to Na with increasing distance from the seacoast (7 kg $Mg \cdot ha^{-1} \cdot yr^{-1}$ in bulk precipitation) to south of Hamburg (2 kg), Göttingen (1.5), Black Forest (1), and Switzerland (Bern, 0.6). Calcium and K in bulk precipitation and cloud droplets originate mainly from soil dust. In small forests embedded in arable land, bulk deposition may exceed 10 kg $Ca \cdot ha^{-1} \cdot yr^{-1}$. The same applies, however, for exposed forests in wooded mountains intercepting cloud water. Low values of bulk deposition are around 4 kg Ca. Bulk deposition of K varies between 2 and 5 $kg \cdot ha^{-1} \cdot yr^{-1}$ (Hauhs et al., 1989).

For sulfate, throughfall data are an appropriate approach to total deposition. For ammonium and nitrate, throughfall data provide minimal estimates of deposition. According to the data compiled by Hauhs and others (1989), sulfate deposition into forests varies between 5 (Norway, Telemark) and 150 (Czechoslovakia, Ore Mountains) $kg \cdot S \cdot ha^{-1} \cdot yr^{-1}$. In central Europe, minimum values are 14 kg S. $NH_4^+$, and $NO_3$ show an irregular spatial distribution with deposition rates in throughfall of spruce stands usually exceeding 8 kg $N \cdot ha^{-1} \cdot yr^{-1}$, respectively. The true rate of N deposition is unknown because in many forests a fraction of deposited N may be directly assimilated in the leaves and therefore cannot be assessed by throughfall measurements (Lindberg et al., 1987). Between the late 1950s and early 1970s, the mean annual increase in nitrate concentration in precipitation was 3% to 6% (Söderlund et al., 1985).

The rate of deposition of nutrients may be compared with the rate of uptake in aboveground biomass (for the beech forest in Solling, ecosystem case study 6 of Table 7–2, in $kg \cdot ha^{-1} \cdot yr^{-1}$: Mg = 4, Ca = 30, K = 40, $SO_4$-S = 7, N = 60) or with the rate of accumulation in the aggrading forest biomass (Mg = 1.7, Ca = 8, K = 7, S = 1.5, N = 13). If these figures are taken as representative of timber trees in central Europe, it follows that the demand for forest increment is covered by deposition for all nutrients listed, except Mg in the southernmost part of Germany and in Switzerland and K in an irregular pattern. Sulfur deposition exceeds the annual uptake by a factor of at least 2, and uptake in forest increment by a factor of at least 9. Sulfate is in such an excess that the uptake and cycling of deposited sulfur is definitely restricted to a negligible fraction of deposition. Deposition rates of N are high enough to cover the demand for forest growth (biomass increment); they can even exceed total annual N uptake in areas with intensive animal husbandry and high $NH_3$ emission.

Clearly, in many forests in central Europe, the deposition of N, S, Ca, K, and Mg is sufficient to allow a productive forest to continue its growth, irrespective of the presence of available soil-borne nutrients, provided its root system is able to utilize the nutrients deposited. There is no doubt that the deposition of nutrients can mask adverse effects of deposition of acidity by leaching soil nutrients. A small but continuous input of dissolved nutrients is the optimal way of fertilization. Nitrogen is, in almost all forest ecosystems, the nutrient that is limiting the growth rate. Its increasing input by deposition, especially since 1950 (cf. Figure

7–1), will (if the root system is able to take it up) most likely increase growth rates (Ulrich, 1975). Therefore, acidic deposition could result in increased forest production. The uncertainty is which soil conditions may limit the uptake of deposited nutrients and negate the effects of nutrient deposition. Magnesium and K deficiency symptoms are the first to be expected. They should first appear in areas where the leaching of cationic nutrients from soil has affected the whole rooting zone, and where deposition of Mg and K is not sufficient to cover the demand of forest increment. These conditions are valid in southern Germany.

## IV. Input–Output Relations of Forest Ecosystems as a Measure of the Deviation from Steady State

Figure 7–2 gives data of the input by deposition (I) and the output by seepage (O) of alkalinity ($HCO_3^-$, as potential alkalinity: $NO_3^-$ and acidity ($\frac{1}{2}Mn^{2+}$, $\frac{1}{3}Al^{3+}$, $H^+$, as potential acidity: $NH_4^+$) for some forest ecosystems in northwestern Germany. Some characteristic features of these ecosystems are compiled in Table 7–2. The chemical soil state is characterized by the buffer range (according to Ulrich, 1981a, 1986). Soil 1 is calcareous, soil 2 has medium base saturation, soil 3 has medium base saturation only in the subsoil, and 4 to 9 have base saturations below 5 to 10% throughout the solum.

The input and output of the ions responsible for acidity and basicity differ

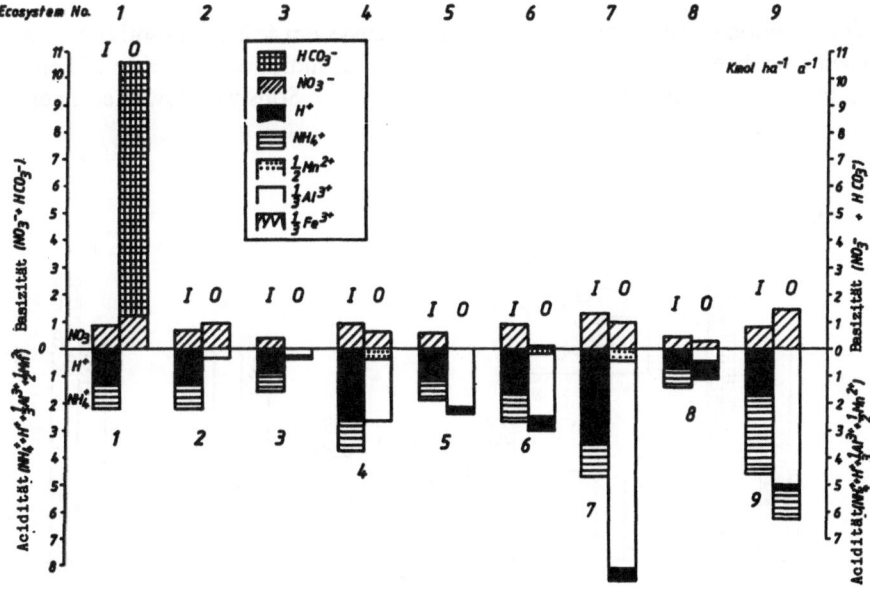

**Figure 7–2.** Input by deposition (I) and output by seepage (O) of acidity and basicity in the case studies listed in Table 7–2.

greatly (Figure 7–2). With the exception of ecosystem 8, the protons deposited are retained in the mineral soil. Ecosystem 8 represents the most strongly acidified soil (Fe/Al buffer range in the whole solum); in this soil, with a mean pH value in seepage water of 3.6, the ability to buffer protons is exhausted. A further acid being deposited, $NH_4^+$, is retained in all ecosystems with the exception of ecosystem 9. Ecosystem 9 lies in an area where the $NH_3$ emission from feedlots is high. This results in a high $NH_4^+$ deposition (3 kmol $\cdot$ ha$^{-1}$ $\cdot$ yr$^{-1}$) and a total nitrogen input of 60 to 80 kg N $\cdot$ ha$^{-1}$ $\cdot$ yr$^{-1}$. Despite the high N input, the ecosystem is characterized by sparse, poor ground vegetation and by slow litter decomposition. The acidity of the soil seems to inhibit the biological activity so that part of the $NH_4$ is not biologically utilized and is therefore leached.

In ecosystems where the subsoil has low base saturation and remains within the aluminum buffer range (ecosystems 4, 5, 6, 7, and 9), the output of acidity is similar to the input. The main acid leached is the cation acid $Al^{3+}$. The deviations between input and output of acidity are correlated with ± equivalent differences in the input and output of sulfate (Figure 7–3): in ecosystems 4, acidity and sulfate are retained; in ecosystems 5, 6, and 7, acidity and sulfate are leached in excess of deposition. In ecosystem 9, the output of $NH_4^+$ and of a small fraction of Al is accompanied by organic anions.

The relation between acidity and sulfate is interconnected with the formation of Al hydroxo sulfates (Prenzel, 1983). As long as the subsoil is in the cation-exchange buffer range, a fraction of $H_2SO_4$ deposited is accumulated as Al hydroxo sulfate in the subsoil:

$$Al(OH)_3 + H_2SO_4 \longrightarrow AlOHSO_4 + 2 H_2O. \qquad (3)$$

This reaction is still occurring in the soil of ecosystem 4. This subsoil is currently switching from exchanger into Al buffer range, as indicated by having still higher storages of exchangeable Ca + Mg (cf. Table 7–2). Soils with low exchangeable

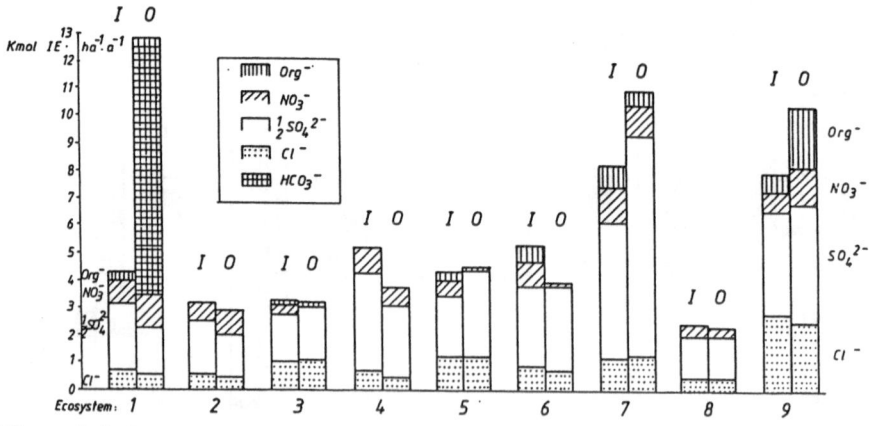

**Figure 7–3.** Input by deposition (I) and output by seepage (O) of anions in the case studies listed in Table 7–2.

Ca + Mg storages (5, 6, 7, and 9) lose more sulfate than they receive. With decreasing pH (i.e., increasing strength of the acids present in soil solution), the Al hydroxo sulfates become mobilized again:

$$AlOHSO_4 + H^+ \longrightarrow Al^{3+} + SO_4^{2+} + H_2O. \tag{4}$$

If the reactive Al compounds are exhausted, input and output of sulfate converge again. An example is ecosystem 8, a young planted spruce stand, in which a low input of acidity and sulfate equals the output. Due to the low height and low degree of roughness of the canopy, the interception (dry and occult) deposition is low, so that total deposition approaches precipitation (wet) deposition (cf. Table 7–2).

The ability to accumulate an acidic sulfate in the solid soil phase as long as the soil is in the cation-exchange buffer range slows down the leaching of base cations and diminishes the appearance of cation acids such as $Mn^{2+}$ and $Al^{3+}$ ions in the soil solution. Thus it decreases the risk of acidic stress to roots and soil organisms. Once the soil is in the Al buffer range, however, soil internal nitrification pulses ("acidification pulses," Ulrich, 1981b) can lead to episodes of high $Al^{3+}$ concentrations in the subsoil (Khanna et al., 1987). Such processes can show a high spatial variation within a stand. The temporal and spatial pattern of the acidification pushes can create a probability distribution of acidic stress to the tree roots located in the subsoil. In Figure 7–4 the time course of Al concentration in soil solution in 20-, 40-, and 60-cm soil depth during 1985 and 1986 is presented for 8 sampling points located in an area of 50 m by 100 m in a Norway spruce timber tree (age 100 years). The data show a remarkable temporal and spatial variability reaching Al concentrations up to 50 mg (2 mmol) Al/L. The highest Al concentrations occur in the subsoil at 60-cm depth. It can be concluded from the ion balance of the soil solution that the high Al concentrations are the result of a climatically (in a dry or warm period) triggered nitrification pulse with subsequent dissolution of Al hydroxo sulfates (Raben, 1988). The dynamics of fine-root concentrations in various soil depths parallels the time course of acidic stress.

The data show that after depletion of the exchangeable base cations the acidity passes through the solum to enrich the seepage water in Al ions. The rate of acidity transfer to the seepage conductor can exceed the rate of acidic deposition considerably, due to the dissolution of Al sulfates accumulated in earlier stages of this development (cf. Figure 7–2, ecosystem 7).

In soils of higher base saturation (cation-exchange buffer range, ecosystems 1 and 2), the output of sulfate is again smaller than the input (as in ecosystem 4; see Figure 7–3). Ecosystem 3 has a sandy soil with a low storage capacity of reactive Al and is also at the borderline between cation exchange and aluminum buffer range. For ecosystems 2 and 3, the anion budget is approximately balanced. This shows that a quantity of base cations from the exchangeable pool almost equivalent to the acidity retained is leached. In these soils, base saturation is decreasing. For the soil of ecosystem 2, the annual loss of base cations is around 2 kmol (+) $ha^{-1} \cdot yr^{-1}$; for the soil of ecosystem 3, around 1 (see input-output relation of acidity in Figure 7–2), compared to a storage of exchangeable Ca + Mg of 200 and 9 kmol $ha^{-1}$, respectively (cf. Table 7–2). For ecosystem 2 the buffer

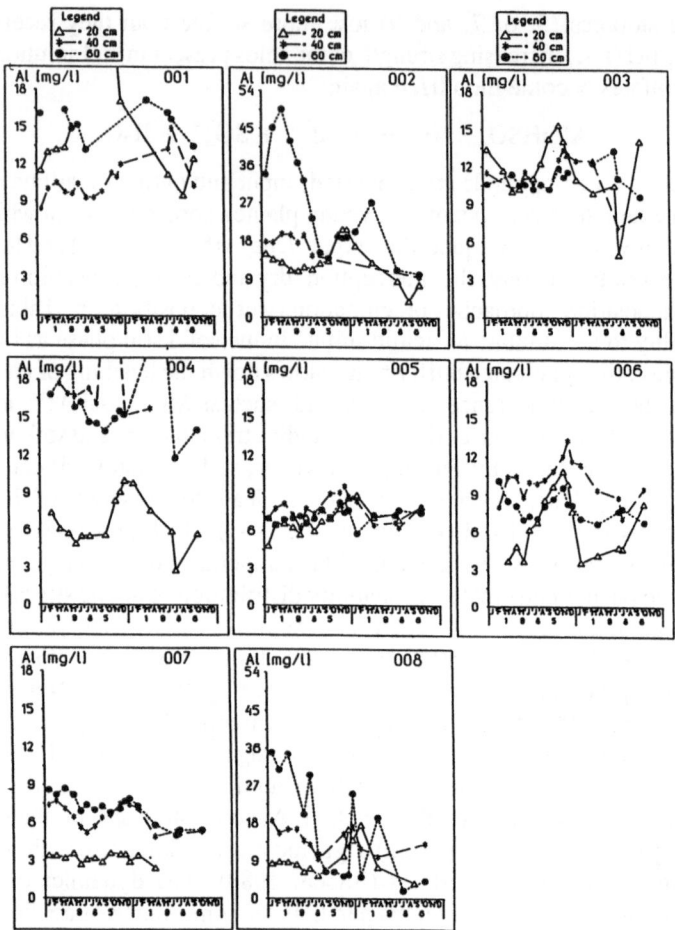

**Figure 7–4.** Time course of Al concentration in soil solution in 20-, 40-, and 60-cm soil depth in a Norway spruce timber tree in forest district Grünenplan no. 79. (From Raben, 1988.)

capacity still existing in the soil is high enough to maintain the cation-exchange buffer range for the next decades.

The calcareous soil of ecosystem 1 is the only one where the seepage water leaving the rooted soil in 1-m depth contains alkalinity. The ion pair for the bicarbonates is calcium, which originates from dissolution of limestone.

As can be seen from Figure 7–2, $NH_4^+$ and $NO_3^-$ are deposited in approximately equivalent amounts with the exception of ecosystem 9. The leaching of $NO_3^-$ may be close to zero (ecosystems 3, 5, and 6), but then again it may exceed $NO_3$ input (ecosystems 1, 2, and 9). Continuous leaching of nitrate or ammonium from an ecosystem indicates "nitrogen saturation": More nitrogen is available than can be utilized for biomass production, for whatever reason. Under the present deposition rates, forest ecosystems with soils in the cation-exchange buffer range (ecosystems 1 and 2) may be generally nitrogen saturated. The same applies to ecosystems with soils in the Al buffer range (ecosystems 4 and 7). Needle analysis shows in northern Germany a trend of increasing nitrogen content in spruce needles (unpublished data from forest administration), but luxury N levels. Increases in tree growth found in healthy trees can therefore be attributed mainly to increased nitrogen deposition. According to soil inventories in ecosystems 6 and 7, the N deposited accumulates mainly as organic bound N in the organic top layer (Matzner, 1989). Most of the deposited N seems to be taken up by plants and microorganisms, being accumulated in soil organic matter due to slow decomposition rates. The absence of apparent mineralization in acidic forest soils in central Europe appears prevalent (see section VII).

The input-output budgets indicate that these ecosystems are not in steady state, as manifested by enhanced leaching of exchangeable base cations or the leaching of cation acids, mainly $Al^{3+}$, at rates even exceeding acidic deposition.

The results of these case studies are confirmed in model ecosystems by open-top chamber experiments: Acidic rain and fumigation with $SO_2$ and $O_3$ showed definite effects on mineral cycling and soil acidification (Seufert and Arndt, 1988).

## V. Changes in Soil Acidity and Present Status of Forest Soil Acidification

The status of soil acidification can be characterized by several parameters: pH values, buffer ranges (Ulrich, 1981a, 1986), base saturation, and storage of the exchangeable nutrient cations Ca, Mg, and K. They are interrelated:

Carbonate and silicate buffer range: pH >5.0, high base saturation, high cation storage (proportional to clay and humus content)

Cation-exchange buffer range: pH usually above 4.2, base saturation above 15%, intermediate cation storage

Al, Al/Fe buffer range: pH measured in salt solution below 4.2, base saturation below 5% (10% in humus-rich soils), low cation storage irrespective of clay and humus content

The time course of acidic emission (Figure 7–1) suggests that changes in the acid-base status of forest soils have occurred over several decades. The earliest pH measurements were made in the late 1920s. Soil data of Hartman and Jahn (1967), collected from 1930 to mid-1950, allow the calculation of base saturation. Ulrich and Meyer (1987) have evaluated these data and suggest that in mixed forests resembling natural vegetation the base saturation of soil was medium to high, even on poor parent material (i.e., soils being in cation exchange and silicate buffer range, at least in the B horizon). The data further indicate strong soil acidification being more prevalent in areas close to industrial activities and to the mountain ridges receiving occult deposition (acidic cloud droplets). Planted spruce stands, which have replaced deciduous or mixed forests, often also show strong soil acidification. In this case, soil acidification can be attributed to ecosystem internal acidification triggered by the change in the vegetation cover (decrease of organic nitrogen stored in mineral soil due to nitrate leaching; Ulrich, 1981b; Kreutzer, 1981).

The resampling of forest soils in central Europe and Scandinavia 20 to 50 years after the first sampling has shown that decreases in the soil pH and base saturation are widespread. Berden and others (1987) critically reviewed these data. As expected, the drop in pH decreases with decreasing pH at the time of the initial measurement. The lower limit of pH with values around 3.0 (Al/Fe and Fe buffer range) is determined by the concentration of strong acids ($H_2SO_4$, $HNO_3$, HCl, organic acids) in the soil solution. Proton buffering approaches zero under these conditions (cf. ecosystem 8, Figure 7–2), and a further pH decrease due to acidic deposition is not to be expected and is indeed not found. Biomass increment, however, cannot be the cause of such low pH values. Although biomass increment can lower the base saturation, it does not produce anions and thus cannot create the high acidic concentration in soil solution necessary for pH values around 3. Zezschwitz (1987) traces back lower pH values in luff compared to lee positions 20 years ago and also ascribes them to acidic deposition.

The fact that a great fraction of the acidic load enters into the soil from the top creates a depth gradient of soil acidity that is shown in Figure 7–5, using data from the ecosystems 5 to 9 and from Ulrich and Malessa (1989). In the leaf litter, the fraction of $M_b$ cations (Ca + Mg + K) exceeds 90%. This indicates a high degree of basicity, which is released during decomposition. The basicity decreases in the fermentation layer (*Of*) strongly in favor of the cation acids $Al^{3+}$ and $Fe^{2+}$ as well as protons ($H^+$). The humus layer (*Oh*) and the top bleached mineral soil horizons (*Ahe*) are in the Fe/Al buffer range, as indicated by pH values between 2.8 and 3.8. The less acidified *Of* is densely rooted; the roots show usually normal branching and are well mycorrhized. In the *Oh* and A horizon, Al and Fe are organically complexed; that is Al exists in nontoxic form. Due to high $H^+$ and low $Ca^{2+}$ concentrations, the acidic stress appears as proton stress. These soil horizons show all features of podzolization. The actual progress of this acidification process is demonstrated in Table 7–3 for the mineral soil of ecosystems 6 and 7 (beech and spruce, Solling). Between 1966 to 68 and 1983, neither the effective cation-exchange capacity ($CEC_e$) nor the exchangeable $Al^{3+}$ showed significant changes. Exchangeable and dissolvable $H^+$ and $Fe^{2+}$ increased, however, indicating a shift to stronger acids. This shift took place mainly in the top 10 cm. A detailed study showed that the transition zone between Fe/Al and Al buffer range is moving downward. Ecosystem 8 represents a soil where this transition zone is below 1-m soil depth. The storage of $M_b$ cations (Ca, Mg, K) was already low in 1966 to 68. Its decrease is small compared with the amount of acidity deposited during the same period (50 and 83 kmol $H^+$ in case of beech and spruce, respectively). As the input-output budgets of acidity in section IV have shown, soil horizons in the Fe/Al and Al buffer range are not able to buffer the acidity deposited, but transfer it with the seepage water to lower layers. Detailed studies (Ulrich and Malessa, 1989) have shown that the acidic buffering in deeper layers (in ecosystems 5 to 9

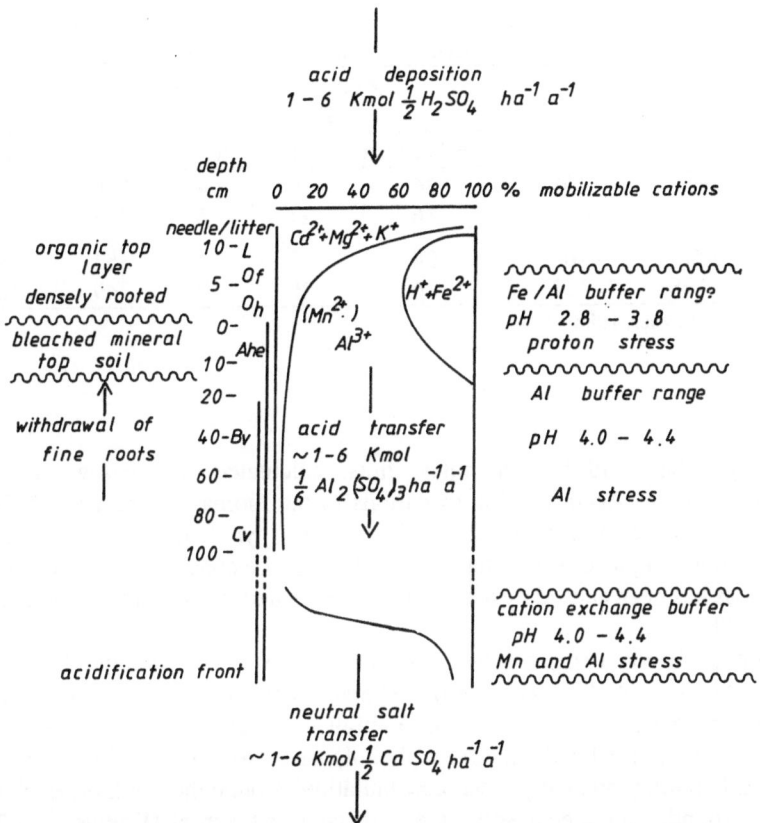

acid    deposition
$1 - 6$ Kmol $\frac{1}{2}H_2SO_4$  $ha^{-1}a^{-1}$

depth
cm        0  20  40  60  80  100 %  mobilizable cations

organic top        needle/litter    $Ca^{2+}+Mg^{2+}+K^+$
layer                 $10-L$
                         $5-Of$
densely rooted       $Oh$                              $H^++Fe^{2+}$          Fe/Al  buffer range
                         $0-$        $(Mn^{2+})$                                   pH   $2.8-3.8$
bleached mineral     $Ahe$           $Al^{3+}$                                    proton  stress
top soil          $10-$
                         $20-$                                                    Al   buffer range

withdrawal of        $40-Bv$      acid  transfer                              pH  $4.0-4.4$
fine  roots                        $\sim 1-6$ Kmol
                         $60-$     $\frac{1}{6}Al_2(SO_4)_3ha^{-1}a^{-1}$              Al   stress

                         $80-$
                    $Cv$
                        $100-$
                                                                              cation exchange buffer
                                                                              pH  $4.0-4.4$
acidification front                                                           Mn and Al stress

neutral  salt
transfer
$\sim 1-6$ Kmol $\frac{1}{2}$ Ca $SO_4$ $ha^{-1}a^{-1}$

**Figure 7–5.** Depth gradient of soil acidity under the influence of acidic deposition.

in soil depth exceeding 2 m; see also Dise and Hauhs, 1987) occurs in a layer of 1 to a few dm thickness in the cation-exchange buffer range. The lower limit of this layer represents the acidification front. Below the acidification front, the base saturation approaches high values (>80%).

A decrease of $M_b$-cations and base saturation is thus limited to soils where at least the root zone initially has been in the cation-exchange buffer range. Decreases in base saturation during the last decades have been reported from Germany (Zezschwitz, 1982, 1985a: exchangeable Ca content halved within 20 years; Hildebrand, 1986a; Rost-Siebert and Jahn, 1988; Gehrmann et al., 1987; Grimm and Rehfuess, 1986) and Sweden (Falkengren-Grerup, 1987: on average 50% decrease in base cation storage within 15 to 35 years). Decreases in exchangeable Ca + Mg storage on the basis of an already low level, comparable to the data shown in Table 7–3, have been reported by Hauhs (1985) and Völker (in Ulrich and Meyer, 1987). If the change in Ca and Mg storage and the time span is known, mean annual acidification rates can be calculated. In most cases in

**Table 7-3.** Changes in storage in 0- to 50-cm depth of $NH_4$ Cl extractable cations in the soils of case studies 6 and 7 (beech and spruce, Solling) after 16 years.

| | | exchangeable cations Kmol (+)/ha | | | | |
|---|---|---|---|---|---|---|
| | $CEC_e$ | K+Mg+Ca | H | Fe(II) | Al | Mn(II) |
| Beech 1966 | 401 | 26 | 0 | 4 | 359 | 10 |
| 0–50 cm 1983 | 416 | 16 | 21 | 6 | 347 | 17 |
| Spruce 1968 | 400 | 28 | 0 | 4 | 351 | 8 |
| 0–50 cm 1983 | 423 | 16 | 20 | 12 | 349 | 15 |

Adopted for Matzner, 1988.

Germany, the acidification rates thus calculated lie below 2 kmol $(+) \cdot ha^{-1} \cdot yr^{-1}$. Accumulation of cations in the biomass increment of timber trees may account for around 0.5 kmol $(+) \cdot ha^{-1} \cdot yr^{-1}$. In agreement with the input-output budgets of ecosystems 2 and 3, it can be concluded that most of the cation loss has been caused by leaching with deposited sulfate and nitrate ions (cf. Section VI).

European beech (*Fagus sylvatica*) concentrates a considerable fraction of precipitation (about 15%) as stemflow, characteristically having lower pH values and higher ion concentrations than throughfall (long-term mean values in ecosystem 6: throughfall pH 4.0, stemflow pH 3.4). In many investigations, increased soil acidification (and heavy metal accumulation) around the trunk of beech trees has been found, compared to soil some distance from the trunk (Koenies, 1982/83; Glatzel et al., 1983, 1986; Wittig and Neite, 1985; Neite and Wittig, 1985; Wittig and Werner, 1986; Glavac and Koenies, 1986; Jochheim, 1985; Schulte and Spiteller, 1987). Beech stands not influenced by acidic deposition, such as those in distinct areas in southern Europe, show in contrast no spatial variation of soil acidity and of heavy metal content in soil (Glavac et al., 1985; Jochheim and Schaefer, 1988). The acidification of the trunk area therefore cannot be attributed to ecosystem internal processes.

The existing data unequivocally show that in general the acidity of forest soils in Germany and Sweden has increased during the last decades.

The evaluation of the present status of forest soil acidification can be based upon investigations made in connection with forest site mapping in several states of northern Germany. In Table 7–4, the storage of exchangeable K, Ca, and Mg in the mineral soil is given for two of the investigations made. The first case study (Hamburg forestry area) is typical for the northwestern German Pleistocene plain. The soils are grouped into buffer ranges; in the table the median values of the cation storage are given. Eighty-five percent of the soils are down to 60-cm soil depth in the aluminum or Al/Fe buffer range. Half of these sites have exchangeable Ca storage below 270 and 170 kg Ca/ha, respectively. The second case study

**Table 7-4.** Storage of exchangeable K, Ca, and Mg in the mineral soil of two areas of northwestern Germany.

| | Median values | | | % of sites |
| --- | --- | --- | --- | --- |
| | K | Ca | Mg | in Al, Al/Fe |
| | | (kg/ha) | | buffer range |
| Hamburg forestry area, 0–60 cm (172 sites)[a] | | | | |
| Buffer ranges | | | | |
|   Carbonate (1% of the sites) | 580 | >10.000 | >500 | |
|   Silicate (2% of the sites) | 160 | 8.000 | 300 | |
|   Cation exchange (12%) | 160 | 1.200 | 64 | |
|   Aluminum (14%) | 160 | 270 | 33 | |
|   Al/Fe (71%) | 120 | 170 | 28 | |
| North Rhine Westphalia, 0–80 cm (150 sites)[b,c] | | | | |
|   Soils on limestone | 1.000 | 10.000 | 550 | 22% |
|   Soils on clay schist | 360 | 300 | 100 | 87% |
|   Loess soils | 320 | 400 | 200 | 81% |
|   Sandy soils | 180 | 220 | 30 | 70% |

[a] From Rastin, 1988.

[b] Higher values given in other publications represent arithmetic mean values.

[c] From Gehrmann, et al., 1987.

(main forest site types of North Rhine–Westphalia) is typical for forests of the medium-altitude landscape. With the exceptions of soils developed on limestones, more than 70% of the soils are again in the Al or Al/Fe buffer range. In these acidic soils, again the median values indicate low storage of exchangeable Ca. These results are confirmed by other studies (see Ulrich and Meyer, 1987).

The finding that the majority of forest soils in areas of central Europe where data exist (mainly northern Germany) are in the aluminum buffer range throughout the whole root zone makes further investigations necessary: How deep does the acidification reach in the weathering mantle or loose sedimentary rocks? How fast is the acidification front moving downward? When will it reach the saturated zone and acidify spring and creek waters?

## VI. Causes of the Present Soil Acidification

Soil acidification can be due to ecosystem internal processes and to acidic deposition (Ulrich et al., 1979; Van Breemen et al., 1983). For the acidification of subsoils, however, ecosystem internal causes can be excluded for the following reasons (cf. Ulrich, 1986; De Vries and Breeuwsma, 1987):

$H_2CO_3$ (from $CO_2$ respired) cannot lead to pH values below 5.0.

Excess cation uptake in the biomass lowers the base saturation in the rooted soil horizons. As no mobile anions are produced, however, there is no effect on the unsaturated zone.

Organic acids, which cause the bleaching of the top mineral soil (podzolization), are arrested in the Bh horizon. They do not reach soil horizons below the Bh and therefore cannot acidify these horizons. As Figure 7–3 shows, the output of organic anions is negligible in the ecosystems studied (exception: ecosystem 9 has sandy soil with extreme deposition of $NH_4^+$).

$HNO_3$ can acidify soil horizons in which organic bound nitrogen has been accumulated. Its net formation, resulting in the leaching of nitrates, can be triggered by changes in land use. It is responsible for the release of strong cation acids within the root zone before the era of acidic deposition. After acidifying the rooted soil, however, the input of N becomes strongly limited due to the absence of $N_2$ fixers (legumes) in the vegetation cover. The ecosystems then become nitrogen limited, and the leaching of nitrate approaches zero.

Ecosystem internal production of $H_2SO_4$ with subsequent leaching of sulfates is negligible. In biomass, the S content is only 7% of the N content. In soil organic matter, the S content is around one tenth that of N. The effects of changes of soil organic matter storage are therefore governed by the fate of N. The effect of sulfides present in the parent material of the soil is also negligible. The proton consumption by the weathering of silicates exceeds by far the proton production by the weathering of sulfides, with the exception of unusual soil parent material.

In Figure 7–6, the proton load of the mineral soil (Bredemeier, 1987a,b) is given for ecosystems 2 through 7 of Table 7–2 and Figures 7–2 and 7–3. The total acidic load varies between 2 and 6 kmol $H^+ \cdot ha^{-1} \cdot yr^{-1}$. The cation excess in biomass increment of the timber trees contributes between 0.25 (pine) and 1 kmol (oak, beech), that is, between 10% and 40% of the total acidic load. The contribution of organic acids, as indicated by the leaching of organic anions, is negligible for the whole solum. However, this may be completely different for the topsoil. Because in these ecosystems $NH_4^+$ is accumulated as organic bound $N_{org}$, the contribution of nitrogen deposition depends on the fate of deposited nitrate. If $NO_3^-$ is also accumulated as $N_{org}$, the contributions of $NH_4^+$ and $NO_3^-$ to the $H^+$ budget tend to balance each other (ecosystems 3, 5, and 6). In all other cases, the leaching output of nitrate is approximately equivalent to the deposition input, leaving $NH_4$ deposition as proton producer. The different behavior of the ecosystems is reflected in the difference of the $NO_3$-N/Cl ratio in deposition and output (see Table 7–2). The leaching of nitrate seems to become a widespread phenomenon, as the $NO_3$-N/Cl ratio in spring and creek waters from forested regions shows (see Table 7–5). In Norwegian lakes, repeated sampling revealed a trend of increasing $NO_3$ concentrations in the water (Hauhs and Wright, 1986; NIVA, 1987). Decreased nitrate utilization by the forests will increase the effective acidic load.

The proton deposition varies between 1 and 4 kmol, which reflects the range given in Table 7–1. The protons buffered in the leaves vary between 0.3 (pine) and 1.6 kmol (spruce, ecosystem 4), and between 20% (spruce, ecosystem 7) and 70% of proton deposition (beech, ecosystem 2). This fraction enters the soil not at the

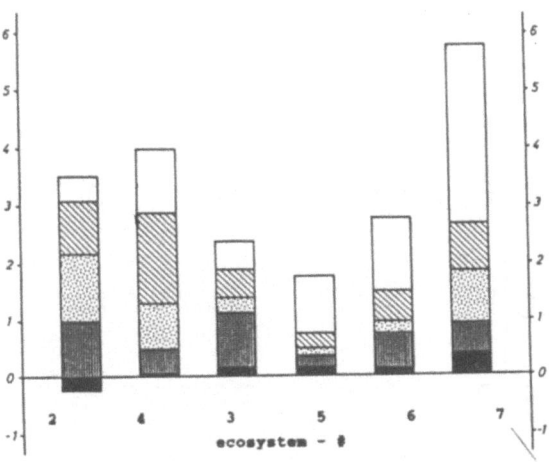

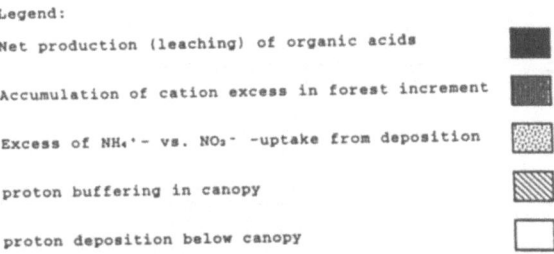

Legend:

Net production (leaching) of organic acids ▪

Accumulation of cation excess in forest increment ▨

Excess of NH₄⁺ - vs. NO₃⁻ -uptake from deposition ▨

proton buffering in canopy ▨

proton deposition below canopy ☐

**Figure 7–6.** Mean annual rates of ecosystem internal proton production (INP) and total proton load (GPM) of different forest ecosystems; values in kmol $H^+$ ha$^{-1}$ yr$^{-1}$. (From Bredemeier, 1987a.)

top surface, but at the surface to the roots. Soil acidification due to acidic deposition proceeds, therefore, not only from the top by percolating water into the subsoil, but may reach the subsoil directly. This is valid especially for beech forests (high stemflow) on soils with higher base saturation (high buffer rate in leaves). Also, for other areas subjected to acid deposition, it has been demonstrated that acidic inputs dominate the acidic load of the soil (Mulder, 1988).

Further evidence for the importance of acidic deposition for soil and seepage conductor acidification is provided by the input-output budgets of acidity in ecosystem case studies: After reaching the Al buffer range, the acidity and the sulfate deposited are leached from the soil, together with deposited acidity accumulated initially as Al sulfates (see Figure 7–2). This is confirmed by comparable data from other areas (Fichtelgebirge: Hantschel, 1987; Hessen: Brechtel, 1988). This result can be transferred to the 80% of forest soils where the condition (soil in Al buffer range) is fulfilled (section V). The sulfate is detectable in spring and creek water by considering the $SO_4$-S/Cl ratio (Ulrich, 1988; cf

**Table 7-5.** Acidity, pH, anion composition, and anion ratios in springs and creeks of forested, unfertilized catchments on rocks free of gypsum.

| | pH | Cation equivalent sum mmol (+)L⁻¹ | $M_a$ cations | HCO₃⁻ | Organic anions | SO₄²⁻ | NO₃⁻ | Cl⁻ | g/g SO₄-S/Cl | NO₃-N/Cl |
|---|---|---|---|---|---|---|---|---|---|---|
| | | | \% of cation equivalent sum | | | | | | | |
| Sea water | | 593 | 0 | 0.4 | 0 | 9 | 0 | 90 | 0.05 | — |
| Springs in the Kaufunger Wald[a] | | | | | | | | | | |
| "Acidic" | 3.7 | 1.820 | 37 | 0 | 8 | 68 | 7 | 17 | 1.9 | 0.2 |
| "Neutral" | 6.4 | 1.560 | <1 | n.d. | n.d. | 38 | 8 | 9 | 1.9 | 0.3 |
| Springs and creeks in the Harz[b] | | | | | | | | | | |
| Quartzite | 4.2 | 0.480 | 52 | 0 | 22 | 47 | 10 | 21 | 1.0 | 0.2 |
| Silica shales | 5.9 | 1.108 | <1 | 11 | 7 | 47 | 20 | 14 | 1.5 | 0.6 |
| Shales + greywacke | 5.5 | 0.869 | 4 | 19 | 6 | 47 | 16 | 12 | 1.8 | 0.5 |
| Greywacke | 5.4 | 0.604 | <1 | 32 | 3 | 39 | 13 | 13 | 1.4 | 0.4 |
| Shales | 5.4 | 1.306 | <1 | 42 | 0 | 38 | 10 | 11 | 1.6 | 0.4 |
| Spilites | 6.1 | 1.540 | <1 | 47 | 5 | 27 | 6 | 15 | 0.8 | 0.2 |
| Creeks in the Black Forest at snow melt[c] | | | | | | | | | | |
| "Acidic" | 4.2 | 0.350 | 42 | 0 | 20 | 44 | 21 | 15 | 1.4 | 0.6 |
| "Medium" | 5.6 | 0.320 | 4 | 16 | 6 | 37 | 28 | 13 | 1.3 | 0.8 |
| "Neutral" | 6.7 | 0.470 | <1 | 37 | 2 | 29 | 21 | 11 | 1.2 | 0.8 |

[a] From Puhe and Ulrich, 1985.
[b] From Heinrichs et al, 1986.
[c] From Zöttl et al, 1985.

Tables 7–2 and 7–5). In areas with rocks free of sulfides, sulfates, and chlorides, or with negligible oxidation rates of sulfides, the $SO_4/Cl$ ratio in soil solutions or seepage, ground, and spring waters can be used to assess the contribution of acidic deposition. An important natural source of sulfate is sea spray. In seawater, the $SO_4$-S/Cl ratio on weight basis equals 0.05. The values for deposition (input) and seepage water (output) for the ecosystem case studies are presented in Table 7–2. As can be seen, the ratio varies depending upon site (proximity to seacoast) and stands between 10- and 50-fold the ratio in seawater. The mean values between input and output differ only slightly. In central Europe, $SO_4$-S/Cl ratios in deposition vary between 0.5 (close to seacoast) to over 2. It can be seen from Figure 7–1 that the $SO_2$ as precursor of sulfate originates to around 90% from anthropogenic $SO_2$ emissions. The increase of the ratio in deposition compared to seawater is therefore almost exclusively due to anthropogenic $SO_2$ emission. In Table 7–5, data characterizing the acid-base status of springs and creeks in forested areas are given. The acidity ("negative alkalinity") is expressed as a fraction of $H^+$ + cation acids (mainly $Mn^{2+}$, $Al^{3+}$) at the cation-equivalent sum. The acidity (as organic bound Al) amounts to less than 4% if the water contains alkalinity ($HCO_3^-$). The $SO_4$-S/Cl ratio varies between 0.8 and 1.9, that is, within the range of deposition and seepage water. The fact that the $SO_4$-S/Cl ratio is independent of the acid-base status of the water shows that all waters are to the same extent subjected to acidic deposition. What varies is the capacity of the cation-exchange buffer in the seepage conductor. In the case of acidified waters, the acidification front has passed through the seepage conductor. In the case of waters containing alkalinity, there is still a cation-exchange buffer system reacting in the seepage conductor before the water is discharged. The pathway of seepage water may differ according to the flow rate (more surface and subsurface flow at high flow rates at times of thaw and snowmelt). The data from the Black Forest show, however, that even under these conditions organic anions account for only half of the total acidity. In most of the waters, $NO_3$ makes a substantial contribution to the anion composition.

The database available allows the generalization presented in Table 7–6. Under A, the acidic load for a rotation period of a European beech (*Fagus sylvatica*) or Norway spruce (*Picea abies*) stand is compiled. The total acidic load of a 100-year-old forest stand varies between 100 and over 400 kmol $H^+ \cdot ha^{-1}$. Acidic deposition amounts to more than 60% of the total acidic load.

Under B, the range of acidic buffering in soil is given. The only process that consumes protons and releases Ca, Mg, and K ions without changes in the acid-base status of the soil is the weathering of silicates. The rate of this process depends upon the types and amounts of silicates present and only to a minor degree on the pH of the soil. The maximum rate of proton consumption reported in soils rich in weatherable silicates is around 2 kmol $H^+ \cdot ha^{-1} \cdot yr^{-1}$ in a soil depth of around 1 m. For most forest soils developed on magmatic and sedimentary rocks with medium silicate content, the rate is around 0.5 kmol $H^+ \cdot ha^{-1} \cdot yr^{-1}$ (Fölster, 1985; Nilsson, 1986). This is in the range of soil acidification due to forest increment (cf. Figure 7–6). If the harvesting is limited to the stemwood

**Table 7-6.** Acidic load and acidic buffering for a rotation period (100 years).

| A. Acidic load | kmol $H^+$/ha |
|---|---|
| 1. Accumulation of cation excess in standing biomass | 20 (beech) to 40 (spruce) |
| 2. Exported timber | Around 30 |
| 3. Decrease in nitrogen storage of soil by nitrate leaching | Between 0 and >100 |
| 4. Acidic deposition | Between 60 and 340 |
| Sum | Between 100 and >400 |
| **B. Acidic buffering in soil (1-m depth)** | |
| 1. Silicate weathering (release of Ca, Mg, K, Na) | Between 20 and 100 (200) |
| 2. Leaching of exchangeable Ca and Mg from the cation exchanger | 20 (sandy soils) to >2000 (clay soils) |

without bark, the acidic production remaining in the soil is around 0.3 kmol $H^+ \cdot ha^{-1} \cdot a^{-1}$. In general, in many forests the soil acidification caused by timber harvesting is balanced by silicate weathering.

This means, however, that any additional acidic load leads to a decrease in exchangeable bases (Ca, Mg, and K). The calculation carried out above leads therefore to the conclusion that an amount of exchangeable bases equivalent to the accumulated acidic deposition (60–340 kmol $H^+$/ha) has been leached from forest soils or subsoils during the last century, half since 1950 and a third since 1970. Assuming 100% base saturation (which is not true, especially for the A horizon of many forest soils), the amount of exchangeable bases in the uppermost meter of forest soils varies between 20 (in sandy soils) and 2000 (in clay soils) kmol ion equivalents (+) per ha (see Table 7–6). If the initial state was 50% base saturation, the values would be 10 to 1000 kmol/ha. The comparison in Table 7–6 between acidic load and acidic buffering during a rotation period of 100 years shows clearly that acidic deposition must have led to extreme low base saturation in great soil depth in many forest soils.

The accumulated acidic deposition of 60 to 340 kmol $H^+$/ha corresponds to 1,200 up to 6,800 kg Ca/ha, which has been (or, in the case of sulfate accumulation, will be) leached together with the sulfate. This can be compared with the median values of Ca storage found today in soils in the Al buffer range (see Tables 7–2 through 7–4). The comparison makes clear that in all these soils, that is, in 80% of the forest soils in the respective area (see Section V), acidic deposition must have had a dominating influence on nutrient stores and the acid-base status of the soils.

# VII. Effects of Acidic Deposition on Litter Decomposition

Relatively little attention has been given to effects of acidic deposition on soil microorganisms and biological soil processes. In the ecosystem case studies 6 and 7 (beech and spruce, Solling, Table 7–2), soil inventories showed a considerable increase in the organic matter and nitrogen storage in the organic top layer during the last two decades (see Matzner, 1989). The decrease in the litter decomposition rate within this period amounts to a third (beech) to two-thirds (spruce) of aboveground litter production. This degree of decomposition inhibition is atypical for mature forests.

From the relationship between deposition of nitrogen (20–80 kg N · ha$^{-1}$ · yr$^{-1}$) and the seepage output of nitrogen (being less than the deposition in all ecosystems studied, approaching zero in many forest ecosystems, for example, case studies 3, 5, and 6), it follows that the majority of forests on acidic soils (Al buffer range) accumulate at present considerable amounts of N. Accumulation in the forest increment of a timber tree can account for only 10 to 15 kg N · ha$^{-1}$ · yr$^{-1}$. In ecosystem 7 (spruce, Solling) denitrification causes losses of <1 kg N · ha$^{-1}$ · yr$^{-1}$ as NO (Gravenhorst and Böttger, 1983) and 5 to 10 kg N · ha$^{-1}$ · yr$^{-1}$ as $N_2O$ (preliminary data, Brumme et al., 1987). Most of the N deposition in soils with low N seepage seems to accumulate in the organic top layer of the soil. In agreement with this is the finding of Zezschwitz (1985b) that the C/N ratio in the humus layer ($O_2$, $Oh$ horizon) of forest soils in North Rhine–Westphalia has become narrower during the last two decades. Wide C/N ratios have been considered to be responsible for retardation of litter decomposition. A nitrogen input exceeding the accumulation in phytomass does not prevent, however, the accumulation of litter and its decomposition residues, indicating that other regulative mechanisms operate.

Bååth and others (1980) found in an experimentally acidified Scots pine forest soil that the amounts of live fungal and bacterial biomasses decreased as did decomposition rates for needle and root litter. In the same ecosystem, the composition of species of microfungi changed after acidic treatment (Bååth et al., 1984). A laboratory study on the fate of $^{14}$C-glucose (Lohm et al., 1984) showed that $^{14}CO_2$ was released more slowly from the acidified soil samples. Also, needle litter from Norway spruce (Hovland et al., 1980; Berg, 1986b) and birch leaves (Hågvar and Kjøndahl, 1981) showed reduced decomposition rates in their respective systems after experimental acidification in forests all over Sweden. Berg (1986a) found a general decrease in rate of needle litter decomposition of 8% after acidic treatment. The decrease in live fungal biomass and respiration rate found by Bååth and others (1984) continued until 5 years after termination of experimental acidification.

In ecosystem case study 1 (beech on limestone, Table 7–2), it was found that additional acidic input may cause the accumulation of litter in an organic top layer (Wolters and Scheu, 1987; Wolters, 1988; Wolters and Schauermann, 1989; Wolters, 1989). Increasing acidic load of the mineral soil results in the disintegration of soil organic matter. In the litter layer, the microbial decomposition rate is slowed down, and the stimulating effect of the micro- and mesofauna on microbial decomposition is decreased. This means that the relationships characteristic for a mor profile (see below) are approached. The acidic input increased the leaching of cations and reduced the activity of the macrofauna. In ecosystem 6 (beech, Solling), the authors found that additional acidic input also decreased microbial decomposition rates in the litter layer. The inhibiting effect of the collembola

*Isotoma tigrina* on microbial decomposition in this acidic soil was increased by the experimental acidification. The authors conclude that even if the soil animals are not directly affected by acidic input, their role within the ecosystem can be changed. This finding is important because until now no change in soil fauna due to acidic deposition could be detected in this ecosystem. The accumulation of easily decomposable organic matter increases the risk of seasonal mineralization and acidification pushes: After the rewetting of the air-dried substrate, the microbial mineralization rate was much higher in the soil with a higher acidic load.

In soils where the A horizon is in the Fe/Al buffer range, the organic fine substance accumulated in the humus layer (O2, *Oh* horizon) of the organic top layer contains predominantly Al and Fe, and the percentage of Ca + Mg of the cation-equivalent sum is usually below 10% (see data in Gehrmann et al., 1987). In Table 7–7, data are compiled about the change in the cation composition in the humus profile of case studies 6 and 7 (beech and spruce in Solling). The percentage of Ca at the cation-equivalent sum increases from leaves to leaf litter to 50%, but drops in the fermentation layer (O1, *Of* horizon) to less than 10% (cf. Figure 7–5). The relative Al and Fe content increases from 1% in leaves to 30% to 50% in the fermentation layer. The cause of this change seems to be the root litter. Dead fine roots contain percentages of Al and Fe comparable to the soil matrix. It is possible that the increase of organic matter in the humus layer (O2, *Oh* horizon) is mainly due to the retardation of root litter decomposition.

The mineralization of organic Al and Fe complexes can be described by the following equation:

$$[(CH_3COO^-)_{\frac{1}{3}} Al^{3+}] + 2\,O_2 \rightarrow 2\,CO_2 + H_2O + \tfrac{1}{3}\,Al(OH)_3 \qquad (5)$$

**Table 7-7.** Change in cation composition in humus profile in the beech and spruce eco-system in Solling (ecosystems 6 and 7 of Table 7–2).

| | Cation sum mmol (+)/kg DM | % of cation sum | | |
| --- | --- | --- | --- | --- |
| | | $\frac{1}{2}Ca^{2+}$ | $\frac{1}{3}Al^{3+}$ | $\frac{1}{2}Fe^{2+}$ |
| Beech (case study 6) | | | | |
| Leaves | 606 | 35 | 1 | 1 |
| Leaf litter | 521 | 49 | 3 | 2 |
| L (litter layer) | 687 | 24 | 39 | 27 |
| Dead fine roots in O2 (*Oh*) | 413 | 6 | 62 | 25 |
| O1 + O2 (*Of* + *Oh*) | 1,150 | 8 | 54 | 32 |
| Spruce (case study 7) | | | | |
| 1-year-old needles | 480 | 36 | 2 | 1 |
| Needle litter | 386 | 49 | 8 | 8 |
| L (litter layer) | 430 | 19 | 33 | 35 |
| Dead fine roots in O1 (*Of*) | 293 | 57 | 16 | 6 |
| O1 (fermentation layer) (*Of*) | 816 | 9 | 48 | 36 |
| Dead fine roots in O2 (*Oh*) | 1,167 | 8 | 61 | 25 |
| O2 (humus layer) (*Oh*) | 805 | 9 | 50 | 34 |

where $CH_3COOH$ represents any organic molecule. In the soil solution of the O horizon, Al and Fe exist predominantly as organic complexes, however (Dietze, 1985; Ares, 1986b; Schierl et al., 1986). In respect to Al, the extreme and moderate labile forms of organic complexes predominate (Dietze and Ulrich, 1989). This shows that reaction 5 is followed by the reaction

$$\tfrac{1}{3} Al(OH)_3 + CH_3COOH \longrightarrow (CH_3COO^-)\tfrac{1}{3} Al^{3+} + H_2O. \tag{6}$$

The mineralization of organic Al and Fe complexes will again lead to new organic Al and Fe complexes. This results only in an increase in the number of acidic groups per unit of organic substance (cf. Table 7–7).

In spruce forests, Funke and others (1986) found no difference in the diversity of soil animal populations between damaged stands (needle loss, dieback) and undamaged stands. This is in agreement with the finding for ecosystem 6. Wolters (1988) considers the formation of a mor profile that is differentiated into a litter layer (O1, L), a fermentation layer (O1, *Of*) and a humus layer (O2, *Oh*), as the result of long-term soil acidification. From this it can be inferred that acidic deposition causes or contributes to the change from a mull to a moder type of humus. Its effect on decomposition rates within these humus types, however, cannot be traced back to the diversity of soil fauna. The deciding role seems to be the change in chemical composition of root litter, the activity of soil microorganisms, and the role of soil animals in regulating microbial activity.

## VIII. Direct Effects of Acidic Deposition on Leaves

When the stomates are open to facilitate photosynthesis, $SO_2$ and $NO_x$ enter leaves together with $CO_2$. The flux is controlled by Fick's law. Within the leaf, the $SO_2$ reacts with water to form sulfurous acid, which may be oxidized or reduced. These reactions can occur in the apoplast (cell wall volume) or, after the passage of $SO_2$ through cell membranes, in the symplast. The passage of $SO_2$ from the apoplast into the cytoplasm of leaf cells should be controlled by the pH of the water films existing at the outer surface of cell membranes. At pH values $>4.0$, the reaction

$$SO_2 \cdot H_2O \rightleftharpoons HSO_3^- + H^+ \; (pK\; 1.81) \tag{7}$$

is strongly shifted to the right. Because the passage of charged species can be controlled by cell membranes, their permeability to $SO_2$ decreases exponentially with increasing pH (Spedding et al., 1980). As long as the proton production due to $SO_2$ dissolution at the cell surface can be buffered and the pH maintained, the reactions should take place mainly in the apoplast. After oxidation, the sulfate is leached by precipitation, together with the cations that have been released in the buffering reactions. The buffering of deposited acidity and the leaching of cationic nutrients from the leaf (see Richardson and Dowding, 1988, for an overview) are thus two aspects of the same process.

Bender and others (1986) found that dry deposition of gaseous $SO_2$ results in a decrease of the acidic buffer capacity of leaves. According to Heber and others

(1987), a pH decrease in the cytoplasm due to the dissociation of $SO_2 \cdot H_2O$ results in the depression of photosynthesis and can lead to light-dependent destruction of pigments. The yellowing of the light-exposed surface of spruce needles is a well-known damage symptom of Mg-deficient trees. A comparison between calculated detoxification rates with calculated rates of $SO_2$ uptake from polluted air suggests that levels of $SO_2$ considered permissible are still too high to avoid damage to forest trees (Heber et al., 1987). In fumigation experiments with 5-year-old Norway spruce growing in sand culture with complete nutrient solution, Ruetze and others (1988) found at $SO_2$ concentrations above 50 $\mu$g $\cdot$ m$^{-3}$ damage to cells in the stomatal area, including the epidermis. Some necrotic cells were present in the substomatal mesophyll, whereas chloroplasts in adjacent cells showed structural alterations decreasing gradually from the parietal face of a chloroplast towards its center. Even after 2 years of exposure, no structural damage occurred in the vascular bundle. After the second exposure period, Ca-oxalate deposition in the needles was reduced. This may be a consequence of the higher proton buffering due to the $SO_2$ load. Cell damage in the stomatal area can be taken as evidence for direct effects of $SO_2$.

For ecosystem case studies 2 through 7 (see Table 7–2), the mean rates of proton buffering and the leaching of cations is given in Table 7–8. The annual variation can be considerable, as the data given by Matzner (1989) show. The cations leached are mainly $K^+$ and $Ca^{2+}$, followed by $Mn^{2+}$ and $Mg^{2+}$. The leaching rate of cations usually exceeds the buffering rate of protons, due to additional leaching of organic anions. Also the uptake of air-borne $NH_4$ results in the leaching of K, Mg, and Ca (Roelofs et al., 1985). The leaching amounts to 22% (Mg, Ca in beech) up to 62% (K in spruce) of the annual uptake rates. For ecosystem 7, as an example, the mean pH value in throughfall is 3.45 with a variation between 4.83 and 2.75. Low pH values are correlated with high concentrations of sulfate (deposited) and K, Mg, Ca, Fe, and Mn ions (mainly leached). This indicates that at low pH values the acidity deposited exceeded the buffer capacity of the needles. The drying of rain and fog droplets or water films on the plant surface can create pH values around 2 (Frevert and Klemm, 1984). The increased leaching of K, Ca, Mg, Zn, Mn, and carbohydrates by acidic mist was demonstrated experimentally (Krause et al., 1983, 1985; Bender et al., 1986; Mengel et al., 1987). The acidic mist also caused damage to the cuticula. The leaching of cations itself does not cause nutrient deficiency as long as the uptake of these cations is not limited. In short-term experiments, the nutrient content of leaves may even increase, despite increased leaching (Skeffington and Roberts, 1985; Guderian, et al., 1987).

Scanning electron microscopy of conifer needle surfaces exposed to air pollutants show fusion and erosion of fibrillar epicuticular waxes. *Pinus* spp. appear to be less sensitive than *Abies alba* or *Picea abies* (Cape, 1988). The effect of acidic mist (pH 3) and/or ozone on the epicuticular wax in the stomatal antechamber of Norway spruce needles was investigated by Magel and Ziegler (1986). In current-year needles, the wax plug formation was severely disturbed. Structural wax outlining the stomatal openings of clear air needles exhibits fused wax rodlets after ozone fumigation (150 $\mu$g $O_3 \cdot$ m$^{-3}$). Cracks were observed in

**Table 7-8.** Proton buffering and leaching of cations in some of the ecosystem case studies (annual mean values in kmol $(+) \cdot ha^{-1} \cdot yr^{-1}$).

| Ecosystem | | Buffering of $H^+$ | Leaching of | | | | | | |
|---|---|---|---|---|---|---|---|---|---|
| | | | $Ca^{2+}$ | $Mg^{2+}$ | $K^+$ | $Mn^{2+}$ | $\Sigma$ cat | Organic anions[a] | Unexplained[b] |
| HE | 2 | 0.92 | 0.43 | 0.13 | 0.37 | 0.06 | 0.99 | 0.19 | -0.12 |
| HDEI | 3 | 0.49 | 0.32 | 0.10 | 0.49 | 0.08 | 0.99 | 0.32 | 0.18 |
| SP | 4 | 1.56 | 0.55 | 0.17 | 0.32 | 0.20 | 1.24 | -0.16 | -0.16 |
| HDKI | 5 | 0.26 | 0.47 | 0.06 | 0.16 | 0.01 | 0.70 | 0.24 | 0.20 |
| SLBU | 6 | 0.66 | 0.35 | 0.09 | 0.54 | 0.12 | 1.10 | 0.28 | 0.16 |
| SLFI | 7 | 0.64 | 0.51 | 0.07 | 0.51 | 0.16 | 1.25 | 0.49 | 0.12 |

[a] Calculated from the cation/anion balance of bulk deposition and throughfall + stemflow.

[b] $\Sigma$ cat (organic anions + buffering of $H^+$) as a measure of plausibility of the data.

Adopted from Bredemeier, 1987a.

wax plugs of acidic mist–treated needles. Ozone and acidic mist in combination caused fusion of wax rodlets and cracks in the wax-filled antechamber. Such changes may be of significance for the leaching of cations from the stomates. Günthardt-Goerg and Keller (1988) found, however, that fumigation with 300 μg $O_3 \cdot m^{-3}$ reduced the leaching of ions from spruce needles due to the formation of a thin "smooth layer," which covered the needle surface, especially the stomatal apertures. Erosion of the epicuticular wax layer has been recorded from needles sampled in the forest (Grill, 1973; Grill et al., 1987; Kazda and Glatzel, 1986; and Kowalkowski et al., 1988, for spruce, and Huttunen and Laine, 1983; Hafner, 1986, for pine). According to Bermadinger and others (1988), $SO_2$ causes severe alterations of the wax structures; with oxidants as the main emission component, the wax structures are nearly unaffected. According to Barnes and others (1988), the effects of ozone on the surface waxes are generally greater on the previous year's needles than on the current year's growth. Elstner and Osswald (1988) conclude that the following chain of effects might cause the bleaching (yellowing) of spruce needles:

1. Trees are subject to certain deficiencies such as magnesium, potassium, or water.
2. Subacute deficiencies increase ethylene formation as a stress response as well as photoactivation of oxygen and thus peroxidation of unsaturated fatty acids
3. Elevated ozone levels damage membrane functions, thus limiting transport of ions and photosynthesis products. The metabolic block thus formed enhances photodynamic reactions as a consequence of limited electron transport.
4. Exogenous ozone and endogenous ethylene react at the needle surface, resulting in reactive aldehydes and peroxide, which in turn damage the wax layer. Damaged surface waxes, especially on top of the stomatal cavities, allow penetration of pathogenic fungi that cause necrotization and finally abscission of needles.

In open-top chambers with nonfiltered air plus 25 ppb ozone, Sutinen and others (1988) observed that ultrastructural changes first occurred in the chloroplasts and that changes always first appeared in the outer cell layer facing the sky. Such results cannot be taken as a contradiction to the conclusions of Elstner and Osswald (1988) as long as it has not been demonstrated that the trees are well supplied with nutrients and water. The importance of mineral deficiencies (Mg, Ca, K) is also stressed by Fink (1988a,b). Mineral deficiencies induce premature collapse of sieve cells within the vascular bundle of spruce needles. This type of damage was found not only in yellow needles but also in green needles from trees with crown thinning. Damage in the vascular bundle and disturbances of phloem transport seem to be an important primary damage symptom that is not caused by the direct effect of air pollutants. These symptoms can be taken as evidence for soil-borne effects weakening the tree.

The harmful effect of gaseous $SO_2$ is not limited to the leaves. Keller (1979) showed that continued fumigation of spruce seedlings with $SO_2$ resulted in root damage. Neighbour (1988) reported an increase in the shoot:root ratio of deciduous trees after exposure to 65 ppb $SO_2$ and 65 ppb $NO_2$. The reduction of root size and, in addition, of stomatal control makes the trees more susceptible to water stress. Jurat and others (1986) found that dry deposition of $SO_2$ (fumigation with 140 $\mu g \cdot m^{-3}$) and of ozone affects the mineral supply of Norway spruce (Jurat and Schaub, 1988). In particular, the contents of Ca and Mg in the roots were drastically reduced after a 7-week exposure period. The combination of acidic soil and sulfur dioxide had a greater effect on both net photosynthetic rates of the plants and chlorophyll content of the needles as compared with the effect of only one of these stresses. With European beech seedlings, a decrease in pH of the applied mist (pH 4.6, 3.6, and 2.6) led to an increased leaching of both $Ca^{2+}$ and $Mg^{2+}$. As a consequence, more cations were taken up by the plants. The concomitant $H^+$ efflux from the roots led to an acidification of the rhizosphere (Leonardi and Flückiger, 1988; Kaupenjohann et al., 1988). The buffering capacity of the leaves decreased with increasing acidic mist treatments, mainly in the extracellular space. Leaf contents of exchanged and leached cations decreased together with decreasing buffer capacities.

The long-term effects of the proton buffering in the leaves on the vitality of the roots is unknown. In soils of low base saturation, acidic precipitation events could exceed rhizosphere buffer rates (Kaupenjohann and Hantschel, 1987). Trees of an age <100 years are subjected to this stress all their lives. As outlined in section III,A, on soils with high base saturation, the direct acidic load of the tree is especially high. The consequences, however, are unknown. Observations show that the dieback of trees is always connected with root damage, even on aerated soil of high base saturation where adverse soil effects seem unlikely. It is unknown whether the root damage precedes the damage in the canopy or whether the opposite is true. Without clarifying the cause of the root damage, little can be said about the cause of tree dieback under these conditions.

# IX. Effects of Soil Acidification on Trees

## A. Assessment of Acidic Stress

Trees that are now around 100 years old have always been subjected to an increasing acidic stress acting on their roots. The acidic stress manifests itself as concentrations of acids in the soil solution. The most important acids are Al ions, protons, Mn ions, Fe ions, and some heavy metal ions. A measure of acidic stress may be either the concentration (or activity) of the ions in the soil solution or ion ratios (e.g., Ca/H, Ca/Al, Mg/Al), depending upon the reaction mechanism at the biological surface. Often the first step in acidic stress may be the competition of a cation acid with a nutrient cation (Ca, Mg, K) at a binding site at the biological surface. Such competition reactions often follow the laws of cation exchange. In this case, ion ratios express the chemical potential of ions in soil solution better than concentrations. Because different stress reactions in the apoplast, at the surface of the symplast, and within the cytoplasm can occur, no simple, general terms describing the dose of acidic stress can be expected.

As in other waters (cf. Table 7–5), soil solutions of pH >5 contain no or negligible concentrations of acids. Above pH 5, soil solutions are mixtures of neutral and alkaline salts (bicarbonates). Below pH 5, soil solutions are mixtures of neutral salts, acidic salts, and mineral acids, as well as organic acids. The fraction of acidic salts and of acids depends mainly upon the presence of anions. A soil with zero base saturation can be in equilibrium with a soil solution of pH close to 5, provided the anion concentration in the soil solution is low. On the contrary, the lower limit of pH in the soil solution is determined by the upper limit of anion concentration. Acidic deposition influences acidic stress not only by the input of acidity but also by the input of mobile conservative anions like $SO_4^{2-}$ and $NO_3^-$. This becomes evident from the data presented in Figure 7–3 and the role of Al sulfates in determining Al concentration in soil solution. In subsoils in the Al buffer range, the acids make up 50% to 70% of the ion composition of the soil solution, the remainder being neutral salts (cf. Table 7–2). The only possibility to assess acidic stress is therefore the analysis of the soil solution. As already shown (see Figure 7–4), the composition of the soil solution may show great variability in space and time. The assessment of acidic stress in soil may therefore require the continuous collecting of soil solution over years. Under the influence of acidic input, the spatial variability of the composition of the soil solution is also of importance of a microscale. Due to the proton production by roots, the soil solution of the rhizosphere may often be much more acidic than that distant from roots. This has been demonstrated by pH measurements close to absorbing roots (Marschner et al., 1985). Also, between the surface and the interior of soil aggregates, chemical gradients can exist (Hildebrand, 1986a, 1986b, 1988a; Horn, 1987; Horn et al., 1988). A consequence can be that the composition of the soil solution in an undisturbed soil under field conditions differs from a disturbed soil sample under laboratory conditions (showing higher acidic concentrations in the undisturbed soil; Hantschel et al., 1987). The assessment of acidic stress under field conditions requires sophisticated methodology. Without using this methodology, however, the contribution of acidic stress to root damage and forest decline cannot be excluded. There are only few field studies in which this methodology has been used.

The fact that the variables indicating stress have to be defined in the solution phase facilitates the transfer of knowledge gained in solution culture studies to the field. Conversely, the knowledge of the composition of the soil solution in the field helps to design the solution culture study. Most of the following results have been gained with nutrient solutions comparable in composition to soil solutions in the aluminum buffer range.

Acidic stress on roots can lead to a reduction in root growth (measurable as reduction in root length), to root damage, to changes in the uptake rates of nutrients, and to changed elemental content in roots. In solution cultures without mycorrhizal infection, the following damage classes can be distinguished

L: Reduction of root *l*ength

D1: Damage class I: The primary root tip becomes light brown and stops growing. It can either regenerate and continue growth, or a lateral root immediately behind the root tip is formed and develops as a new main root.

D2: Damage class II: The root tip dies back (meristematic abortion; Metzler, 1985). The cortex becomes brown and can easily be separated from the vascular cylinder (especially in spruce). Scanning electron microscopy shows a necrosis in the region between the cortex and the central cylinder (Hüttermann and Ulrich, 1984). The tips of the lateral roots show also browning and dieback.

D3: Damage class III: The roots are necrotic and show no length growth or formation of lateral roots.

UCa, UMg, etc: Strong decrease in *u*ptake rates of Ca and Mg. The uptake rate can be decreased by more than 50% (Junga, 1984; Stienen and Bauch, 1988; Godbold et al., 1987, 1988b, 1988c). This results in a decrease of the nutrient contents in roots, bark, and shoots (Rost-Siebert, 1985). In Al-treated roots, the Al content is high. Cellular element analytical techniques of x-ray microprobe (EDAX) and laser-micro-mass spectroscopy (Godbold et al., 1988b; Stienen and Bauch, 1988) reveal that Ca is displaced in the cortex cell walls by Al. As Ca is an important element for maintaining cell wall structure, the exchange of Ca and Mg by Al may inhibit cell elongation. In the primary xylem, the Ca values are not lowered, and Al is detectable only in traces. In case of reduction of Mg uptake, the Mg content in the whole cortex and endodermis, as well as in the inner central cylinder, can fall to very low values.

Under field conditions, the following features have been used to classify root damage:

1. Distinction between fine root (<2 mm) biomass (live, vital roots) and necromass (dead, subvital roots). The necromass is characterized by missing root tips and high fragility; the central cylinder is missing or dark brown (Murach, 1984).
2. The number of root tips per unit length (root tip frequency, ramification density; Blasius et al., 1985; Meyer, 1987a, 1987b; Eichhorn et al., 1988).
3. The number of lateral roots continuing axial root growth after damage to the apical meristem, per unit root length (number of root derangement points; Eichhorn, 1987).
4. Histological features like degeneration in the cortex of the fine roots (Stienen et al., 1984; Nienhaus, 1988).
5. Decrease in mycorrhiza frequency (Blasius et al., 1985; Blaschke, 1986), changes in mycorrhizal type, change from ectomycorrhizae to ect-endomycorrhizae and parasitic behavior of the fungi (Meyer, 1984, 1987b), microscopic and ultrastructure (Haug et al., 1986; Kottke et al., 1986), and FDA fluorescence (Ritter et al., 1986). Effects on mycorrhizae are, however, still controversial. Kottke (1986) and Kottke and Oberwinkler (1986) conclude that growth of mycelia of mycorrhizal fungi appears to be not so much influenced by

soil pH and Ca/Al ratio as by sand content, aeration, and humus supply. However, the determination of exchangeble cations is not sufficient to exclude acidic and nutrient stress, and the aeration has to be quantified by measurements to draw such conclusions.

6. Chemical analysis of live (vital) and dead (subvital) fine roots by ash analysis and cellular element analytical techniques in order to quantify the effect of acidic and nutrient stress in the roots themselves (Ulrich 1981a; Bauch and Schröder, 1982).

7. Assessing the depth functions of fine-root biomass, fine-root necromass, and the woody root system in order to assess long-term changes in the depth gradient of the fine-root system (Ulrich et al., 1984).

8. Following the changes in fine-root biomass and necromass as a depth function with time, together with the time pattern of growth factors (temperature, water suction, soil solution composition) in order to identify the relevant growth or stress factors (Murach, 1984).

Roots of all damage classes, even of D3, may be able to regenerate after being transferred to solutions free of acidic stress. Even if the apical meristem is destroyed, new lateral roots can be formed (Junga, 1984; Rost-Siebert, 1985). Stienen and Bauch (1988) and Schröder and others (1988) showed that the Al in the cortex of young roots can be reexchanged by Ca in Ca-enriched media with a higher pH. In the older roots Al is not reexchanged completely, and they may never regain their full apoplast function.

According to Rost-Siebert (1983, 1985), the damage caused by protons and monomeric Al ions depends—within the limits of soil solution composition—strongly on the Ca concentration in the solution: Increased Ca concentrations can depress the adverse effects of Al or H ions. This observation led to the use of ion ratios as stress variables.

## B. Proton Stress

For the acid-tolerant species Norway spruce and European beech, proton stress, according to the data compiled in Table 7–9, seems to play no role at pH >3.9. It is thus bound to soil horizons in the Al/Fe or Fe buffer range, that is, to the top soil horizons (with the exception of case study 8; cf. Table 7–2) or to periods of acidification pulses. Beech is much less tolerant to proton stress than spruce, showing root damage already at Ca/H ratios <1 (spruce at ratios <0.1). Below pH 3, the Ca concentration has no influence on proton stress. The data from the ecosystem case studies 6 and 7 of Table 7–2, presented in Figure 7–7, show Ca/H ratios between 1 and 0.2 in the topsoil of both sites. Minimum values occur after the warm and dry years 1982–83 in 1984–85 (climatic acidification pulse). Beech reacts to the proton stress with shortened lifetimes for fine roots, as indicated by high percentages of fine-root necromass in the topsoil (see Figure 7–8). Spruce should not be under proton stress in the O and A horizons. Acidification pulses may, however, decrease pH periodically and thus create acidic stress even in the O

**Table 7–9.** Ca/H and pH relations in nutrient solutions and soil solutions as criteria of proton stress.

| Author | pH | mol Ca/ mol H | Effects or remarks |
|---|---|---|---|
| A. Norway Spruce (*Picea abies*) | | | |
| Rost-Siebert, 1985 | >3.9 | — | None |
| | 3.5–3.9 | <0.1 | L, D2, D3 |
| Stienen and Bauch, 1988 | 3.0 | 1 | L, UCa, UMg |
| Stienen, 1986 | | | Reduction in transpiration |
| Simon and Rothe, 1985 | 3.0 | 0.1 | D1, D2 |
| Metzler and Oberwinkler, 1986 | 2.6–3.0 | 0.09–0.23 | D1, D2 |
| Vogelei and Rothe, 1988 | 3.5;2.5 | 0.41; 0.04 | L, D1, D2, D3 |
| | | | Lysis of cell wall |
| Case study 7 (spruce Solling) | 3.52 | 0.55 | x̄ (1970–1979) of input into mineral soil |
| Case study 8 (young spruce, Hils) | 3.56 | 0.27 | x̄ (1984–1986) of seepage water in 1 m soil depth |
| B. European Beech (Fagus sylvatica) | | | |
| Rost-Siebert, 1985 | 3.6–5.7 | >1 to 1.5 | None |
| | | <0.7 to 1 | L, D2, D3, UCa, UMg |
| Case study 6 (beech Solling) | 3.76 | 0.47 | x̄ (1969–1979) of input into mineral soil |

horizon (Murach, 1984). This coincides with the observation that the main fine-root mass of spruce is located in the topsoil (see Figure 7–8). As indicated by the ratio of fine-root biomass to necromass, the best rooting conditions seem to exist in the fermentation layer (O2, *Of* horizon) and in 5- to 20-cm mineral soil depth (cf. also Figure 7–5).

Factors influencing the Ca/H ratio are the pH of throughfall, the deposition of $Ca^{2+}$, nitrification or acidification pulses in warm and/or dry years, and the

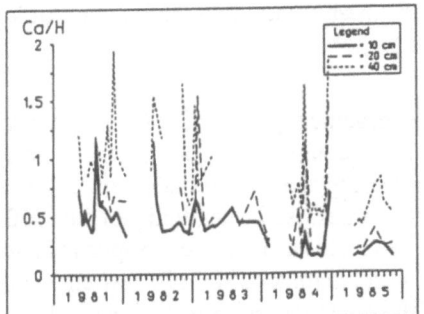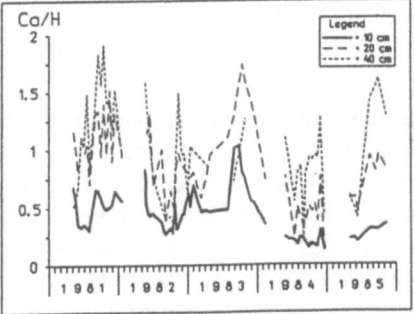

**Figure 7–7.** Time course of Ca/H ratio in soil solution in ecosystem 6 (beech–left, Solling) and 7 (spruce–right, Solling); cf. Table 7–2. (From Murach and Ulrich, 1988.)

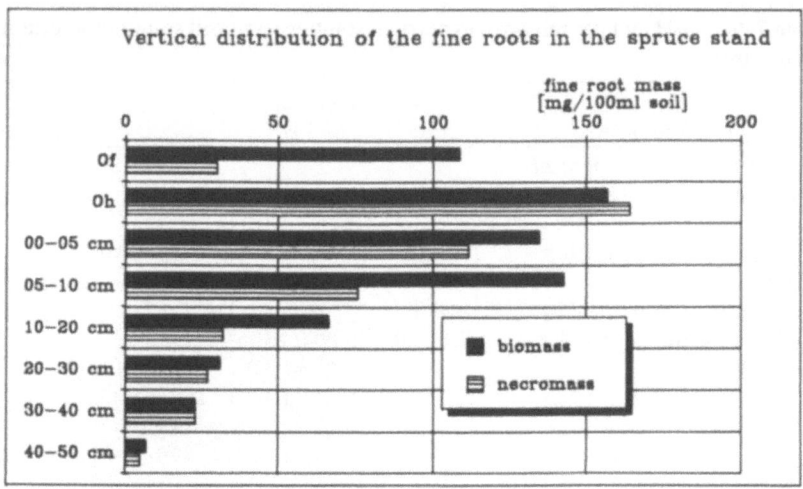

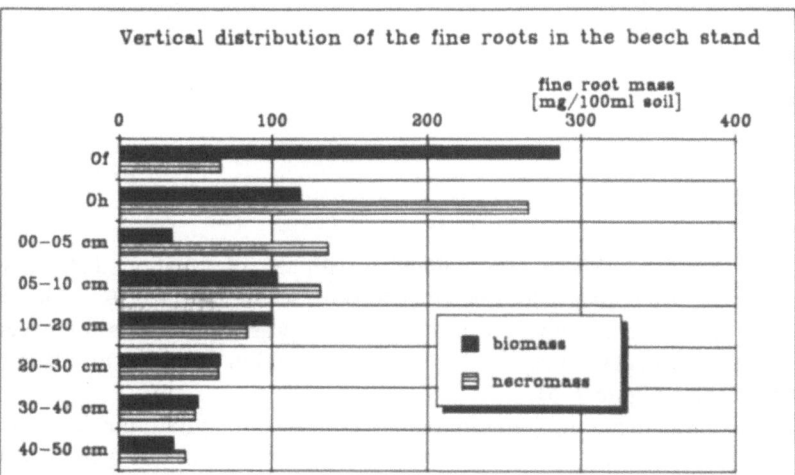

**Figure 7–8.** Vertical distribution of fine roots in ecosystem 6 (beech, Solling) and 7 (spruce, Solling; cf. Table 7–2) (several years' mean values, from Murach and Ulrich, 1988).

leaching of $Ca^{2+}$ accumulated earlier in the organic top layer. The Ca stored in the organic top layer and slowly leached as a consequence of buffering acidic deposition has delayed the effect of increasing soil acidification on roots by maintaining Ca/H and Ca/Al ratios at higher values. At the beginning of the 1960s, difficulties in beech regeneration led to silvicultural measures: Superficial soil cultivation, bringing the fruits in contact with the mineral soil of a few cm depth, was found to enable the establishment of the regeneration (Burschel et al., 1964). As discussed in section V, since the 1960s, the depth of soil in the Al/Fe buffer range has increased considerably. This development upsets more and more the effect of superficial soil cultivation in respect to natural regeneration of beech. The strong acidification of the topsoil is now a great obstacle in natural beech regeneration (Gehrmann, 1984; Hüttermann and Ulrich, 1984; Glatzel and Kazda, 1985).

## C. Aluminum Stress

Again, experimental data for European tree species from solution culture experiments are almost entirely restricted to Norway spruce and European beech. For the acid-tolerant species spruce and beech, Al stress is restricted to pH values below 4.2 (4.7 in soils, cf. section IX,A) and the presence of monomeric Al species. Under the conditions of acidic soil solutions regarding pH, Ca, Mg, and Al concentrations, the Al stress is, according to the data compiled in Table 7–10, strongly governed by the ratios $Ca^{2+}/Al_{inorg}$ and $Mg^{2+}/Al_{inorg}$. Such ratios represent the relation in the chemical potential of the ions involved to be adsorbed at negatively charged surfaces like the cell wall and cell membranes. This adsorption is the first step in a series of chemical and biochemical reactions resulting in an effect. This makes the usefulness of such ratios understandable but shows also the limitations to define threshold values for specific cause-and-effect relationships (cf. Table 7–11).

Eldhurst and others (1987) used a spraying technique to supply the nutrient solution to the roots. The authors found for Norway spruce at 0.17 mmol Al/L (lowest concentration used) swelling of root tips and the development of grey and brown zones behind the tip. The Ca and Mg content of roots and shoots decreased, indicating a marked decrease of nutrient uptake. At >0.3 mmol Al/L the growth rates decreased, and an increase in the dry weight/fresh weight ratio of seedlings indicated a decrease in water uptake. At >1 mmol Al/L the needle colour turned to yellow or yellowish red. At >3 mmol Al/L the roots died from the tip inwards. The Ca/Al ratio in the spraying solution was in all experiments <0.06. As demonstrated in Table 7–10 and Figure 7–4, Al concentrations up to or even exceeding 2 mmol/L are realistic values in acidic forest soils subjected to acidic deposition.

The data compiled in Table 7–10 show that Al can reduce the uptake of Ca and Mg. Junga (1984) reported a depression of K uptake. Reduction in root growth and root damage seems to be regulated mainly by the Ca/Al ratio, whereas a depression of Mg uptake is regulated by the Mg/Al ratio and can be important even in the case of high Ca/Al ratio and low root damage (Jorns and Hecht-Buchholz, 1985; Jorns, 1988). Apart from the roots, nutrient deficiencies are greatest in the bark (Stienen and Bauch, 1988) and extend to the leaves (Murach and Wiedemann 1988).

A generalization of the results of the solution culture experiments is given in Table 7–11. As a criteria for Al stress, the Ca/Al and Mg/Al ratios are used. The meaning of the ratios is not independent of the actual concentrations, but their variation is limited in acidic soils. These ratios may be considered as an approximation to threshold values in order to evaluate the Al stress acting in the field. The values imply that all ion species exist in monomeric form in the soil solution. This is not the case in O and A horizons. In soil horizons of low organic

**Table 7-10.** Al concentrations and Ca/Al and Mg/Al ratios in nutrient solutions and soil solutions as criteria of aluminum stress.

| Author | pH | Ca mmol/L | Al mmol/L | Mg mmol/L | Ca/Al range | (mol/mol) threshold | Mg/Al range | Duration of experiment (days) | Effects and remarks |
|---|---|---|---|---|---|---|---|---|---|
| A. Norway spruce (*Picea abies*) | | | | | | | | | |
| Rost-Siebert, 1985 | 3.8 | 0–3.2 | 0–0.37 | 0.08 | 5.4–0.008 | 1 | 0.21–2.1 | 14 | L, D2, D3, Uca, UMg |
| Junga, 1984 | 3.8 | 0.1 | 0–0.37 | 0.08 | 0.3–0.1 | 0.3 | 0.22 | 56, 94 | D2, D3, UCa, UMg, UK |
| Metzler and Oberwinkler | 2.6–3.7 | 0.23 | 1 | — | 0.23 | 0.23 | — | 180 | D1, D2 |
| Godbold et al., 1988b,c | 3.8 | 0.13 | 0–1.2 | 0.08 | 0.19–0.1 | 0.19 | 0.07–0.12 | 1237 | UCa, UMg |
| Tischner et al., 1983 | <4.0 | 1.0 | 0–2.0 | — | 0.5–2.0 | 1 | — | 42 | D2, D3 |
| Simon and Rothe, 1985 | 4.0 | 0.1 | 0–1 | — | 0.1–1 | 0.1 | — | 154 | D2 |
| Stienen and Bauch, 1988 | 3.0 | 1.0 | 1.5 | 0.19 | 0.67 | 0.67 | 0.13 | — | L, D2, D3, UCa, UMg |
| Jorns et al., 1985, 1988 | 3.8 | 0.13 | 0–1.7 | 0.03 | 0.75–0.075 | 0.25 | 0.02–0.18 | 42 | D1, D2, UCa, UMg needle yellowing |
| Asp et al., 1988 | 4.2 | 0.1 | 0.1;1;10 | 0.1 | 2.5;0.19;0.016 | 0.19 | like Ca/Al | 63 | L, D2, D3, UCa, UMg, UK |

| | | | | | | | | |
|---|---|---|---|---|---|---|---|---|
| *Data from ecosystem case studies* | | | | | | | | |
| No. 4 (Spanbeck) | 3.96 | 0.52 | 1.2 | 0.27 | 0.43 | — | 0.22 | — | $\bar{x}$ (1982–1985) |
| No. 7 (Solling) | 4.01 | 0.086 | 0.45 | 0.057 | 0.19 | — | 0.12 | — | $\bar{x}$ (1973–1983) |
| No. 8 (young stand) | 3.5 | 0.08 | 0.11 | 0.031 | 0.7 | — | 0.28 | — | $\bar{x}$ (1984–1986) |
| No. 9 (Wingst) | 4.0 | 0.15 | 0.81 | 0.11 | 0.19 | — | 0.14 | — | $\bar{x}$ (1984–1986) |
| **B. European beech (*Fagus sylvatica*)** | | | | | | | | |
| Rost-Siebert, 1985 | 3.8 | 0–0.8 | 0–0.59 | 0.08 | 5.4–0.005 | 0.1 | 0.21–2.1 | 20 | D2, D3, UCa, UMg |
| Neitzke and Runge, 1985 | 2.3–5.5 | — | — | — | 2.5–0.4 | 1 | — | 200 | L, D2 effects time dependent |
| Neitzke and Runge, 1987 | 4.0 | 0.125; 1 | 0–1.29 | — | 9–0.1 | 0.3 | — | 140 | L, D2, D3 |
| *Data from ecosystem case studies* | | | | | | | | |
| No. 2 (Harste) | 4.45 | 0.42 | 0.04 | 0.08 | 10 | — | 1.9 | — | $\bar{x}$ (1982–1985) |
| No. 6 (Solling) | 4.09 | 0.044 | 0.11 | 0.023 | 0.4 | — | 0.21 | — | $\bar{x}$ (1969–1983) |

**Table 7-11.** Threshold values for aluminum stress, derived from solution culture experiments.[a]

| Spruce | Beech | Effects |
|--------|-------|---------|
| $Ca^{2+}/Al_{inorg}$ mol/mol in solution | | |
| >1 | >1 | Negligible Al stress |
| 1–0.3 | 1–0.3 | Decrease in axial root growth. |
|  |  | Spruce: shortened lifetime and increased turnover of fine roots, resulting in an increase in fine-root necromass; decrease in uptake of Ca and basicity; increase in ramification index? |
| 0.3–0.1 | <0.3 | Increase in damages to the apical meristem, resulting in increased frequency of root derangement points, reduced axial root growth, shortened lifetime of fine roots, and further increase in fine-root necromass. |
|  |  | Spruce: reduction in root tip frequency. In the long run, the reduced axial growth may result in withdrawal of the fine roots from such soil horizons. Strong decrease in Ca uptake. |
| <(0.2–0.1) | ? | A fine-root system can no longer be maintained or established |
| $Mg^{2+}/Al_{inorg}$ | | |
| <0.2–0.3 | ? | Strong reduction of Mg uptake, resulting in decreasing Mg content of leaves. After years, development of Mg deficiency and needle yellowing |

[a] Cf. Table 7-10.

matter content, however, the Al in soil solution exists predominantly in monomeric inorganic form (Dietze, 1985; Schierl et al., 1986; Ares, 1986a, 1986b). Godbold and others (1988c) found that nitrate does not modify the toxicity of Al to spruce seedlings.

In Figure 7–9, the time course of the Ca/Al ratio in 1-m soil depth in the ecosystems 6 and 7 (Solling) is shown. After a period of great variation from 1969 to 1975, the Ca/Al ratio stayed more or less at values below 0.3 and slowly approached 0.1. There is no great difference between the beech and the spruce ecosystem. At 40-cm soil depth, the ratio varies still between 0.1 and 0.3 (data not given). The vertical fine-root distribution (Figure 7–8) shows an exponential decrease approaching zero below 50-cm depth. This is in agreement with the Al stress expressed by the Ca/Al ratio. It is not in agreement, however, with the depth growth of the large-diameter woody roots. They reach as far as 1-m soil depth. From this it can be deduced that during the last 15 years the fine roots in the subsoil could not be maintained, and the rooting became more superficial. In damaged Norway spruce stands in southern Sweden, Puhe and others (1986) found that in mineral soil horizons that were strongly acidified and depleted of basic cations, the large-diameter roots were to a great extent dying off. The same observations have been made by Gehrmann and others (1984) and Ulrich and others (1984).

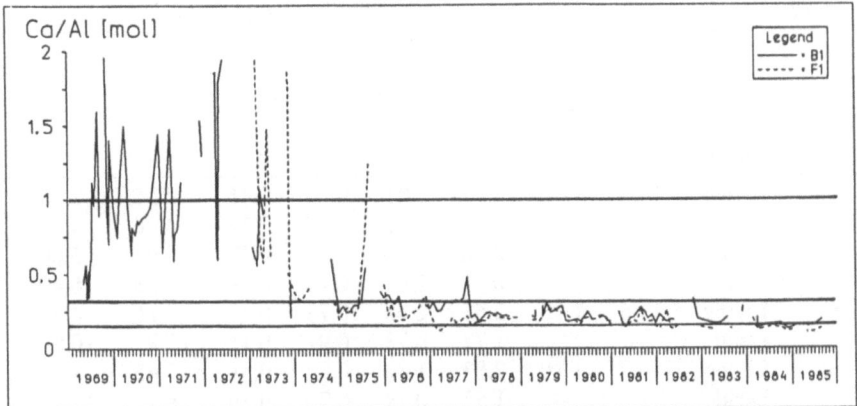

**Figure 7-9.** Time course of Ca/Al ratio in soil solution in 1-m depth of ecosystem 6 and 7 (*B1* = beech, Solling; F1 = spruce, Solling; cf. Table 7-2). The horizontal lines represent the threshold values given in Table 7-11. (From Matzner, 1988.)

As in nutrient solution cultures, the acidity of the soil solution is reflected in the elemental composition of the fine roots. X-ray microanalysis showed high Al contents in walls of root cortex cells (Godbold et al., 1988b). Walls of the mycorrhizal hyphae contained much less Al and Fe, with a minimum content in the walls of the outer layer of the mycorrhizal hyphae surrounding the roots. In walls of the central cylinder, Al and Fe could not be found, but the Ca content of the stele was high. Very small Ca amounts could be detected in mycorrhizal hyphae. Magnesium was detectable only in the stele in very small amounts. These data show that, under the existing conditions at the Solling, the mycorrhizal mantle does not prevent the transport of Al ions into the root cortex. Liming increased Ca and Mg and decreased Al content in the root cortex (Bauch et al., 1985).

Data from dry or wet ashing (see Table 7-12) reveal that the chemical composition of fine roots differs significantly according to soil depth and vitality (Murach, 1988). Especially the transition from the organic top layer (O horizon) to the top mineral soil (0-5 cm) is reflected in changes of Al and Fe content. The highest Fe concentrations are found in the dead roots of the 0 to 5-cm layer, which is in the Fe buffer range. The Al concentrations of live and dead roots increase with depth, with dead roots always reaching higher values. Mobile cations like K and Mn are obviously leached from dead roots. Their low Ca content may be due to a decreased uptake as well as leaching. In the fermentation layer (O1, *Of*), where acidic stress is at its lowest, Ca content is at its highest and does not differ according to vitality. These depth gradients are consistent with results of x-ray microanalysis (Bauch et al., 1985).

The elemental composition clearly reflects the acidic stress and its increase with soil depth. In dead roots, the Ca/Al ratio approaches that of the soil solution. This indicates that the chemical state in the root apoplast is controlled by the soil and not by the symplast. The loss of control of the chemical state of the apoplast by the

Acidic Precipitation

**Table 7-12.** Chemical composition of fine roots in ecosystem No. 6 and 7.

### No. 7 Norway spruce

| Depth | Vitality | N | P | K | Ca | Mg | Mn | Fe | Al | Molar Ca/Al |
|-------|----------|-----|------|------|------|------|------|------|------|-------|
| | | | | | | mg/g | | | | |
| Of | Living | 14.6 | 1.38 | 2.38 | 3.41 | 0.58 | 0.46 | 2.36 | 2.59 | 0.88 |
| Of | Dead | — | — | 0.58 | 3.31 | 0.45 | 0.27 | 4.96 | 4.20 | 0.53 |
| Oh | Living | 15.0 | 1.40 | 2.90 | 2.70 | 0.58 | 0.37 | 4.03 | 3.60 | 0.51 |
| Oh | Dead | — | — | 0.93 | 1.91 | 0.43 | 0.19 | 8.23 | 6.40 | 0.20 |
| 0–5 cm | Living | 11.5 | 1.18 | 2.65 | 2.45 | 0.73 | 0.42 | 8.16 | 4.84 | 0.34 |
| | Dead | — | — | 1.77 | 0.73 | 0.91 | 0.11 | 17.9 | 10.5 | 0.05 |
| 5–10 cm | Living | 10.7 | 1.10 | 2.83 | 1.93 | 0.87 | 0.51 | 7.45 | 7.02 | 0.18 |
| | Dead | — | — | 1.53 | 0.74 | 0.70 | 0.20 | 14.1 | 12.7 | 0.04 |
| 10–40 cm | Living | 9.8 | 0.98 | 2.54 | 1.70 | 0.78 | 1.07 | 4.30 | 10.6 | 0.11 |
| | Dead | — | — | 1.00 | 0.54 | 0.53 | 0.20 | 5.53 | 18.5 | 0.02 |

### No. 6 European beech

| Depth | Vitality | N | P | K | Ca | Mg | Mn | Fe | Al | Molar Ca/Al |
|-------|----------|-----|------|------|------|------|------|------|------|-------|
| | | | | | | mg/g | | | | |
| Of | Living | 16.5 | 0.97 | 1.66 | 3.03 | 0.59 | 0.18 | 0.81 | 0.55 | 3.72 |
| Oh | Living | 13.9 | 0.80 | 1.37 | 1.80 | 0.32 | 0.16 | 1.67 | 1.45 | 0.83 |
| | Dead | — | — | 0.47 | 0.19 | 0.15 | 0.05 | 2.94 | 2.32 | 0.14 |
| 0–5 cm | Living | 17.7 | 1.06 | 1.57 | 0.96 | 0.37 | 0.12 | 5.85 | 3.55 | 0.18 |
| | Dead | 11.1 | — | 1.43 | 0.47 | 0.45 | 0.07 | 17.4 | 7.79 | 0.04 |
| 5–10 cm | Living | — | — | 1.91 | 1.02 | 0.56 | 0.21 | 5.67 | 4.58 | 0.15 |
| | Dead | — | — | 1.40 | 0.45 | 0.55 | 0.13 | 9.99 | 10.0 | 0.03 |
| 10–20 cm | Living | 9.9 | 0.69 | 2.04 | 0.90 | 0.65 | 0.31 | 3.76 | 6.80 | 0.09 |
| | Dead | 6.7 | 0.70 | 1.23 | 0.38 | 0.47 | 0.22 | 5.86 | 13.53 | 0.02 |
| 20–30 cm | Living | — | — | 2.17 | 0.74 | 0.58 | 0.35 | 1.93 | 7.97 | 0.06 |
| | Dead | — | — | 1.14 | 0.34 | 0.42 | 0.28 | 3.34 | 18.2 | 0.01 |
| 30–40 cm | Living | 9.3 | 0.55 | 2.17 | 0.86 | 0.57 | 0.35 | 1.58 | 8.40 | 0.07 |
| | Dead | 8.5 | 0.67 | 1.25 | 0.44 | 0.47 | 0.25 | 3.54 | 19.1 | 0.03 |
| 40–50 cm | Living | — | — | 2.10 | 0.88 | 0.52 | 0.32 | 1.59 | 7.64 | 0.08 |
| | Dead | — | — | 0.75 | 0.27 | 0.29 | 0.15 | 2.18 | 19.4 | 0.01 |

From Murach and Ulrich, 1988.

symplast as a consequence of acidic load exceeding buffer ability seems to be the cause of fine-root injury and fine-root necromass.

Similar results have been found in other studies. With x-ray microanalysis, Bauch (1983) and Stienen and others (1984) showed that in the fine roots of spruce trees exhibiting needle loss, Al and Fe had accumulated, and Ca and especially Mg had decreased. In Figure 7–10, the Ca/Al molar ratio of live fine roots of various spruce decline case studies in Germany is shown. With the exception of the sites Odenwaldstetten, Sauerlach, and Selb, the soils are in the Al buffer range

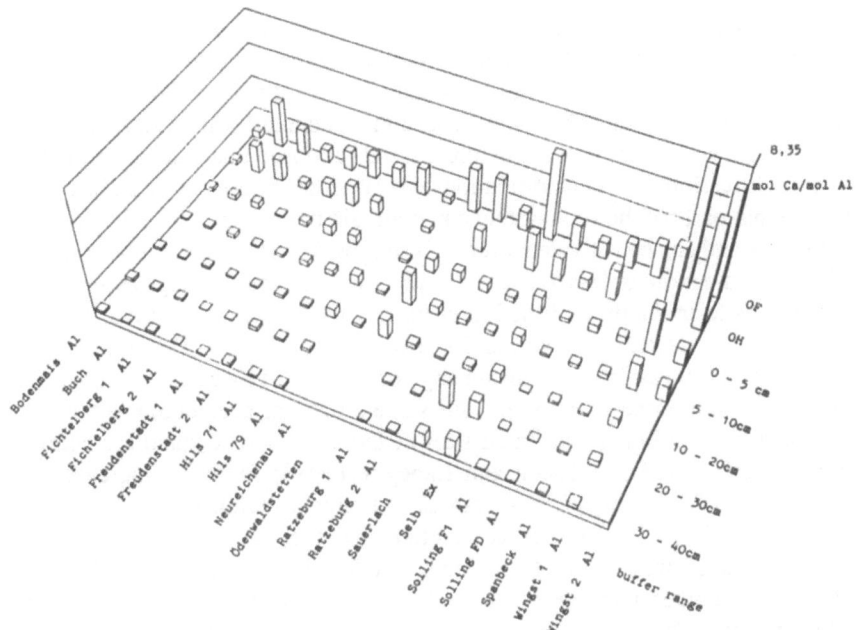

**Figure 7–10.** Ratios of Ca/Al in live fine roots of spruce-decline case studies in Germany. The names represent the case studies. *Ex* = soil below A horizon in cation exchange buffer range; *Al* = soil in Al buffer range. (From Murach and Wiedemann, 1988.)

throughout the whole rooting zone. This is reflected in the low values of the Ca/Al ratio, indicating the same form of stress as in the Solling. In the topsoils, where Al is complexed, the Al stress seems in most cases much lower. Of the dead fine roots of the same sites, 90% have Ca/Al ratios between 0.05 and 0.14 (Murach, 1988). These data show that Al stress seems to be a common feature throughout Germany. For sites in the Fichtelgebirge, Meyer and others (1988) found a significant correlation between the Ca/Al ratio in the soil solution of the top 20 cm of soil (range 0.01 to 0.6) and the number of root tips, as well as the number of mycorrhizal root tips of Norway spruce. In damaged Norway spruce stands in southwestern Sweden, it was suggested that the proportion of live and dead roots was dependent on acidification variables like base saturation and Ca/Al ratio in the roots (Puhe et al., 1986).

Cronan and others (1987) compared the solution chemistry of 13 watersheds in America and Europe. The ratio of maximum monomeric Al to Ca ($Al_{monomeric}$/Ca) varied between 31 and 0.27, the three European case studies (Solling, Lange Bramke, Gardsjon) being <0.33. They assume that Al toxicity to trees results not so much from short-term exposure to elevated levels of aqueous Al as it does from the integrated effect of chronic exposure to soluble Al. As such, root tissue chemistry might provide a better-integrated index of Al toxicity potential than instantaneous measurements of monomeric Al in the rooting zone. Regardless of species or soil horizon sampled, very fine roots (<1 mm) had two to three times higher Al concentrations than slender roots (1–3 mm). Furthermore, fine roots from the organic-rich surface horizons had significantly lower Al concentrations than those from mineral B horizons. The authors conclude that Al toxicity may be an important stress factor in some forest stands exposed to acidic deposition.

In acidification and deacidification experiments, Stienen and Bauch (1988) showed that parts of the root system can maintain their vitality under the Al stress for months and can be activated after Al is removed from the rhizosphere solution. Other roots die more or less quickly. Despite this fact, the mucilage as a protective layer seems to maintain its function at least partially for a long time, so certain parts of the roots can survive acidification pulses in soils. Deacidification by raising pH to 6.0 resulted in a reexchange of aluminum by calcium, magnesium, and sodium in the fine-root cortex. The same effect has been demonstrated when fine roots from unlimed and limed sites of the same spruce stand are compared (Bauch et al., 1985). LeTacon and Lapeyrie (1988) found an increase in the mycorrhizal infection of spruce seedlings by liming an acidic soil (pH(H$_2$O) 3.5). In earlier liming trials on spruce, Kottke and Oberwinkler (1988) found an increased rate of production of fine roots, a reduced lifespan of mycorrhizae, and a decreased number of mycorrhizal species (Müller and Oberwinkler, 1988) on the unlimed plots. The rapid aging of the mycorrhizae facilitates root infection by pathogenic soil fungi. This is in agreement with the findings of Schlechte (1986, 1987) that soil acidification by acidic deposition has caused a dramatic decrease of efficient mycorrhizal species never observed in Central Europe before 1970. In the Netherlands, conifers on sites of higher acidic deposition show a lower fruiting body production, lower number of species fruiting, lower mycorrhizal infection, a lower number of mycorrhizal types, and a higher number of root tips without mycorrhizal mantle (Arnolds, 1985; Arnolds and Jansen, 1987; Jansen, 1988; Roelofs et al., 1988). The cause can be either enhanced aluminum or enhanced ammonium concentrations (Boxman et al., 1986).

The view of great variability in a root system under acidic stress in respect to injury and tolerance is in agreement with attempts to relate the short-term (seasonal) dynamics of the fine-root biomass and necromass in the spruce ecosystems 4 and 7 to the variation of the growth and stress factors (soil temperature, soil humidity, starch content of roots as a measure of assimilate supply, ion concentrations in soil solution as a measure of nutrient supply and acidic stress). Decreases in fine-root biomass and corresponding increases in necromass could be related to the acidic stress parameters (Murach, 1984). Due to ecosystem internal acidification pulses, which are mainly caused by a higher rate of nitrification compared to the rate of nitrate uptake (Ulrich, 1981b; Matzner and Ulrich, 1985), the acidic stress parameters in the soil solution can greatly vary, even in already strongly acidified soils (cf. section IV). Under field conditions, roots pass through phases of injury and recovery. In acidic soils, high fluctuations of the fine-root concentration in the soil are related to the fluctuations of the proton and metal concentrations in the soil solution (Murach, 1984; Raben, 1988). This creates a high temporal variation that makes it difficult to assess long-term changes from short-term measurements. According to Bartsch (1987), the regeneration ability of spruce roots after a drought period is slowed down by preceding acidic stress.

The adverse conditions for root growth in the acidic soil of ecosystem 7 have been tested by ingrowth core studies, using the same soil after liming as well as nonacidic soil. The amount of roots that had grown into the limed ingrowth cores were much higher, and the percentage of dead roots was less when compared to unlimed treatment. The fertilized peat-sand soil provided optimal conditions for root growth as indicated by maximum root biomass and zero percent root mortality (Matzner et al., 1986).

Fine roots of declining spruce trees with an unfavorable base status (low Ca + Mg, high Al content, symptom: needle loss) developed cortex cells with a reduced diameter and thicker cell walls; in addition, accessory compounds were accumulated in this presumably protective tissue. Tannins were deposited in the parenchyma of the vascular cylinder, and many pit membranes of the primary xylem often did not differentiate fully (Stienen et al., 1984). Severely damaged trees possess more fine roots with small, opaque tip zones than do slightly damaged trees. Meristematic cells of these tips show an intensive vacuolization; this probably indicates a lower root vitality (Schmitt and Liese, 1987).

## D. Long-term Consequences of Nutrient Cation Losses from Soil

Forest ecosystems accumulate considerable amounts of nutrients in their standing biomass. Data are given in Table 7–13 for ecosystems 6 and 7 (beech and spruce in Solling). These data can be compared with the storage of exchangeable K, Ca, and Mg (see Tables 7–2 and 7–4). The comparison shows that in the majority of forest soils the cation storage is now comparable or even smaller than the amount of nutrients accumulated in the biomass of a timber tree. Therefore, nutrient deficiencies can be expected. The nutrient demand of timber trees, however, is covered by the increased nutrient deposition due to air pollution (Ulrich et al., 1979). Aggrading stands are confronted with a different situation: The nutrient input by deposition is less, whereas the amount of nutrients fixed in the aggrading biomass is higher. On soils in the Al/Fe buffer range, they develop only a

**Table 7-13.** Nutrient storage in the aboveground biomass in a European beech and a Norway spruce stand in the Solling (ecosystems 6 and 7 of Table 7-2).

|        | N    | P    | S   | K   | Ca  | Mg  |
|--------|------|------|-----|-----|-----|-----|
|        | kg/ha | | | | | |
| Beech  | 600  | 90   | 60  | 270 | 310 | 70  |
| Spruce | 730  | 70   | 80  | 400 | 410 | 60  |
|        | kmol (+)/ha | | | | | |
| Beech  | 42   | 2.9  | 3.7 | 7.0 | 15  | 5.6 |
| Spruce | 52   | 2.4  | 4.8 | 10  | 21  | 4.6 |

superficial root system, so the leaching of deposited nutrients with soil water out of the rooted soil is increased.

The rooting pattern of young Norway spruce stands in soils in the Al/Fe and Al buffer range is shown in Table 7–14. The years 1982 and 1983 were warm and dry with water stress (drought) and high acidic stress (climatic acidification pulse). Then 1986 was the third successive cool and/or wet year, enabling recovery of the root system. The vertical distribution pattern, however, is not changed: The mass of the root system remains restricted to the organic top layer, where the trees develop horizontal roots extending over more than 10 m in length. Only in the O horizons is the lifetime of live roots long enough for them to become the dominant fraction. The time course of the concentrations of live and dead fine roots reveals that the lifetime of fine roots in the mineral soil is very short. The superficial rooting is caused by the acidic stress affecting the roots in the humus layer (*Oh* horizon, proton stress) and especially in the mineral soil. There is no chance that these trees will ever develop a deep root system as they would have done in old existing spruce stands.

The superficial rooting also prevents the uptake of nutrient cations released by weathering. Nevertheless, N input from deposition at least in western central Europe is high enough to stimulate growth as long as the acidic stress and the cationic nutrients deposited and released by litter decomposition provide a sufficient nutrient supply. In ecosystem case study 8, the nutrient that limits growth is potassium: The trees show very little K contents in all plant parts (Schulte-Bisping and Murach, 1984).

The generalization of these results leads to the prognosis that many aggrading forest stands in central Europe will, during the next years and decades, run into increasing nutrient deficiencies and acidic stress. Which nutrient first becomes deficient depends upon the local deposition and soil conditions. The needle yellowing in pole stands and young spruce timber trees (see section XI,A) is the first widespread symptom of this kind.

## X. Heavy Metals: Input–Output Relations and Possible Effects

Heavy metals in cationic form belong to the group of cation acids. They are constituents of acidic deposition and may be a component of acidic stress. The fluxes of heavy metals have been assessed for a few forest ecosystems, including ecosystem case studies 2 through 7 and 9 (Schultz, 1987).

As an example, the fluxes of Cr, Co, Ni, Cu, Zn, Cd, and Pb in a Norway spruce stand (*Picea abies*) in the Solling (ecosystem 7, Table 7–2) are shown in Figure 7–11 (from Schultz, 1987). In the following discussion of the behavior of heavy metals, additional data from other forests on acidic soils are taken into consideration.

Deposited Cr is in general accumulated in the ecosystem. The intercepted Cr is deposited mainly as water-insoluble compounds in aerosols (Schmidt et al.,

**Table 7-14.** Fine-root distribution in a young spruce stand on soil in Al/Fe buffer range (ecosystem study 8 of table 7-2).[a]

| | Fine root mass kg DM/ha | | Distribution in % of fine-root mass | | | | | | |
| | | | Total | Of | Oh | 0-5[b] | 5-10[b] | 10-20[b] | 20-40[b] |
|---|---|---|---|---|---|---|---|---|---|
| 1983 (age 13) | 2,575 | Living | 31 | 20 |  | 9 | 2 | 0 | 0 |
| | | Dead | 69 | 7 |  | 20 | 21 |  | 21 |
| 1986 (age 16) | 5,370 | Living | 55 | 21 | 20 | 9 | 3 | 2 | 0 |
| | | Dead | 45 | 4 | 20 | 11 | 4 | 4 | 2 |

[a] Adopted from Schulte-Bisping and Murach, 1984; Gerriets and Ulrich, unpublished.
[b] Soil depth in cm.

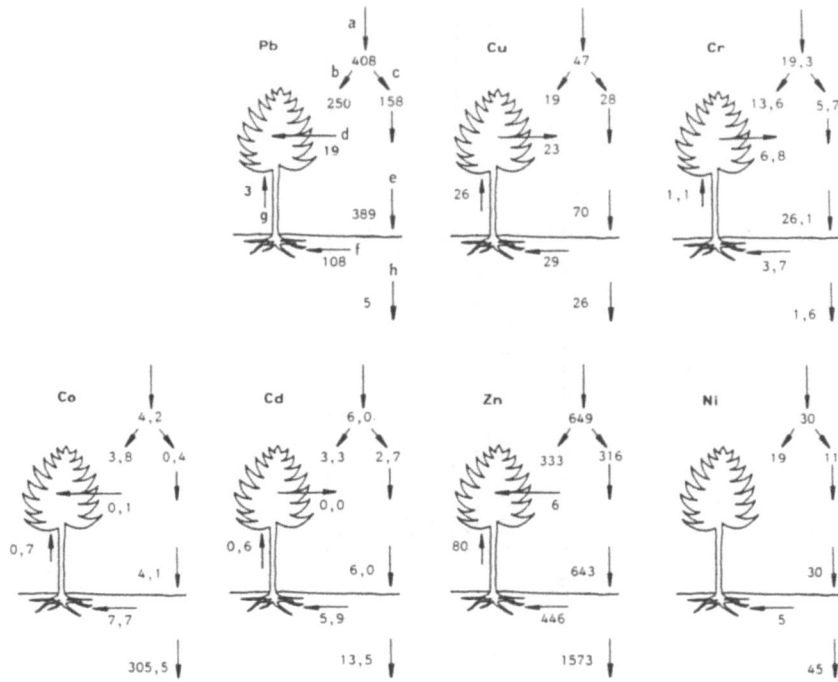

**Figure 7–11.** Fluxes of Pb, Cu, Cr, Co, Cd, Zn, and Ni in Norway spruce ecosystem 7 from Table 7–2. Values given in $g \cdot ha^{-1} \cdot yr^{-1}$. $a$ = total deposition; $b$ = interception deposition; $c$ = precipitation deposition; $d$ = canopy leaching ($\rightarrow$) and adsorption in canopy ($\leftarrow$); $e$ = soil input; $f$ = root uptake; $g$ = translocation in tree; $h$ = seepage output. (From Schultz, 1987.)

1985). A small fraction of the deposited Cr is washed off; leaching is negligible. The main fraction of deposited Cr is adsorbed in the canopy and reaches the soil with litter fall; part of it accumulates in the canopy, however. The output of Cr from the ecosystem with seepage water is minimal on loamy soils but reaches the level of deposition on sandy soils.

Cobalt shows a negative ecosystem balance in most forests with acidic soils. Deposition is small and contributes little to the turnover in the ecosystem. The soils continuously lose Co with the seepage water. In the forest shown in Figure 7–12, the mobilizable Co storage in soil compensates the net loss from soil for a period of only 16 years.

Deposited Ni is adsorbed in the canopy and accumulates in the bark. The balance of the soil is usually negative (output exceeds input). Compared to the turnover, however, the loss is minimal.

In all forests investigated, Cu accumulates in the mineral soil and the top organic layer. The accumulation in biomass is small; most of the Cu taken up is either leached or reaches the soil with litter fall. Whereas the Cu content in spruce

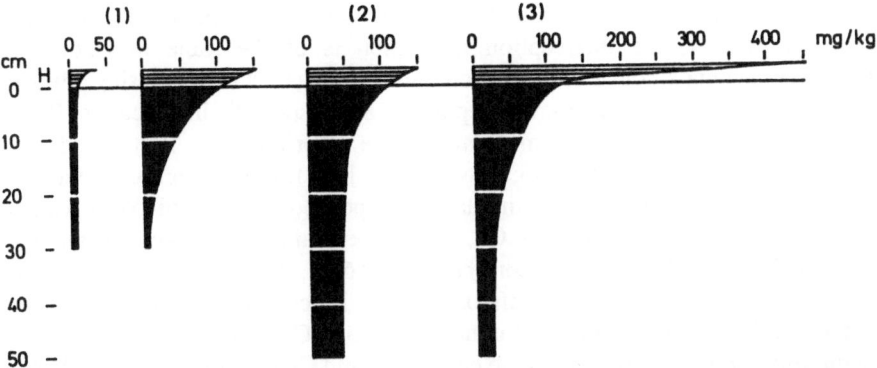

**Figure 7–12.** Depth gradient of Pb concentration in European forest soils. $H$ = top organic layer; *cm* = depth of mineral soil. (1) Central Norway (*left*) and South Norway (*right*) (Solberg and Steinnes, 1983); (2) South Germany (Black Forest, Bärhalde) (Keilen, 1978; Raisch, 1983); (3) Central Germany (ecosystem 7 of Table 7–2) (Mayer, 1981, 1985).

needles approaches the level of deficiency (Schultz et al., 1987, 1988), the Cu content in the top organic layer (O horizon) of the soil approaches the level of toxicity (Tyler, 1972).

The fluxes of Zn vary considerably between different forests, depending upon the acidic strength as indicated by pH. In ecosystems 2 and 3 (cf. Table 7–2), there is no net loss from the soil. In the other ecosystems, however, the Zn budget of the soil is negative. This is most pronounced in ecosystem 7 (Figure 7–11). The deposition increases with rainfall and therefore with altitude. The interception deposition seems to depend strongly on the exposure of the forest. The content in needles is generally low and approaches the level of deficiency. The balance of the soil is generally negative. Soil acidification may thus in the long term lead to Zn deficiency in the vegetation.

In acidic forest soils, Cd is mobilized and leached. The input-output balances of the ecosystems as a whole as well as of the soil are negative. Spruce forests show a considerable rate of Cd uptake from soil in contrast to deciduous forests. The different uptake pattern is reflected in different Cd contents in the wood and may be related to higher Cd concentrations in the soil solution. The Cd now mobilized in the soil probably stems from deposition and had been accumulated before soil acidification reached the Al buffer range.

Almost the total Pb that is deposited is accumulated within the ecosystem. The seepage output of Pb is always very low and shows no trend (for the ecosystem depicted in Figure 7–11, seepage outputs are measured since 1972). Most of the lead accumulates in the top organic layer (see Figure 7–12). The deposition of Pb has been decreasing since 1974, according to the decrease of the lead content in fuel. In Figure 7–12, the decrease of deposition from central Germany (spruce Solling, ecosystem 7) to southern Germany (Bärhalde) and Scandinavia (Norway) can also be seen.

The deposition rates of Pb, Cd, Cu, and Zn increase with increasing precipitation. In general, the interception deposition is of the same magnitude as precipitation deposition. Because precipitation increases with increasing altitude, this creates a strong dependency of deposition and, especially in the case of Pb, of accumulation in the ecosystem from altitude (Korff et al., 1980; Godt and Lunkenbein, 1983; Glavac, 1986; Glavac et al., 1988): The higher the altitude, the greater are the deposition and accumulation. Deposited Cd, Zn, and Co are washed off from the canopy, whereas Pb, Cu, and Cr are retained. This is connected with increased content in the bark (Lötschert and Köhm, 1978; Schreiber and Rentschler, 1988). Clear air and washoff experiments in ecosystems 6 and 7 (beech and spruce, Solling) have shown that uptake of Pb and Cr from soil into the plant is negligible and is insignificant in the case of Cd and Cu (Godt et al., 1986, 1988).

With increasing soil acidification, the mobility of heavy metals increases. This applies particularly to those metals, like Co, Cd, and Be, that do not form strong complexes.

Lead, Cr, As, Hg, Ag, and Cu form stable complexes with organic matter and thus accumulate in the top organic layer of forest soils. A mobilization and downward movement of these elements is governed by the mobilization of soil organic matter. The presence of dissolved humic substances in the soil solution is a characteristic feature of soil horizons in the Al/Fe and Fe buffer range. This applies to the topsoils of ecosystem case studies 4 through 9 (cf. Table 7–2). Soil internal acidification pulses by net nitrification therefore usually increase the heavy metal concentrations in the soil solution and the leaching rates (Lamersdorf, 1989).

For central Europe, Table 7–15 gives the range of deposition, accumulation, and concentrations of some heavy metals in forest floor organic matter and fine roots. High rates of deposition are reached in areas close to emission sources, as well as in the more distant high altitudes. The maximum values given for the accumulation in forest floor represent beech timber trees in highly industrialized areas (Gehrmann, 1984; Rastin et al., 1985). Zezschwitz (1986) has shown that in northwestern German forest top soils, the content of Pb and Cu has increased since the mid-1960s, whereas Zn has decreased on spruce sites in the stronger acidified soils. This is reflected in the distribution of metals in annual rings of beech trees: Ca, Mg, Mn, and Zn decrease in concentrations, whereas Fe and Al increase in younger rings (Meisch et al., 1986).

With respect to adverse effects, there are no hints to direct effects of deposited heavy metals on leaves. If their accumulation in the forest floor hinders litter decomposition, one would expect that a threshold concentration is reached. This is indeed the case (Mayer, 1985). A concentration of 500 mg Pb/kg dry matter is exceeded only in areas close to industrial activities emitting Pb dust (Koenies, 1982/85). Comparable data are reported from North America (Parker et al., 1978; Smith and Siccama, 1981; Friedland et al., 1984).

From Table 7–15, it can be seen that the concentrations in fine roots approach or exceed the maximum concentrations in the forest floor organic matter. In the case of the easier acidic soluble elements such as Cd and Zn, the dead fine roots show

**Table 7-15.** Heavy metals: Range of deposition and accumulation and concentration in forest floor organic matter and fine roots. [a]

| | Pb | Cd | Cu | Zn | Cr |
|---|---|---|---|---|---|
| Deposition on open land (g · ha⁻¹ · yr⁻¹) | 70-140 | 2-4 | 10-50 | 70-300 | 2-4 |
| Deposition in forests (g · ha⁻¹ · yr⁻¹) | 150-400 | 3-6 | 40-90 | 300-700 | 5-50 |
| Accumulation in forest floor (kg · ha⁻¹) (x) | 0.7-60 (20) | 0-0.2 (0.05) | 0.2-8 (0.3) | 0.5-40 (15) | 1-20 (3) |
| Concentration in Oh (mg · kg⁻¹)[b] | 100-500 | 0.2-3 | 10-100 | 50-200 | 10-50 |
| Concentration in live fine roots[b] (mg · kg⁻¹) | 50-400 | 4-12 | 2-20 | 50-240 | 0-70 |
| Concentration in dead fine roots[b] (mg · kg⁻¹) | 50-800 | 1-4 | 2-35 | 10-120 | 0-120 |

[a] Adopted from Schultz, 1987; Lamersdorf, 1988; and Schultz et al., 1988.
[b] Data from ecosystems 7 and 9 of Table 7-2.

lower contents than the live ones. Live fine roots can accumulate Cd: $Cd^{2+}$ in acidic soil solution seems to be transferred to insoluble compounds at the higher pH existing in the free space of vital fine roots. If these roots become injured due to acidic stress, pH may decrease and Cd as well as Zn will be leached. This implies that inside the root cortex much higher Cd and Zn solution concentrations can build up in case of acidic stress to a vital fine root. According to Godbold and others (1985, 1986) and Godbold and Hüttermann (1985), Godbold et al., (1988a) a comparison of the levels of Hg, Cd, Pb, and Zn shown to inhibit root elongation in nutrient solution with those in forest soils under stands showing symptoms of dieback suggests that the metal levels in such soils are sufficiently high to influence root growth. In Pb-treated Norway spruce seedlings, Pb was found mainly in cell walls and electron-dense deposits; in field samples, in cell walls and in vacuolar deposits (Godbold et al., 1988a). The toxicity of Pb is dependent upon the composition of the nutrient solution, a greater inhibition of root elongation was found in a nutrient-poor artificial soil solution than in a full nutrient solution (Godbold et al., 1988c).

The available data suggest that direct adverse effects of deposited heavy metals on leaves are not to be expected. However, effects on soil organisms in the top organic layer of forest soils and on roots cannot be ignored. The probability for such effects to occur increases with increasing soil acidity. Heavy metal toxicity is thus one aspect of acidic stress that may not be distinguishable from Al and proton stress in the ecosystem case studies.

# XI. Cause-and-Effect Relationships in Tree Damage on the Basis of Case Studies

Tree dieback and forest decline in Europe may show symptoms of

1. Pathogen and insect attack
2. Leaf (needle) loss (crown thinning) without participation of pathogens, insects, nutrient deficiency (according to leaf analysis), and mechanical effects
3. Nutrient deficiency, which may be followed by leaf (needle) loss. On acidic soils, nutrient deficiency has been demonstrated in respect to Ca, Mg, K, Mn, Zn (Zöttl and Mies, 1983; Zöttl and Hüttl, 1985; Hüttl, 1985; Gehrmann et al., 1984; Evers and Schöpfer, 1988; Fiedler, 1988) and other micronutrients. Micronutrient deficiency, especially boron deficiency, has been found in Finland (Huikari, 1977) and Sweden (see Kolari, 1983) to cause growth disturbances. In the central and southern parts of Germany, Mg deficiency in spruce causing needle yellowing is widespread (see below). From calcareous sites in the Alps, K and Mg deficiency are reported (Forschungsbeirat, 1986). On deciduous trees and shrubs, intercostal chlorosis due to Fe and Mn deficiency has been reported from Germany (Rayermann and Schlote, 1986).
4. Changes in branching (*Fagus sylvatica* and other deciduous species: Roloff 1986, 1988, 1989; Norway spruce: Gruber 1987a, 1987b). The changes in

branching in beech trees now showing signs of dieback have often started 30 to 40 years ago. This indicates that we are confronted with a long-term dieback process that proceeds slowly. This renders it difficult to investigate causal dose-effect relationships within a few years of research.

5. Decline in radial growth of trees (tree-ring width). The narrowing of tree rings can have started decades or only a few years before the process of dieback sets in.

As described in the introduction, a cause-and-effect relationship in the long-lived forest ecosystem cannot be deduced from simple correlations, for instance, between nutrient concentration in needles and appearance of the tree. It has to be shown how the input vector, which may be modified by air pollutants, changes the state vector, which in turn changes the transfer functions. As acidic deposition is a widespread phenomenon in the whole of Europe, the assessment of acidic stress in soil and roots is indispensable for the evaluation of possible cause-and-effect relationships. If such data are missing, the contribution of acidic stress, either acting directly through leaves or indirectly through soil, cannot be rejected. Such data do exist, however, only for case studies showing leaf (needle) loss (without participation of pathogens, insects, and nutrient deficiency) and Mg deficiency (yellowing of spruce needles with subsequent needle loss). The discussion of possible cause-and-effect relationships is therefore restricted to both of these cases.

## A. Needle Yellowing of Norway Spruce Connected with Magnesium Deficiency

### 1. Symptoms and Development of the Damage

The typical symptom is the yellowing of older needles (Forschungsbeirat, 1986). In the initial stage only the upper faces of branches that are exposed to direct sunlight turn yellow. Later the yellowing may spread from the interior and the basis of the crown to its outer face and the top. The discoloring starts at the needle tip and continues to the middle part of the needle, whereas the base of the needle may remain green. The transition from the yellow to the green part is sudden. Usually the yellow needles turn brown and then shed. It has also been observed, however, that needles that are not completely discolored turn green again. This type of yellowing is always connected with Mg deficiency, as indicated by low Mg content in the needles.

Histological and histochemical evaluations (Fink, 1983, 1987; Cape et al., 1988; Paramesvaran et al., 1985) have shown no necrosis of mesophyll cells, but a specific necrosis and collapse of the sieve cells occur in the phloem of the vascular bundle. Cambial cells and adjacent parenchyma cells, however, show hypertrophy and hyperplasy. The mesophyll exhibits a general and uniform accumulation of starch. This symptom is always found in combination with yellowing symptoms of the older needles, though it also occurs in needles that are still green. The symptom found in needles collected in the forest agree with those from Mg-deficiency experiments. With the present knowledge, these pathological changes can be attributed to disturbances in mineral nutrition (especially lack of magnesium) rather than to a direct impact of gaseous air pollutants. The phloem necrosis blocks the translocation of assimilates from the needles to the stem and thus induces a pathological starch accumulation in the chloroplasts of the needles. This starch accumulation also occurs in needles that are still green but may be on the verge of turning yellow.

Especially in needles from higher elevations, a flecking type of damage can appear. Originally they were considered to be caused by ozone; this, however, is now being called into question (Senser et al., 1987). Histologically (Cape et al., 1988), these flecks differ from those occurring in ozone-fumigated needles: Whereas the latter are always related to stomata and consist of totally collapsed cells, the former are mainly found along the edges of the upper side of the needle; the cells are dead, although not collapsed, and thus appear to be freeze-dried. The dead cells are in most cases full of starch. It seems highly likely that at least some of these flecks are due to climate (winter fleck injury) and not necessarily related to the direct impact of air pollutants (Cape et al., 1988).

As a widespread phenomenon occurring not only in plantations but also in spruce stands of different ages, the needle yellowing was first observed in the early 1970s in the Giant Mountains (USSR, Kandler et al., 1987) and in the late 1970s in the Fichtelgebirge (FRG, Zech and Popp, 1983). It spread out in the medium altitudes in the warm, dry years of 1982 and 1983 (Prinz et al., 1982; Bosch et al., 1983, Zöttl and Mies, 1983; Hauhs, 1985; Nebe et al., 1987). In the cool, wet years of 1985 and 1986, partial regreening was observed (in the Bavarian forest, Kandler et al., 1987), whereas in other areas (Harz, Hartmann et al., 1986) the extent of yellowing increased steadily. A decrease in the magnesium content in spruce needles has been observed for 1 to 2 decades (Reemtsma, 1986). Between 1986 and 1987 (again a wet and cool year) in the FRG, the percentage of spruce area showing needle yellowing on more then 10% of the needles decreased from 7% to 5.4% (Bundesministerium f. Ernährung, 1987). This shows that recovery is possible if the stress is lowered.


## 2. Performance of Spruce at Different Stages of Decline: The Case Study Fichtelgebirge

The Fichtelgebirge is located in the eastern part of the FRG at the Czechoslovakian border. Since 1980, *Picea abies, Abies alba*, and *Larix decidua* have exhibited needle-tip chlorosis and necrosis in the area. Affected trees usually occur on acidic soils above an altitude of 700 m. Needles of affected spruce trees had ample N and P concentrations. However, needle concentrations of Mg were very low, and those of Ca, K, and Zn of affected trees were relatively low. Visible leaf injury in *Abies alba, Larix decidua, Fagus sylvatica*, and *Acer platanoides* was connected with low concentrations of Mg in the leaves (Zech et al., 1985).

In these areas, two spruce stands aged 30 to 35 years in 675 to 755-m altitude (precipitation $1,100–1,400 \ 1 \cdot m^{-2} \cdot yr^{-1}$) have been selected for the investigations: site Oberwarmensteinach as a severely damaged stand and site Wülfersreuth showing minor damage. Most results reported in the following originate from these two sites. In addition results from comparable spruce stands in the Black Forest (Zöttl, 1987; Fink, 1988b; Hüttl, 1985) are included.

According to Lange and others (1985, 1986), in yellow needles the photosynthetic capacity is reduced by about 50%. This applies to the comparison between undamaged (100% green needles) and damaged trees as well as between green and yellow needles from the same branch of a damaged tree. The photosynthetic capacity of the green needles shows a slight decrease with increasing needle age. The stomates respond to air humidity irrespective of needle age (Zimmermann et al., 1988). The photosynthetic capacity of 1-year-old yellow needles on damaged trees is reduced, as stated above, but the age trend is no longer noticeable. The lack of age trend in photosynthesis rules out a direct effect of air pollutants, as in fumigation experiments with 150 $\mu$g $O_3 \cdot m^{-3}$ and 90 $\mu$g $SO_2 \cdot m^{-3}$ a decrease in photosynthesis and an accumulation of starch was found in the needles of the second year's class (Küppers and Klumpp, 1988). The yellowing is accompanied by low Mg content in the needles, and other biomass components (Oren et al., 1988a). According to Küppers and others (1985) and Schulze and others (1987), photosynthesis declines at a threshold of needle magnesium content with less than 0.3 mg $\cdot$ g$^{-1}$. The authors conclude that the damage of the phtotsynthetic ability of needles is not directly caused by air pollutants but by systemic damage of the tree. The results presented by Benner and Wild (1987) on Norway spruce on the Hunsrück Mountain showing needle yellowing do not contradict these findings. By comparing damaged trees and trees without symptoms, these authors found a reduced rate of photosynthesis and photosynthetic capacity, whereas the transpiration rates differed only slightly. Distinct damage to photosynthesis was observed only when the chlorophyll content of the needles was markedly lowered. These findings are in agreement with those from the Fichtelgebirge and Black Forest. According to Dietz and others (1988), the photosynthetic electron transport chains in the thylakoid membranes of the chloroplasts are the sites of early injury. A lower chlorophyll content and this membrane damage were apparent even in green needles from the damaged spruce trees. In this case the primary cause could also be a systemic damage of the tree because the lowering of the chlorophyll content should precede the yellowing.

The fact that the symptoms appear first on older needles is explained by the transport of Mg from the older needles in the new flush (Mies and Zöttl, 1985; Oren et al., 1988b). An experimental elimination of the new flush indeed prevented the outbreak of the symptoms as well as the reduction in photosynthetic capacity (Lange et al., 1987). The authors conclude that the bleaching and the loss in photosynthetic capacity result from nutrient deficiencies through soil environment changes and/or root damage. Magnesium fertilization increases the Mg content in needles and results in a revitalization of 40-year-old spruce trees (Zöttl, 1987; Kaupenjohann et al., 1987; Hüttl and Fink, 1988; Fiedler et al., 1988). Throughout a 4-month measuring period, the daily rates of $CO_2$ uptake of the unfertilized tree exhibiting symptoms were markedly lower than that of the

Mg-fertilized tree, whereas the transpiration rates for both trees were very similar (Beyschlag et al., 1987). Also Werk and others (1988) found no significant difference in transpiration between "green" and "yellow" stands. Thus, the water use efficiency of the unfertilized tree was noticeably lower. As the stomatal behavior of both needle types was found to be almost identical, the observed difference must be caused by the reduction in photosynthetic capacity, which is at least partly due to a 50% lower chlorophyll content. In addition, a lower carboxylation efficiency was found in the damaged needles of the unfertilized tree. This indicates damage of the photosynthetic apparatus (loss of the activity of the $CO_2$ binding enzyme).

The fact that an increase of Mg in the soil by fertilization leads to a marked recovery of damaged trees supports the hypothesis that the decline phenomena in the Fichtelgebirge are the expression of a systemic damage of the trees due to changes in the root zone. The soils are in the aluminum buffer range throughout the whole rooting zone. The actual acidic stress can be assessed using the threshold values for the Ca/Al and Mg/Al molar ratio in the soil solution of soil horizons low in organic matter content (see Table 7–11). In the unfertilized plot showing damage, the two-year mean value of the Ca/Al ratio is 0.25 and the Mg/Al ratio is 0.09 (Hantschel, 1987). This indicates damage in the apical root meristem and strong reduction of Mg uptake. Due to Mg fertilization, the Mg/Al ratio changed drastically to values between 0.3 and >1.0. In the undamaged stand (site Wülfersreuth), both ratios are >1.0 and thus beyond the stress range. The appearance or nonexistence of the Mg deficiency and yellowing is thus in agreement with the results of solution culture studies presented in section IX,C. Meyer and others (1988) found a highly significant decrease of the number of root tips in 0 to 20-cm soil depth with decreasing Ca/Al ratio (range 0.6 to 0.3), as is to be expected (cf. section IX,C). There is no effect on the frequency of mycorrhizae. The Ca and Mg content in the needles is, however, significantly positively related with the number of ectomycorrhizae per leaf area. The nutrient relationship between root and shoot becomes obvious through analysis of the xylem sap. The Ca and Mg content in the 1-year-old needle increases proportionally to the concentration of both ions in the xylem sap (Osonubi et al., 1988).

The yellowing symptom may appear in trees with excellent growth. Its appearance is usually not preceded by long-term growth reduction. This is also valid for the Oberwarmensteinach site. The growth is driven by nitrogen deposition (22 kg $N \cdot ha^{-1} \cdot yr^{-1}$), whereas the accumulating biomass of the aggrading stand needs only 13 kg $N \cdot ha^{-1} \cdot yr^{-1}$. Eleven kg $N \cdot ha^{-1} \cdot yr^{-1}$ are leached as nitrate with the seepage water, in spite of a well-developed fine-root system in the mineral soil (68% of the fine-root mass of 1.640 kg $\cdot ha^{-1}$ is located in the mineral soil; Schulze et al., 1987). This indicates that growth is not limited by nitrogen.

Acidic deposition amounts to a mean annual rate of 2.2 kmol $H^+ \cdot ha^{-1}$ + 0.9 kmol $NH_4^+ \cdot ha^{-1}$ = 3.1 kmol $ha^{-1} \cdot yr^{-1}$. This corresponds to 63% of the acidic load in the Solling (cf. section III,A). Assuming the same percentage, the acidic deposition accumulated at the Oberwarmensteinach site since the beginning of industrialization is estimated at 160 kmol $H^+ \cdot ha^{-1}$ (63% of the value calculated for ecosystem 7). If one compares this figure with the storage of exchangeable Ca + Mg + K + Na in the mineral soil (25–35 kmol (+) per ha and 80-cm depth, calculated from data given by Süsser, 1987), it becomes evident that acidic deposition must have caused a drastic change in the acid-base status of the soil, resulting in very low storage of exchangeable bound nutrients, and, finally, in Mg/Al ratios in the soil solution, preventing Mg uptake. The development of the spruce stand under these conditions shows that Mg is not a growth-regulating factor. Growth is regulated by the nitrogen supply. If Mg is not sufficiently available or its uptake is impossible, the tree utilizes the Mg stored in old needles for new growth. Once this is depleted, growth continues until Mg deficiency in the needles causes a decrease in the chlorophyll content and damage of the photosynthetic apparatus. Acidic deposition is the driving force (Ulrich, 1983a) for this development.

## 3. Driving Force of the Soil Changes: The Case Study Lange Bramke (Harz)

In the Harz Mountains, the appearance of needle yellowing is correlated with the parent material of the soil (the geology): On poor sandstone and granite, 30% to 40% of the spruce stands show the yellowing symptom; on richer parent rocks, the yellowing is less widespread (Stock, 1988). Sometimes damaged (yellow) and undamaged (green) parts of spruce stands are clearly separated. Investigations (Blanck et al., 1988) showed that the soil of the undamaged parts is influenced by spring and creek waters. These waters passed through deeper soil layers and contain alkalinity. The Ca/Al and Mg/Al ratios in the soil solution differ significantly between the undamaged and damaged parts and are in agreement with the threshold values given in section IX,C.

In the Harz Mountains, the watershed Lange Bramke has been monitored since 1977 with respect to deposition, seepage output from the soil, creek output from the valley, and storage of exchangeable cations in the soil (Hauhs, 1985, 1989). The watershed is afforested with an even-aged spruce stand planted in 1949. The first yellowing symptoms were observed in 1982. Figure 7–13 shows the trends of Ca/Al and Mg/Al molar ratios in the seepage water in 1-m soil depth from 1977 to 1986. The horizontal lines represent the threshold values given in Table 7–11. There is a clear trend toward increasing stress (Ca/Al) or to reach values below the threshold of severe root damage (Mg/Al). The time course of the Mg/Al ratio indicates that Mg uptake from the mineral soil was increasingly inhibited. The new flush, therefore, had to be supplied with Mg from the older needles, which then became Mg deficient. No significant change in leaching of Mg from the canopy was observed. Thus, not the leaching, but the decrease in Mg uptake is responsible for the Mg deficiency.

In Table 7–16, the mean annual flux balance of the watershed is given for calcium and magnesium. There have been considerable decreases of exchangeable Ca and Mg storage (A). The flux balance of the root zone for both ions is negative (B). As the main anion in seepage water is sulfate, it can be concluded that the net leaching of Ca and Mg from the soil is a consequence of acidic deposition. The

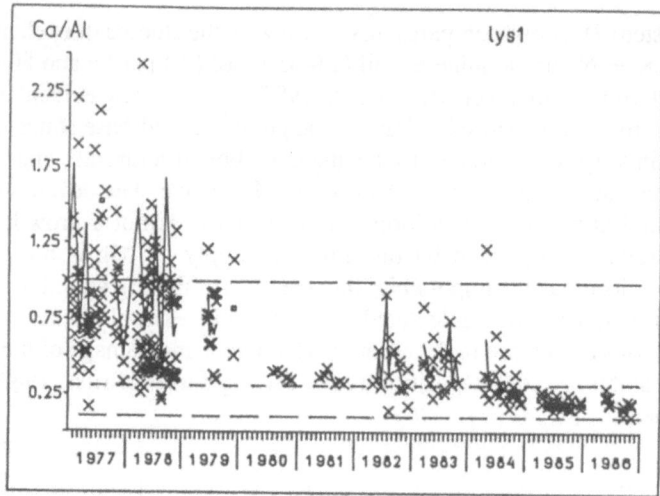

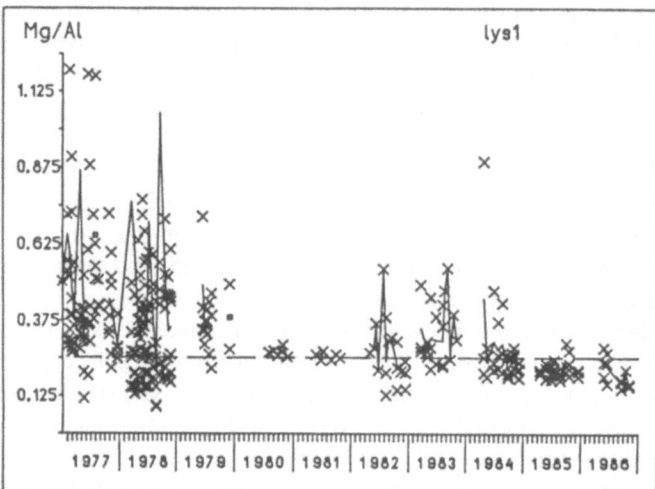

**Figure 7–13.** Time course of Ca/Al and Mg/Al molar ratios in the seepage water in 1-m soil depth in the Lange Bramke watershed, Harz. The horizontal lines represent the threshold values given in Table 7–11.

difference $A-B$ represents the sum of the fluxes that have not been measured: the weathering rate as input, and the accumulation of Ca and Mg in organic matter as output from the mineral soil. This balance is again negative, that is, the weathering rate cannot compensate for the accumulation of Ca and Mg in the increasing biomass and organic top layer of this aggrading stand. The mean annual accumulation rate in an aggrading spruce stand of this yield class amounts to about $8 \text{ kg Ca} \cdot \text{ha}^{-1}$ and $1 \text{ kg Mg} \cdot \text{ha}^{-1}$. During the passage through the deeper layers

**Table 7-16.** Mean annual flux balance of Ca and Mg of the watershed Lange Bramke.

|                                                                        | Ca    | Mg   |
|------------------------------------------------------------------------|-------|------|
| A. Changes in exchangeable cation storage in root zone                 |       |      |
| Storage 1974 (kg · ha$^{-1}$)                                           | 98    | 24   |
| Storage 1984 (kg · ha$^{-1}$)                                           | 42    | 9    |
| Mean annual storage change (kg · ha$^{-1}$ · yr$^{-1}$)                 | −5.6  | −1.5 |
| B. Flux balance:                                                       |       |      |
| Deposition I (kg · ha$^{-1}$ · yr$^{-1}$)                               | 7.9   | 1.8  |
| Seepage from root zone 0                                                | 8.7   | 2.8  |
| I-0                                                                     | −0.8  | −1.0 |
| C. A − B (kg · ha$^{-1}$ · yr$^{-1}$)                                   | −4.8  | −0.5 |
| Represents the sum of not measured fluxes: weathering minus accumulation in organic matter |  |  |
| D. Flux balance (kg · ha$^{-1}$ · yr$^{-1}$):                          |       |      |
| Infiltration below root zone I                                          | 8.7   | 2.8  |
| Output with creek water 0                                               | 19.7  | 10.1 |
| I-0                                                                     | −11   | −7   |

of the weathering mantle, the seepage water is considerably enriched with Ca and Mg in connection with an increase in pH to values above 5.0 and the appearance of bicarbonate.

These data clearly show that acidic deposition has been the driving force for the leaching of exchangeable Ca and Mg and the shifting of the acid-base status of the soil solution to the acidic side, causing acidic stress to the trees. An additional driving force is the input of nitrogen (around 30 kg N · ha$^{-1}$ · yr$^{-1}$), which triggers excellent growth with the consequence of absorbing the last reserves of exchangeable cations.

The application of Mg fertilizers can temporarily shift the Mg/Al ratio in the soil solution above the threshold level and thus increase the Mg uptake. The deficiency symptoms may disappear (Hüttl and Fink, 1988). The Al-saturated soil cannot retain the Mg ions in exchangeable form, however, so that the fertilizer applied is leached from the root zone within a few years. Fertilization can therefore be helpful, but it is no solution to the problem of soil acidification caused by acidic deposition. The addition of soluble fertilizer salts may even trigger the transfer of acidity from the organic top layer to the mineral soil (Hildebrand, 1988b) and thus increase acidic stress in the mineral soil. The amelioration of the soil requires the addition of a base, that is, liming.

## B. Crown Thinning of Norway Spruce

Throughout middle and northern Europe, a widespread symptom in Norway spruce is crown thinning. It results from both above-average loss of the older sets of needles and insufficient branching, especially proventitious branching inside the crown (Gruber, 1987b). The typical pattern of thinning is a loss of needles from the inside to the outside and from the base to the top of the crown (Schütt, 1984). Other well-known symptoms are the *Lamettasyndrom* (silver tinsel syndrome) (Schütt, 1984; Rehfuess and Rodenkirchen, 1984; Gruber, 1987a; Magel and Ziegler, 1987) and the *Fenstereffekt* (window effect) (Schröter and Aldinger, 1985; Gruber, 1988a). Older needles can be discolored (yellow to red-brown). The symptoms can be well distinguished from damage symptoms caused by other abiotic or biotic factors (Hartmann et al., 1988).

## 1. The Mechanism of Needle Abscission

Most of the research on abscission has been performed with herbaceous species, which are very different from trees. Especially the separation process in the conifers are totally different, as they seem to depend primarily upon the water status of the needle and are apparently independent of hormone-regulated cell divisions and cell differentiations (Fink, 1988a). Gruber (1987a) has examined the mechanism by which needle shedding occurs. Figure 7–14a shows a longitudinal section through the abscission zone of the needle. The green needle develops on a footlike cushion on the twig. Between the thick-walled axially oriented cells of the hyaline layer (shrinkage tissue) at the base of the needle and the lignified sclerenchyma layer of the cushion (resistance tissue), which is oriented at right angles to the needle axis, is the cutin layer (separation tissue), which is one cell

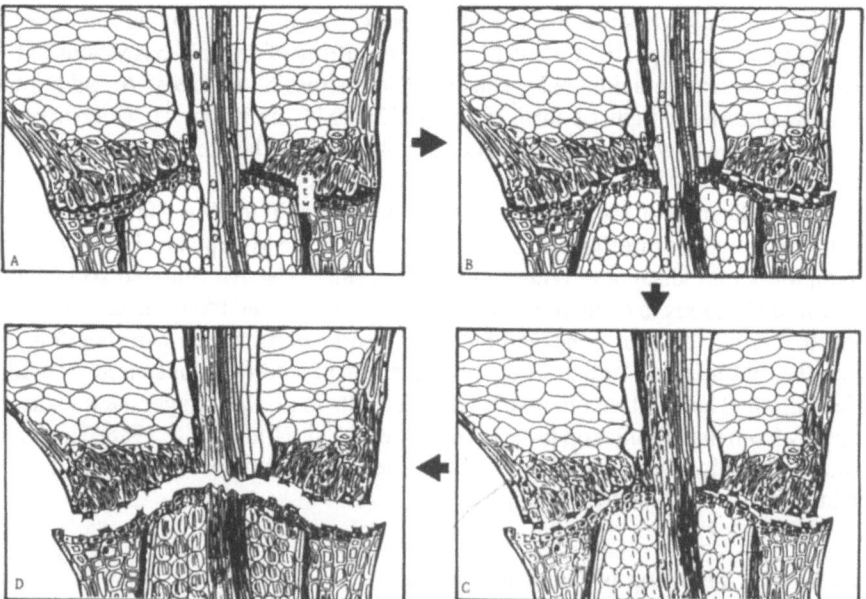

**Figure 7–14.** Operating mode of the abscission zone on Norway spruce needle; *A* = resting phase; *B* = stretching phase; *C* = tearing phase; *D* = breaking of the leaf trace. (From Gruber, 1987a.)

thick and in which the separation takes place. The cutin cells have a thin primary wall and a thick secondary wall, which consists of cutin or cutinlike material. Needle and cushion are connected by the primary walls of the separation layer cells, the leaf trace, and the cuticle.

The separation of the needle is triggered by the shrinkage of the hyaline layer, whose capacity will reach over 20% of the cross-sectional area. The shrinkage, which occurs only on drying needles, causes a tearing of the primary walls and the cuticle. The needle drops off when the leaf trace is broken. This takes place by the slightest touch (Figure 7–14b and 7–14c). Thus, the abscission of the spruce needles is a simple physical process, a "hygroscopic rhexolyse." Because of the physical nature of the abscission, the falling needle is a highly sensitive sensor indicating and reacting to water stress on a twig, branch, or part of the crown.

As litter fall data show, the needle fall happens throughout the whole year with maxima in autumn and at the end of winter or at the beginning of spring. These times are after the dry periods in summer and winter.

## 2. The Branching of Crowns of Norway Spruce

GRUBER (1986, 1987a, 1988b, 1988c) has investigated and described the branching system of young and old Norway spruce trees. He found that a tree undergoes three modes and phases of shoot formation during its life cycle, primarily as a result of adaptation to the environment. These phases are the proleptic, the regular, and the proventitious shoot formations (Figure 7–15).

The first mode, which only appears in youth and in stands with optimal growth conditions, develops from embryonic buds. It allows fast crown expansion and quick regeneration of directly damaged or lost crown parts.

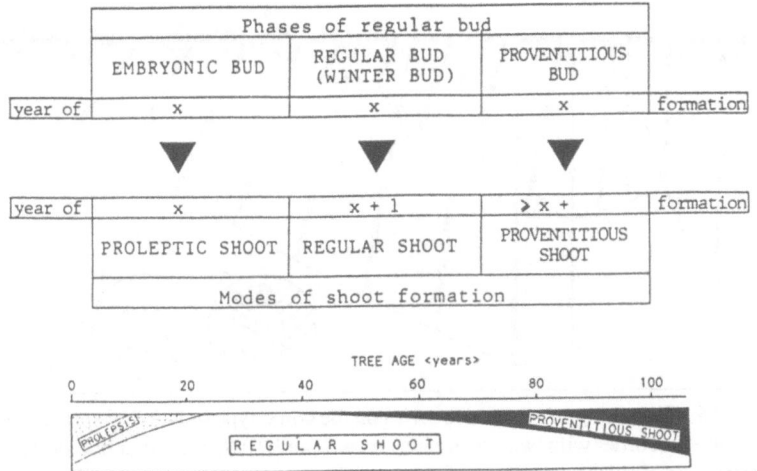

**Figure 7–15.** Modes of shoot formation through the life cycle of Norway spruce trees. (Adopted from Gruber, 1988c.)

The regular shoot formation takes place every year from 1-year-old buds (winter buds). It is the foundation of the tree architecture and can be observed in the spruce crown in all phases of the tree's life. The regular branching has the function of maximizing the crown expansion and defending the conquered crown space against crown expansion from neighboring trees. Therefore, the regular shoot formation makes up the periphery of the crown.

With the aging of a tree the proventitious shoot formation becomes more and more important. It emerges from dormant buds at the base of annual shoots and on the upper site along horizontally oriented axes. Thus, older, ineffective twigs and branches can be replaced by younger, effective ones (Figure 7–16). This permanent substitution of ineffective branches by proventitious shoots and twigs, which is called "branching in successive series" (Gruber, 1987a), is quite a normal process in vital crown (see also Rehfuess, 1983; Rehfuess and Rehfuess, 1988; Mohr, 1984). This process makes it possible for the tree to develop a highly effective needle mass even in the inner crown. The proventitious shoot formation represents a highly developed repair system for substituting damaged or lost needles and branches. Of further importance is the great flexibility of the crown structure and the enormous architectural adjustment to environmental conditions. Altogether, the plasticity and regenerative capacity of the spruce crown is very high.

## 3. Thinning and Regeneration of Spruce Crowns

Crown thinning shows very different patterns. Thinning may occur suddenly or gradually, may be acute or chronic, and may result from a high loss of needles or

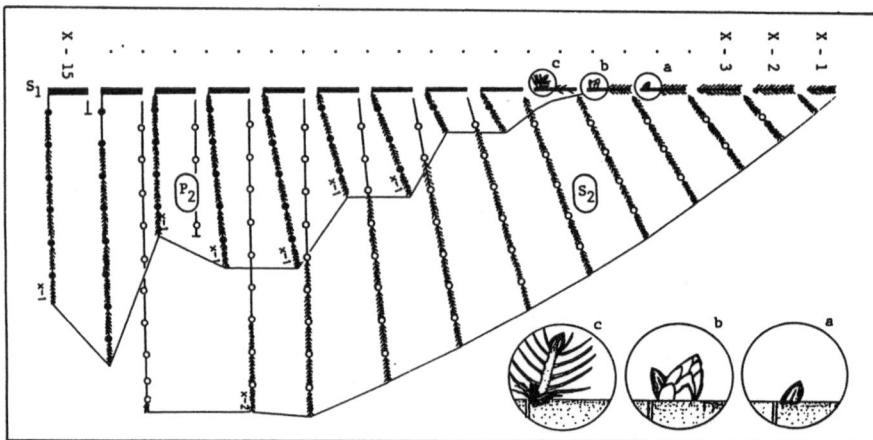

**Figure 7–16.** Branching in successive series on the comb type of Norway spruce. This is a branch of the first order with two successive series $S_2$ and $P_2$: $S_1$ = regular branch axis of the first order; $S_2$ = regular branch axis of the second order; $P_2$ = proventitious branch axis of the second order; $a$ = proventitious bud; $b$ = activated proventitious bud; $c$ = proventitious start shoot. (From Gruber, 1987a.)

from insufficient branching. Although high needle losses are a sign of acute damage or stress, poor branching generally indicates chronic stress or disease. Therefore, important conclusions can be drawn from the regenerative behavior of the canopy.

In Figure 7–17, the development of needle mass and branching of the crown under acute and chronic stress with and without the capacity of regeneration is illustrated on a comb type of spruce.

In describing the effect of acute stress on the crown, it must be distinguished between crowns with and without a regenerative capacity, which is dependent upon the extent of the initial damage. This again is regulated by differences in environment and genotype. Due to the high regenerative capacity of the crowns, the spruce is normally able to withstand and overcome extreme climatic and biotic damage under natural conditions. Pechmann (1958) showed that only needle losses above 60% are lethal.

A strong growth in proventitious shoots, as shown in the example of the comb type of spruce where the older branch parts are covered with a large number of green shoots, is an indication of high turnover in the crown. This is dependent upon an intact uptake and transport system for water and nutrients. It is therefore unlikely that spruce trees with this branching pattern are suffering from stronger or chronic stress. On the contrary, this indicates acute damage from abiotic or biotic stress factors. The number of proventitious shoots can be taken as a measure of exogenous stress, and, when the shoot length is taken into account, the number of proventitious shoots may be used to estimate the regenerative capacity of the tree.

Under chronic stress, crown thinning develops not only as a gradual decrease in the number of annual sets of needles but also as a decrease in branching, especially in the inner crown. This is determined by the insufficient formation or lack of formation of proventitious shoots. Consequently, the needle mass and the number of new shoots decrease mainly on the older parts of the branches. Providing this is not caused by shading due to the branching pattern or to the position of the tree in the stand, it must be assumed that this is due to a reduced capacity of regeneration.

The symptom of crown thinning from the inside to the outside of the crown, which is typical for the present spruce damage, may be attributed to reduced proventitious shoot formation. This poor branching in the crown, apart from genetic differences, is always a result of an insufficient supply of water (Gruber, 1988a).

## 4. The Water Stress Hypothesis of Crown Thinning

The mechanism of needle abscission and the analysis of branching patterns leads to the conclusion that the physiological cause of crown thinning of Norway spruce could be water stress if other stresses (nutrient deficiency, pathogens, mechanical action) can be excluded. This hypothesis is supported by the following observations.

Rademacher and others (1986) reported that the water content of the outer and inner sapwood of spruce trees with needle loss >60% is significantly lower than in

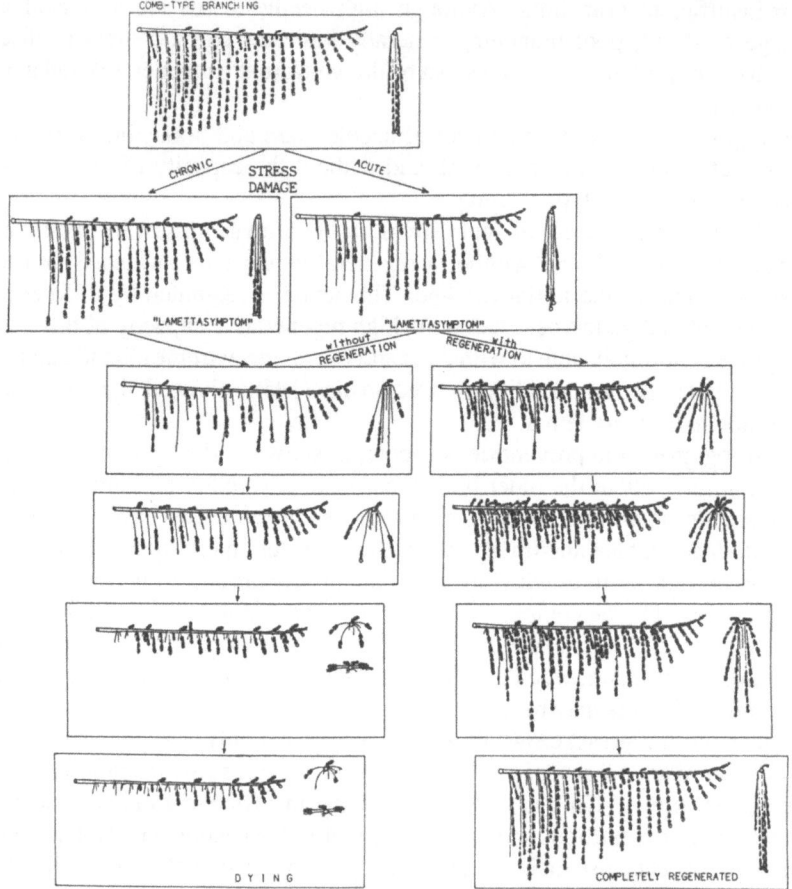

**Figure 7–17.** Diagram of the alternation of the needle mass and branching due to chronic and acute stress with complete regeneration and without regeneration. (From Gruber, unpublished.)

trees with needle loss <10%. This water deficit is intensified by the reduction of their sapwood portion by 20% to 25%. These results are confirmed by other authors (Knigge et al., 1985). Bode and others (1985) and Cape and others (1988) report that the water potential in the needles of diseased trees is more negative than in healthy ones; the branches show no water deficit, however. Strack and Unger (1988) investigated the water flow system using tritiated water. In a spruce tree showing needle loss and yellowing, a severely altered water flow system was observed, due to a reduced water conducting cross section. The tissue water of needles from the healthy tree showed a distinct gradation of tritium concentration in relation to age which was missing in the diseased tree. Water potential measurements also provided evidence for a tighter water budget but not for an acute water stress situation.

Schütt (1984) proposed that the starting point of the novel tree dieback is a disturbance of photosynthesis due to the long-term load of low concentrations of air pollutants. This should limit the root development and subsequently water and nutrient uptake. The tree would enter a downward life spiral with reduced renewal of needles and fine roots. Persson (1988) has summarized the evidence against this hypothesis of a gradual strangulation of the life processes of the tree. Investigations on diseased trees showed that the photosynthetic apparatus is working perfectly (Siefermann-Harms and Hampp, 1987). Fine roots of diseased trees have, despite showing damage symptoms, a fairly high starch content (Murach, 1984; Puhe et al., 1986; Eichhorn et al., 1988). This is comparable to the results found for the yellowing symptom (see section XI,A). A green needle on a diseased tree is physiologically largely healthy (Hüttermann and Godbold, 1985). Koch (1988) measured photosynthesis and transpiration on two branches of the same tree in the forest, comparing the uncleaned air at the site and clean air. He found a depression of the rates of photosynthesis and transpiration only during a frost period at high $SO_2$ concentrations (up to 380 $\mu g \cdot m^{-3}$). In spring, with low $SO_2$ concentrations, the depression disappeared again. Measurements of the osmotic potential of green needles of healthy and diseased spruce trees showed no significant difference (Koch, 1985).

The present state of knowledge can be summarized by saying that there seems to be no significant difference between the functioning of the green, visibly undamaged needles of trees subjected to crown thinning and those that show no or negligible (<10%) needle loss in respect to photosynthesis and water status.

Acute water stress does not become apparent in increased water potentials (as in visibly damaged needles) and leads to the closure of the stomata. This, however, causes a strong decrease in photosynthesis. Persson (1985) has shown that water deficiency, caused by the drying out of the soil, depressed photosynthesis to 5% of the control plots. If the situation of water stress continues and becomes chronic, the tree has to find ways to discouple older, shaded, and therefore ineffective needles from the transpiration stream in order to allow younger, light-exposed needles of high photosynthetic effectivity and water-use efficiency to open the stomata. Gruber (1987a) has observed that the needle abscission can be triggered by the incrustation of the xylem in the needle cushion. He also observed that the leaf trace can be interrupted in connection with the annual wood formation. Thus spruce has mechanisms to separate ineffective needles from the transpiration stream in order to increase the water-use efficiency of the whole trees. By this, the transpiring needle mass can be adjusted to the amount of water transported to the canopy. If this mechanism operates effectively, the water status of the undamaged needles should indeed show no signs of stress as long as there is no acute stress situation (as, for example, a drought or frost-drought period).

A chronic water stress can be caused by

A change in climate, for example, a lasting series of warm and/or dry years.

An increase in transpiration rate. This could be caused by ineffective stomata regulation or an increase in cuticular transpiration. Such effects could be the

result of a direct influence of air pollutants, especially of a combination of acidic mist and ozone on the needle (cf. section VII). As discussed above, however, the exising data do not show a dramatic influence on the water status of the needle.
A decrease in the uptake rate of water.
A decrease in the area of the conductive tissue, connecting the sites of water uptake with those of transpiration.

A chronic decrease in water uptake can be caused by a decrease in the water-absorption ability of the roots, by a decrease in the amount of absorbing roots, or by a decrease in soil volume utilized for water uptake. Klein (1985) and Stienen (1986) found in solution culture experiments decreases in transpiration as a consequence of acidic stress (protons, Al, and heavy metal ions). This indicates that a decrease in the water-absorption ability of fine roots due to acidic stress may contribute to the chronic water stress.

Decreases in fine-root biomass or in root-tip frequency have been related to acidic stress (Ulrich, 1980; Murach, 1984; Raben, 1988). These processes represent an acute stress, a top load superimposed on the long-term chronic load. Depending upon the physical and chemical soil conditions and the available carbohydrates, the fine-root system may be renewed within weeks, months, or at the latest in the following year. If the restoration of the fine-root system is limited, the acute stress turns into chronic stress, raising the already existing chronic stress to a higher level. This situation may occur in trees with higher needle losses. In this case one would expect a decreasing fine-root biomass and root-tip frequency, an increasing percentage of fine-root necromass, increasing degeneration symptoms of the fine-root cortex, and increasing invasion of root cells by mycorrhizal fungi with increasing needle loss, as shown by Ulrich and others (1984), Gehrmann and others (1984), Puhe and others (1986), Eichhorn and others (1988), and Nienhaus (1988). From a dynamic point of view, this situation may be characterized by an increasing demand of photosynthates for fine-root regeneration and a decreasing rate of photosynthesis due to needle loss and decreased water uptake (stomate closure). If this situation continues, tree death is inevitable.

Recovery is possible if the acidic stress in soil is decreasing (e.g., the Ca/Al ratio in the soil solution increasing) and fine-root lifetime is increasing, which could decrease the demand of carbohydrates in order to maintain an efficient fine-root system within the limits of the carbohydrate supply. This view is supported by the relationship between fine-root biomass and necromass. Decreases in fine-root biomass are usually accompanied by increases in fine-root necromass. The decreasing fine-root biomass with increasing needle loss is accompanied by an increasing fine-root necromass. These relationships show that the fine-root turnover is increased.

Even in undamaged trees, the consumption of photosynthates can account for around half of the net photosynthesis, so that an increased demand of assimilates in the root system can cause a considerable reduction in diameter growth of the aboveground part of the tree at constant rates of photosynthesis. Praag and others (1988) assessed in a Norway spruce stand a turnover rate of 7,000 kg root dry weight $\cdot$ ha$^{-1}$ $\cdot$ yr$^{-1}$. This can be compared with the aboveground biomass production of 9,000 to 13,000 kg dry mass in ecosystem 7, which contains 3,400 kg needle litter production.

For the development of a chronic water stress, a change in the depth distribution of fine roots seems to be of great importance for the following two reasons: The development of a superficial root system (1) decreases the amount of water that can be utilized by the tree (this simulates a change to a dryer climate) and (2) makes ineffective the water-conductive area in woody roots expanding deeper into the soil. Whereas the fine-root mass can be renewed rapidly, the renewal of water-conductive area in the roots transporting the water to the stem needs many years. The existing discrepancy between the depth distribution of woody coarse roots and that of fine roots in spruce timber trees is often very striking. Whereas the coarse root system reaches 0.8- to 1.3-m depth, the fine roots are, as a consequence of Al stress in the subsoil, accumulated in the top soil (Figure 7–8; Ulrich et al., 1984; Hauhs, 1985; Puhe et al., 1986; see also section IX,C). Rademacher and others (1986) found in a comparison of the total biomass of needles, branches, thick roots, and fine roots that these proportions in the healthy tree and the severely diseased tree (>60% needle loss) correlate well with the stem wood with the exception of thick roots (in the diseased tree, biomass reduced to one quarter). The drastic deficit in coarse-root biomass is underlined by a decrease of the root-plate volume from 24 m$^3$ for healthy trees to 9 m$^3$ for severely diseased trees.

The existence of a functional relationship between leaf area and the hydroactive xylem (sapwood) area was first suggested by Huber (1928) and later confirmed for many diverse plant species. Variations in this relationship that have been found in experimental work have been resolved by Oren and others (1986). These authors conclude that the slope of the regression relating needle biomass (or area) to the conductive area is a decreasing function of the transpirational demand and possibly an increasing function of the hydraulic conductance. The intercept is related to what degree the morphologically identifiable sapwood area is actually involved in conductance. The conducting system continues from the stem base through thick roots and ending in the fine roots. The change from a deep to a superficial fine-root system reduced primarily the conductive area below the stem base. This should precede the reduction of the stem sapwood area, which has been described by Rademacher and others (1986). The hypothesis proposed in this chapter is in full agreement with the Huber theory.

Crown thinning can, but need not be, correlated with nutrient deficiency, as indicated by needle analysis, especially of Ca and Mg, but also of others (Forschungsbeirat, 1986; Siefermann-Harms and Hampp, 1987; Puhe and Aronsson, 1986). The plant organ where deficiency manifests itself most clearly is thus the root: The dead fine root, the living fine root (see Figure 7–10 and Table 7–12), and the coarse root (Ulrich et al., 1984). The root analysis shows that the change in elemental composition reflects a change in the acid-base status of the plant that has severe consequences for its nutrient status. The term *nutrient deficiency* restricts scientific concepts that are not in agreement with experimental data (see section IX), and this has already led to false conclusions.

The change from a deep to a superficial fine-root system has a further consequence. In order to overcome a drought period without a decrease in transpiration, a superficial root system needs a higher-absorbing root mass compared to a deep-reaching and homogeneous distributed fine-root system (Schlichter et al., 1983; for example, 6,000 instead of 2,500 kg/ha). The data on fine-root mass show indeed higher values in stands with a superficial root system as long as the crown thinning has not proceeded far.

Crown thinning of Norway spruce can be traced back to chronic water stress. The chronic water stress can be traced back mainly to a change from a deep-reaching to a superficial fine-root system. Such a change can be the result of an internal ecosystem development that results in the acidification of the rhizosphere after decades or centuries of a tree's life. The ecosystem internal acidification of the rhizosphere will be accelerated if a stand opens and soil temperature increases, due to either natural events, to grazing, or to cutting. Crown thinning is therefore neither a new symptom nor is it specific for air pollutants. Acidic deposition, however, has acidified either the root or the soil on a large scale and to such an extent that the symptom of crown thinning is now widespread and taken as a measure of tree vitality.

## XII. Future Research Needs

The manifold linkages between the components of a forest ecosystem render it highly improbable to ascertain simple linear relationships between environmental stress and tree vitality. Anthropogenic air pollutants are superimposed on the natural environmental stress to which forest ecosystems are continuously subjected. In order to reveal the cause-and-effect relationships of air pollutants in forest ecosystems, the cause-and-effect relationships of natural environmental stress factors have to be known. This knowledge, however, was and still is very limited. Tree dieback and forest decline are not new phenomena; they can happen for natural reasons. Research on forest decline till now resulted in a description of the symptoms and a discussion of possible causes but did not give a conclusive explanation of the processes involved. A scientific explanation requires a knowledge of how a change in the environment of the forest ecosystem (e.g., climate or air pollutants) or in the ecosystem itself (e.g., by management) affects the components of the ecosystem (soil, plants, animals, microorganisms) and how these factors affect each other. We need not be deterred by the complexity of forest ecosystems; we need, however, much more basic ecosystem research to elucidate this complexity.

A central aspect is the effects of climatic variation where the effects of drought periods on the water status have been assumed to be of primary importance. Effects on soil chemistry (acid-base and nutrient status) due to a discoupling of the turnover processes by primary and secondary producers have been neglected. For medium-term effects (a decade), they may be of greater importance than the short-term effects on water status.

Another central aspect is nutrient and acidic stress and the acid tolerance of trees. How should nutrient and acidic stress be defined? The research of the last years has shown that one needs to know the composition of the soil solution in respect to ion species. It has also become evident that, especially in soils subjected to acidic deposition, the stress parameters in the soil solution can reach extreme values and show a great spatial and temporal variability. How can the effects of acidic stress be tested? The approaches presented in section IX seem neither well developed nor comprehensive. How is the development of tree damage in the forest related to the spatial and temporal variability of air pollutants, on the one hand, and of stress parameters in the soil solution on the other hand? How can long-term effects (decades) of acidic stress be assessed? Research on the factors determining root and mycorrhizal status and growth is of primary importance in order to increase our understanding of the relationships between a tree and its belowground and aboveground environment. It requires not only an extended application of the present methodology but also the development of new methodical approaches.

The results presented have shown that acidic deposition as well as dry deposition of $SO_2$, $NO_x$, and ozone on leaves affects the roots even in the short term. However, it has been demonstrated that adverse soil conditions, affecting roots and tree nutrition, lower tolerance against direct effects of air pollutants on the leaves. This shows that the reaction of the tree to a stress cannot be assessed by analyzing one organ. That the tree reacts as one organism with rapid transfer of matter and information between roots and shoots has to be taken into account in designing experiments and control measurements. This makes experimentation complex and expensive, but it is unavoidable if reliable results are to be achieved.

Results have been presented that relate ultrastructural and histochemical symptoms in leaves and roots to the stressor that caused them. These results should also be confirmed in other areas. The approach could be used to differentiate between the direct and indirect effects of air pollutants and to identify the cause-and-effect relationship of tree damage on an areal scale.

It is evident that nutrient deficiencies and the acid-base status of the soil play an important role is predisposing trees to the direct effects of air pollutants. Optimal nutrition in the sense of Ingestad (1987) and Ingestad and Agren (1988) is therefore a measure not only to increase growth and production but also to maintain the tree in a state of high vitality and to minimize adverse effects of air pollutants. What are the prerequisites in respect to chemical and biological soil state and to the state of the ecosystem (deviation from the steady state) to approach optimal nutrition without the continuous addition of fertilizers? Which silvicultural measures should be taken in order to achieve this state?

# References

Ares, J. 1986a. Analytica Chimica Acta 187:195–211.

Ares, J. 1986b. Soil Sci 142:13–19.

Arndt, U., G. Seufert, J. Bender, and H.J. Jaeger. 1985. VDI-Ber 560:783–803.

Arnolds, E.J.M., ed. 1985. Wetenschappelijke Mededelingen K N N V 167:1–101.

Arnolds, E.J.M., and A. E. Jansen. 1987. Biological Station Wijster, The Netherlands, Rep 334:1–29.

Asche, N., and F. Beese. 1987. Allgem Forstzeitschr 42:868–870.

Asp, H., B. Bengtsson, and P. Jensen. 1988. Plant Soil 111:127–133.

Bååth, E., B. Berg, U. Lohm, B. Lundgren, H. Lundkvist, T. Rosswell, B. Söderström, and A. Wirén. 1980. Pedobiologia 20:85–100.

Bååth, E., B. Lundgren, and B. Söderström. 1984. Microbial Ecology 10:197–202.

Barnes, J. D., A. W. Davison, and T.A. Booth. 1988. *In* J. N. Cape and P. Mathy, eds. *Scientific basis of forest decline symptomatology* (Air Poll Rep Ser 15), 245–251. Comm Eur Communities, Brussels.

Bartsch, N. 1987. Can J For Res 17:805–812.

Bauch, J. 1983. *In* B. Ulrich and J. Pankrath. eds. *Effects of accumulation of air pollutants in forest ecosystems*, 377–386. Reidel, Dordrecht.

Bauch, J., and W. Schröder. 1982. Forstwiss Centralbl 101:285–294.

Bauch, J., H. Stienen, B. Ulrich, and E. Mazner. 1985. Allgem Forstzeitschr 40: 1448–1451.

Baumbach, G., and M. Käss. 1985. Staub Reinhaltung der Luft 45:274–278.

Bender, J., H.-J. Jäger, G. Seufert, and U. Arndt. 1986. Angew Botanik 60:461–479.

Benner, P., and A. Wild. 1987. J Plant Physiol 129:59–72.

Berdén, M., S.I. Nilsson, K. Rosen, and G. Tyler. 1987. *Soil acidification: Extent, Causes, Consequences.* Nat Swed Environ Prot Board Report 3292. Solna. 164 p.

Berg, B. 1986a. Forest Ecology Management 15:195–213.

Berg, B. 1986b. Scand J For Res 1:317–322.

Bermadinger, E., D. Grill, and P. Golob. 1988. GeoJournal 17:289–293.

Beyschlag, W., M. Wedler, O.L. Lange, and U. Heber. 1987. Allgem Forstzeitschr 42:738–741.

Blanck, K., E. Matzner, R. Stock, and G. Hartmann. 1988. Forst u Holz 43:288–292.

Blaschke, H. 1986. Forstwiss Centralbl 105:477–487.

Blasius, D., I. Kottke, and F. Oberwinkler. 1985. Fortwiss Centralbl 104:318–325.

Bode, J., H.P. Kühn, and A. Wild. 1985. Forstwiss Centralbl 104:353–360.

Bosch, C., E. Pfannkuch, U. Baum, and K.E. Rehfuess. 1983. Forstwiss Centralbl 102:167–183.

Boxman, A.W., R.J. Sinke, and J.G.M. Roelofs. 1986. Water Air Soil Poll 31:517–522.

Brechtel, M. 1988. Forst und Holz 43:298–302.

Bredemeier, M. 1987a. Stoffbilanzen, interne Protonenproduktion und Gesamtsäurebelastung des Bodens in verschiedenen Waldökosystemen Norddeutschlands. Diss Univ Göttingen, Ber Forschungszentr Waldökosys Univ Göttingen A33. 183 p.

Bredemeier, M. 1987b. Plant Soil 101:273–280.

Bredemeier, M. 1988. Water Air Soil Pollut, 40:121–138.

Breemen, N. van, and E.R. Jordens. 1983. *In* B. Ulrich and J. Pankrath, eds. *Effects of accumulation of air pollutants in forest ecosystems*, 171–182. Reidel, Dordrecht.

Breemen, N. van, J. Mulder, and C.T. Driscoll. 1983. Plant Soil 75:283–308.

Brumme, R., N. Loftfield, and F. Beese. 1987. Mitt Dtsch Bodenkdl Ges 55:585–586.

Bundesministerium für Ernährung, Landwirtschaft und Forsten, Bonn. 1987. Waldschadenserhebung 1987. 85 p.

Burschel, P., J. Huss, and R. Kalbhenn. 1964. *Die natürliche Verjüngung der Buche.* Schriftenr Forstl Fak, Univ Göttingen. Vol. 34.

Büttner, G., N. Lamersdorf, R. Schultz, and B. Ulrich. 1986. *Deposition und Verteilung chemischer Elemente in küstennahen Waldstandorten.* Ber Forschungszentr Waldökosys, Uni Göttingen B 1. 172 p.

Cape, J.N. 1988. *In* J. N. Cape and P. Mathy. *Scientific basis of forest decline symptomatology.* (Air Poll Rep Ser 15), 133–148, Comm Eur Communities, Brussels.

Cape, J.N., J. Wolfenden, H. Mehlhorn, P.H. Frier-Smith, I.S. Paterson, A.R. Welbonn, and S. Fink. 1988. *Early diagnosis of forest decline.* Inst of Terrestr Ecology, Grange-over-Sands. 68 p.

Cassens-Sasse, E. (1987) *Witterungsbedingte saisonale Versauerungsschübe im Boden zweier Waldökosysteme.* Diss Univ Göttingen, Ber Forschungszentr Waldökosyst Univ Göttingen A30. 163 p.

Cronan, C.S., J.M. Kelly, C.L. Schofield, and R. A. Goldstein. 1987. *In* R. Perry, R.M. Harrison, J.N.B. Bell, and J.N. Lester, eds. *Acid Rain, Scientific and Technical Advances,* 649–656. Selper Ltd, London.

Dietz, B., I. Moors, U. Flammersfeld, W. Rühle, and A. Wild. 1988. Z Naturforsch 43c:581–588.

Dietze, G. 1985. Bindungsformen und Gleichgewichte von Aluminium im Sickerwasser saurer Böden. Diss Univ Göttingen, Ber Forschungszentr Univ Goettingen 16. 118 p.

Dietze, G., and B. Ulrich. 1989. *In* G. Matthess, ed. *Hydrogeochemical processes in the water cycle in the unsaturated and saturated zone.* Springer, Berlin. In press.

Dise, N., and M. Hauhs. 1987. *In* B. Moldau and T. Pàces, eds. *Geomon,* 58–61. Geological Survey, Prague.

Eichhorn, J. 1987. Vergleichende Untersuchungen von Feinwurzelsystemen bei unterschiedlich geschädigten Altfichten. Diss Univ Göttingen. 154 p.

Eichhorn, J., E. Gärtner, and A. Hüttermann. 1988. Allgem Forst- u Jagdztng 159:37–42.

Eldhuset, T., A. Göransson, and T. Ingestad. 1987. *In* T. C. Hutchinson and K. M. Meema, eds. *Effects of atmospheric pollutants on forests, wetlands and agricultural ecosystems* NATO ASI Ser G16), 401–409. Springer-Verlag, Berlin.

Ellenberg, H., R. Mayer, and J. Schauermann, eds. 1986. Ökosystemforschung: Ergebnisse des Sollingprojekts. Ulmer, Stuttgart.

Elstner, E.F., and W. Osswald. 1988. *In* J. N. Cape and P. Mathy. *Scientific basis of forest decline symptomatology* (Air Poll Rep Ser 15), 126–131. Comm Eur Communities, Brussels.

Evers, F.H., W. Schöpfer. 1988. Allgem Forst- u Jagdztng 159:146–154.

Falkengren-Grerup, U. 1987. Environmental Pollution 43:79–90.

Fiedler, H.J. 1988. Beitr Forstwirtschaft 22:61–66.

Fiedler, H.J., F. Leube, and W. Nebe. 1988. Forst u Holz 43:398–400.

Fink, S. 1983. Allgem Forstzeitschr 38:660–663.

Fink, S. 1987. 3. Statuskolloq PEF, March 1987, Karlsruhe. *Kernforschungszentr Karlsruhe,* 113–122.

Fink, S. 1988a. *In* J.N. Cape and P. Mathy, eds. *Scientific basis of forest decline symptomatology* (Air Poll Rep Ser 15), 39–47. Comm Eur Communities, Brussels.

Fink, S. 1988b. *In* F. Horsch, ed. 4. Statuskoll PEF 35, *Kernforschungszentrum Karlsruhe* 1:209–218.

Fölster, H. 1985. *In* J.I. Driever, ed. *The chemistry of weathering*, 197–209. Reidel, Dordrecht.

Forschungsbeirat Waldschäden/Luftverunreinigungen der Bundesregierung und der Länder. 1986. 2. Bericht, *Kernforschungszentrum Karlsruhe*. 229 p.

Frevert, T., and O. Klemm. 1984. Arch Met Geoph Biocl Ser B 34:75–81.

Friedland, A.J., A.H. Johnson, and Th.G. Siccama. 1984. Water Air Soil Pollut 21:161–170.

Funke, W. et al. 1986. 2. Statuskolloq PEF. 1986. *Kernforschungszentr Karlsruhe*, Kfk-PEF 4 1:337–346.

Gehrmann, J. 1984. Einfluβ von Bodenversauerung und Kalkung auf die Entwicklung von Buchenverjüngungen (*Fagus sylvatica* L.) im Wald. Diss Univ Göttingen, Ber Forschungszentrum Waldökosysteme Univ Göttingen 1. 213 p.

Gehrmann, J., G. Büttner, and B. Ulrich. 1987. Ber Forschungszentr Waldökosys Univ Göttingen B4:69–74.

Gehrmann, J., M. Gerriets, J. Puhe, and B. Ulrich. 1984. Ber Forschungszentr Waldökosys 2:169–206(202–206).

Glatzel, G., and M. Kazda. 1985. Z Pflanzenernaehr Bodenk 148:429–438.

Glatzel, G., M. Kazda, and L. Lindebner. 1986. Düsseldorfer Geobot Kolloq 3:15–32.

Glatzel, G., E. Sonderegger, M. Kazda, and H. Puxbaum. 1983. Allgem Forstzeitschr 38:693–694.

Glavac, V. 1986. Natur und Landschaft 61:43–47.

Glavac, V., H. Jochheim, H. Koenies, R. Rheinstädter, and H. Schäfer. 1985. Allgem Forstzeitschr 40:1397–1398.

Glavac, V., and H. Koenies. 1986. Düsseldorfer Geobot Kolloq 3:3–13, Verh Ges f Okologie 14:293–298.

Glavac V., H. Koenies, H. Jochheim, and R. Heimerich. 1988. *In* E.P. Mathy, ed. *Air pollution and ecosystems*, 520–529. Reidel, Dordrecht.

Godbold, D.L., K. Dictus, and A. Hüttermann. 1988c. Can J For Res 18:1167–1171.

Godbold, D.L., E. Fritz, and A. Hüttermann. 1988a. *In* E. P. Matthy, ed. *Air pollution and ecosystems*. Reidel, Dordrecht, pp. 864–869.

Godbold, D.L., E. Fritz, and A. Hüttermann. 1988b. Proc Natl Acad Sci USA 85:3888–3892.

Godbold, D.L., and A. Hüttermann. 1985. Environ Pollut Ser A 38:375–381.

Godbold, D.L., and A. Hüttermann. 1986. Water Air Soil Pollut 31:509–515.

Godbold, D.L., H. Schlegel, and A. Hüttermann. 1985. VDI Ber 560:703–716; Int conf on heavy metals in the environment, Athens, 500–502.

Godbold, D.L., R. Tischner, and A. Hüttermann. 1987. *In* T.C. Hutchinson, ed. *Effects of acid deposition on forests, wetlands and agricultural ecosystem* (NATO ASI Series, Springer) G16:387–400.

Godt, J., and H. Lunkenbein. 1983. Mitt Dtsch Bodenkdl Ges 38:203–208.

Godt, J., M. Schmidt, and R. Mayer. 1986. *In* H.W. Georgi, ed. *Atmospheric pollutants in forest areas*, 263–274. Reidel, Dordrecht.

Godt, J., M. Schmidt, and R. Mayer. 1988. *In* J.N. Cape and P. Mathy, eds. *Scientific basis of forest decline symptomatology* (Air Poll Rep Ser 15), 307–315. Comm Eur Communities, Brussels.

Gravenhorst, G., and A. Böttger. 1983. *In* S. Beilke and A.J. Elshout, eds. *Acid deposition*, 172–184. Reidel, Dordrecht.

Grill, D. 1973. Phytopath Z 78:75–80.

Grill, D., H. Pfeifhofer, G. Halbwachs, and H. Waltinger. 1987. Europ J Forest Pathol 17:246–255.

Grimm, R., and K.E. Rehfuess. 1986. Allgem Forst- u Jagdztng 157:205–213.

Gruber, F. 1986. Beiträge zum morphogenetischen Zyklus der Knospe, zur Phyllotaxis und zum Triebwachstum der Fichte (*Picea abies* (L) Karst) auf unterschiedlichen Standorten. Diss Univ Göttingen, Ber Forschungszentr Waldökosys A25. 215 p.

Gruber, F. 1987a. Das Verzweiggungssystem und der Nadelfall der Fichte (*Picea abies* (L) Karst) als Grundlage zur Beurteilung von Waldschäden. Ber Forschungszentr Waldökosys A26. 214 p.

Gruber, F. 1987b. Allg Forstzschr 43:1285–1288.

Gruber, F. 1988a. Forst und Holz 43:58–60.

Gruber, F. 1988b. Schweiz Zschr f Forstw 139:173–201.

Gruber, F. 1988c. Flora 181: 205–242.

Guderian, R., A. Klumpp, and K. Küppers. 1987. Verhandl Ges f Ökologie (Giessen, 1986) 16:311–322.

Günthardt-Goerg, M.S., and T. Keller. 1988. *In* J.N. Cape and P. Mathy, eds. *Scientific basis of forest decline symptomatology* (Air Poll Rep Ser 15), 316–322. Comm Eur Communities, Brussels.

Hafner, L. 1986. Allgem Forstzeitschr 41:1119–1121.

Hågvar, S., and B. KjØndahl. 1981. Pedobiologia 22:232–245.

Hantschel, R. 1987. Wasser- und Elementbilanz von geschädigten, gedüngten Fichtenökosystemen im Fichtelgebirge unter Berücksichtigung von physikalischer und chemischer Bodenheterogenität. Diss Univ Bayreuth, Bayreuther Bodenkundl Ber 3:219 p.

Hantschel, R., M. Kaupenjohann, R. Horn, and W. Zech. 1987. Z Pflanzenernaehr Bodenk 150:13–16.

Hartmann, F.K., and G. Jahn. 1967. *Waldgesellschaften des mitteleuropäischen Gebirgsraumes nördlich der Alpen.* Fischer Verlag, Stuttgart.

Hartmann, G., F. Nienhaus, and H. Butin. 1988. *Farbatlas Waldschäden.Ulmer Verlag, Stuttgart. 256 p.*

Hartmann, G., J. Saborowski, R. Nebel, and A. Voretzsch. 1986. Forst- u Holzwirt 41:413–420.

Haug, I., I. Kottke, and F. Oberwinkler. 1986. Z Mykol 52:373–391.

Hauhs, M. 1985. Wasser- und Stoffhaushalt im Einzugsgebiet der Langen Bramke (Harz). Diss Univ Göttingen, Ber Forschungszentr Waldoekosys 17. 206 pp

Hauhs, M. 1986. Geoderma 38:97– 113.

Hauhs, M. 1989. Lange Bramke, an ecosystem study of a forested watershed. Adv Environm Sci (in press).

Hauhs, M., T. Paces, H.M. Seip, B. Vigernst, G.H. Raben, and K. Rost-Siebert. 1989. Evaluation of nitrogen and sulfur fluxes (in press).

Hauhs, M., and R.F. Wright. 1986. *In* B.G. Blackmon and R.S. Beasley, ed. *Proc Mid-South Symposium on Acid Deposition*, 15–26. Univ of Arkansas, Little Rock.

Heber, U., A. Laisk, H. Pfanz, and O.L. Lange. 1987. Allgem Forstzeitschr 42:700–705.

Heinrichs, H., B. Wachtendorf, K.H. Wedepohl, G. Rössner, and G. Schwedt. 1986. Neues Jahrbuch Miner Abh 156:23–62.

Hildebrand, E.E. 1986a. Forstwiss Centralbl 105:60–76.

Hildebrand, E.E. 1986b. Z Pflanzenernaehr Bodenk 149:340–346.

Hildebrand, E.E. 1988a. Bulletin Bodenk Ges Schweiz 12:67–86.

Hildebrand, E.E. 1988b. Forst und Holz 43:51–56.

Horn, R. 1987. Z Pflanzenernaehr Bodenk 150:13–16.

Horn, R., R. Hantschel, and H. Taubner. 1988. *In* J. Bervaes, P. Mathy, and P. Evers, eds. *Relationships between above and below ground influences of air pollutants on forest trees* (Air Poll Rep Ser 16), 119–131. Comm Eur Communities, Brussels.

Hovland, J., G. Abrahamsen, and G. Ogner. 1980. Plant Soil 56:365–378.

Huber, B. 1928. Jb Wiss Bot 67:877–959.

Huikari, O. 1977. Silva Fenica 11:251–254.

Hüttermann, A., and D.L. Godbold. 1985. *In* E. Niesslein and G. Voss, eds. *Was wir über das Waldsterben wissen*, 197–215. Deutscher Institutsverlag, Köln.

Hüttermann, A., and B. Ulrich. 1984. Phil Trans R Soc Lond B 305:353–368.

Hüttl, R. 1985. Freiburger Bodenkdl Abh 16. 195 p.

Hüttl, R., and S. Fink. 1988. Forstwiss Centralbl 107:173–183.

Huttunen, S., and K. Laine. 1983. Ann Bot Fennici 20:79–86.

Ingestad, T. 1987. Geoderma 40:237–252.

Ingestad, T., and G.I. Agren. 1988. Physiol Plant 72:450–459.

Irens, W.P.M.F., G.P.J. Draaijers, M.M. Bos, and W. Blenten. 1988. Dutch Priority Programme on Acidification, Report 37–09.

Jansen, A.E. 1988. *In* J.N. Cape and P. Mathy, eds. *Scientific basis of forest decline symptomatology*. (Air Poll Rep Ser 15), 182–189. Comm Eur Communities, Brussels.

Jochheim, H. 1985. Der Einfluss des Stammablaufwassers auf den chemischen Boden-zustand und die Vegetationsdecke in Altbuchenbeständen. Diss GH Kassel, Ber Forschungszentr Waldoekosys Univ Göttingen 13. 225 p.

Jochheim, H., and H. Schaefer. 1988. Z Pflanzenernaehr Bodenk 151:81–85.

Jorns, A.C. Hecht-Buchholz. 1985. Allgem Forstzeitschr 41:1248–1252.

Jorns, A.C. 1988. Aluminiumtoxizität bei Sämlingen der Fichte (*Picea abies (L) Karst)* in Nährlösungskultur. Ber Forschungszentr Waldökosyst Univ Göttingen A 42: 86 p.

Junga, U. 1984. Sterilkultur als Modellsystem zur Untersuchung des Mechanismus der Aluminium-Toxizität bei Fichtenkeimlingen (*Picea abies* Karst). Diss Univ Göttingen, Ber Forschungszentr Waldoekosyst 5, 1–173.

Jurat, R., and H. Schaub. 1988. Z Pflanzenernaehr Bodenk 151:379–384.

Jurat, R., H. Schaub, H. Stienen, and J. Bauch. 1986. Forstwiss Centralbl 105:105–115.

Kandler, O., W. Miller, and R. Ostner. 1987. Allgem Forstzeitschr 43:715–723.

Kaupenjohann, M., and R. Hantschel. 1987. Z Pflanzenernaehr Bodenk 150:156–160.

Kaupenjohann, M., B.U. Schneider, R. Hantschel, W. Zech, and R. Horn. 1988. Z Pflanzenernaehr Bodenk 151:123–126.

Kaupenjohann, M., W. Zech, R. Hantschel, and R. Horn. 1987. Forstwiss Centralbl 106:78–84.

Kazda, M., and G. Glatzel. 1986. Allgem Forstzeitschr 41:436–438.

Keilen, K. 1978. Spurenelementverteilung und Bodenentwicklung im Bärhaldegranit gebiet (Südschwarzwald). Freiburger Bodenkdl Abh 8. 278 p.

Keller, Th. 1979. Schweiz Z Forstwes 130:429–435.

Khanna, P.K., J. Prenzel, K.J. Meiwes, B. Ulrich, and E. Matzner. 1987. Soil Sci Soc Am J 51:446–452.

Klein, R.M. 1985. *In* D.D. Adams and W.P. Page, eds. *Acid deposition*, 303–322. Plenum, New York.

Knigge, W., H. Aszmutat, and W.J. Weiss. 1985. Forstarchiv 56:65–70.

Koch, W. 1985. Eur J For Path 15:207–216.

Koch, W. 1988. Trees 1988:213–222.

Koenies, H. 1982/85. Über die Eigenart der Mikrostandorte im Fussbereich der Altbuchen unter besonderer Berücksichtigung der Schwermetallgehalte in der organischen Auflage und im Oberboden. Diss GH Kassel, Ber Forschungszentr Waldoekosys Univ Göttingen 9. 288 p.

Kolari, K.K. ed. 1983. Commun Inst For Fenn 116:111–170.

Korff, H.C., E. Schrimpff, I. Brandner, and F. Janocha. 1980. Bayreuther Geowiss Arbeiten 1:39–56.

Kottke, I. 1986. In G. Einsele, ed. Das landschaftsökologische Forschungsprojekt Naturpark Schönbuch, 443–462. VCH, Weinheim.

Kottke, I., and F. Oberwinkler. 1986. Trees 1:1–24.

Kottke, I., and F. Oberwinkler. 1988. KfK PEF 39. 19 p. Kernforschungszentrum Karlsruhe.

Kottke, I., C. Rapp, and F. Oberwinkler. 1986. Eur J For Pathol 16:159–171.

Kowalkowski, A., Z. Brogowski, and J. Kocon. 1988. Forst und Holz 43:293–295.

Krause, G.H.M., K.D. Jung, and B. Prinz. 1983, 1985. VDI Ber 500:257–266; 560:627–656.

Kreutzer, K. 1981. Mitt Dtsch Bodenkdl Ges 32:272–286.

Küppers, K., and G. Klumpp. 1988. GeoJournal 17:271–275.

Küppers, M., W. Zech, E.-D. Schulze, and E. Beck. 1985. Forstwiss Centralbl 104:23–36.

Lakhani, J.H., and H.G. Miller. 1978/1980. In T.C. Hutchinson and M. Havas, eds. Effects of acid precipitation on terrestrial ecosystems, 161–172. Plenum Press, New York.

Lamersdorf, N. 1989.Toxicol Environm Chem 18:239–247.

Lamersdorf, N. 1988. Verteilung und Akkumulation von Spurenstoffen in Waldökosystemen. Diss Univ Göttingen, Ber Forschungszentr Waldoekosys 36. 205 p.

Lange, O.L., J. Gebel, E.-D. Schulze, and H. Walz. 1985. Forstwiss Centralbl 104:186–198.

Lange, O.L., J. Gebel, H. Zellner, and P. Schramel. 1986. Tagungsber Statusseminar Ursachenforschung zu Waldschäden, Jülich 1986, 127–147.

Lange, O.L., H. Zellner, J. Gebel, P. Schramel, B. Köstner, and F.-C. Czygan. 1987. Oecologia (Berlin) 73:351–357.

Leonardi, S., and W. Flückiger. 1987. Tree Physiology 3:137–145.

Leonardi, S., and W. Flückiger. 1988. Forstwiss Centralbl 107:160–172.

LeTacon, F., and F. Lapeyrie. 1988. In J. Bervaes, P. Mathy, and P. Evers, eds. Relationships between above and below ground influences of air pollutants on forest trees (Air Poll Rep Ser 16), 44–59. Comm Eur Communities, Brussels.

Lindberg, S.E., G.M. Lovett, and K.-J. Meiwes. 1987. In T.C. Hutchinson and K.M. Meema, eds. Effects of atmospheric pollutants on forests, wetlands and agricultural ecosystems, 117–130. Springer, Berlin.

Lohm, U., K. Larssen, and H. Nömmik. 1984. Soil Biology Biochemistry 16:343–346.

Lötschert, W., and H.-J. Köhm. 1978. Oecologia (Berl) 37:121–132.

Ludyck, G. 1977. Theorie dynamischer Systeme. Elitera Verlag.

Magel, E., and H. Ziegler. 1986. Forstwiss Centralbl 105:234–238.

Magel, E., and H. Ziegler. 1987. Allgem Forstzeitschr 43:731–733.

Marschner, H., N. Haeussling, and E. Leisen. 1985. In Root-rhizosphere interaction

Commission of the European Communities (Proc EC-COST Workshop, Juelich, 1985), 113–118.

Matzner, E. 1988. Der Stoffumsatz zweier Waldökosysteme im Solling. Habilitation thesis, Forstl Fak, Univ Göttingen. 254 p.

Matzner, E. 1989. *In Acidic precipitation: case study Solling Vol. I*. Springer-Verlag, New York (in press).

Matzner, E., D. Murach, and H. Fortmann. 1986. Water, Air Soil Pollut 31:273–282.

Matzner, E., and B. Ulrich. 1985. Experientia 41:578–584.

Mayer, R. 1981. *Natürliche und anthropogene Komponenten des Schwermetallhaushalts von Waldökosystemen* Goettinger Bodenkdl Ber 70. 292 p.

Mayer, R. 1985. *Im Nationalpark Bayer Wald 5* 343–351. Tagungsber, Grafenau.

Mayer, R., and B. Ulrich. 1974. Ecol Plant 9:157–168.

Meisch, H.V., M. Vessler, W. Reinle, and A. Wagner. 1986. Experientia 42:537–542.

Meiwes, K.-J. 1983. Mitt Dtsch Bodenkdl Ges 38:257–262.

Meiwes, K.-J. 1985. Mitt Dtsch Bodenkdl Ges 43(II):981–986.

Mengel, K., H.J. Lutz, and M.Th. Breininger. 1987. Z Pflanzenernaehr Bodenk 150:61–68.

Metzler, B. 1985. *In Root-rhizosphere interaction* Commission of the European Communities (Proc EC-COST workshop, Juelich, 1985), 92–94.

Metzler, B., and F. Oberwinkler. 1986. Allgem Forstzeitschr 42:649–651.

Meyer, F.H. 1984. Allgem Forstzeitschr 40:212–228.

Meyer, F.H. 1987a. Forstwiss Centralbl 106:84–92.

Meyer, F.H. 1987b. Allgem Forstzeitschr 43:754–757.

Meyer, J., B.V. Schneider, K. Werk, R. Oren, and E.-D. Schulze. 1988. Oecologia, 77:7–13.

Mies, E., H.W. Zöttl. 1985. Forstwiss Centralbl 104:1–8.

Mohr, H. 1984. Biologie in unserer Zeit 14:103–110.

Mulder, J. 1988. Impact of acid atmospheric deposition on soils: Field monitoring and aluminum chemistry. Diss Univ Wageningen. 163 p.

Müller, Ch. and F. Oberwinkler. 1988. *In* F. Horsch. 4. Statuskoll PEF 35, Kernforschungszentrum Karlsruhe, Vol 1:113–121.

Murach, D. 1984. Die Reaktion der Feinwurzeln von Fichten (*Picea abies* Karst.) auf zunehmende Bodenversauerung. Diss Univ Göttingen, Göttinger Bodenkdl Ber 77. 126 p.

Murach, D. 1988. *In* J. Bervaes, P. Mathy, and P. Evers, eds. *Relationships between above and below ground influences of air pollutants on forest trees* (Air Poll Rep Ser 16), 148–167. Comm Eur Communities, Brussels.

Murach D. and H. Wiedemann. 1988. Dynamik und chemische Zusammensetzung der Feinwurzeln von Waldbäumen als Maß für die Gefährdung von Waldökosystemen durch toxische Luftverunreinigungen. Ber Forschungszentr Waldökosyst Univ Göttingen B 10: 287 p.

Murach, D., and B. Ulrich. 1988. GeoJournal 17:253–260.

Nebe, W., H.J. Fiedler, G. Ilgen, and W. Hofmann. 1987. Flora 179:453–462.

Neighbour, E.A. 1988. *In* J. Bervaes, P. Mathy, and P. Evers, eds. *Relationships between above and below ground influences of air pollutants on forest trees.* (Air Poll Rep Ser 16), 168–174. Comm Eur Communities, Brussels.

Neite, H., and R. Wittig. 1985. Acta Oecologica 6:375–385.

Neitzke, M., and M. Runge. 1985. Flora 177:237–249.

Neitzke, M., and M. Runge. 1987. Bot Jahrb Syst 108:403–415.

Nienhaus, F. 1988. *In* Minister f Umwelt NRW, ed. Forschungsber Luftverunreinigungen u Waldschäden 1:355–380.

Nilsson, J., ed. 1986. *Critical loads for nitrogen and sulphur.* Nordisk Ministerråd Miljo Rapport 11. Stockholm. 232 p.

NIVA. 1987. *1000 lake survey 1986.* Norweg State Poll Control Authority Rep 283/87. Oslo.

Oren, R., E.D. Schulze, K.S. Werk, and J. Meyer 1988b Oecologia (Berlin). 77:163–173.

Oren, R., K.S. Werk, and E.D. Schulze, 1986. Trees 1:61–69.

Oren, R., K.S. Werk, E.D. Schulze, J. Meyer, B.U. Schneider, and P. Schramel 1988a Oecologia (Berlin) 77:151–162.

Osonubi, O., R. Oren, K.S. Werk, E.-D. Schulze, and H. Heilmeier. 1988. Oecologia (Berlin) 77: 1–6.

Paramesvaran, N., S. Fink, and W. Liese. 1985. Eur J For Path 15:168–182.

Parker, G.R., W.W. McFee, and J.M. Kelly. 1978. J Environ Qual 7:337–342.

Pechmann, H.V. 1958. Forstwiss Centralbl 77:357–373.

Persson, H. 1985. *In* H. Persson, ed. Vad händer med skogen: shogsdöd pa väg? 117–131. Liber Förlag, Stockholm.

Persson, H. 1988. *In* J. Bervaes, P. Mathy, and P. Evers, eds. *Relationships between above and below ground influences of air pollutants on forest trees* (Air Poll Rep Ser 16) 180–186. Comm Eur Communities, Brussels.

Praag, H.J. van, S. Sougnez-Remy, F. Weissen, and G. Carlett. 1988. Plant Soil 105:87–103.

Prenzel, J. 1983. *In* B. Ulrich and J. Pankrath, eds. *Effects of accumulation of air pollutants in forest ecosystems,* 157–170. Reidel, Dordrecht.

Prinz, B., G.H.M. Krause, and H. Stratmann. 1982. *Bericht Nr 28.* Landesamt f Immissionsschutz, Essen.

Puhe, J., and A. Aronsson. 1986. Forst- und Holzwirt 41:464–470.

Puhe, J., and B. Ulrich. 1985. Arch Hydrobiol 102:331–342.

Puhe, J., H. Persson, and I. Börgesson. 1986. Allgem Forstzeitschr 42:488–492.

Raben, G. 1988. Untersuchungen zur raumzeitlichen Entwicklung boden- und wurzelchemischer Stressparameter und deren Einfluss auf die Feinwurzelentwicklung in bodensauren Waldgesellschaften des Hils. Diss Forstl Fak, Univ Göttingen, Ber Forschungszentr Waldökosys, Univ Göttingen A 38: 252 p.

Rademacher, P., J. Bauch, and J. Puls. 1986. Holzforschung 40:331–338.

Raisch, W. 1983. Bioelementverteilung in Fichtenökosystemen der Bärhalde (Südschwarzwald). Freiburger Bodenkundl Abhandl 11:239 p.

Rastin, N., B. Ulrich. 1988. Z Pflanzeneraehr Bodenk 151:229–235.

Rastin, N., H. Heinrichs, B. Ulrich, and R. Mayer. 1985. Ber Forschungszentr Waldoekosys, Univ Göttingen 10:234–321.

Rayermann, M., and F.-W. Schlote. 1986. Allgem Forstzeitschr 42:1264–1265.

Reemtsma, J.B. 1986. Allgem Forst- u Jagdzeitung 157:196–200.

Rehfuess, K.E. 1983. Allgem Forstzeitschr 39:1111.

Rehfuess, K.E., and Chr. Rehfuess. 1988. Allgem Forst- u Jagdztng 1559:20–26.

Rehfuess, K.E., and H. Rodenkirchen. 1984. Forstwiss Centralbl 103:248–262.

Reiter, H., J. Bittersohl, R. Schierl, and K. Kreutzer. 1986. Forstwiss Centralbl 105:300–309.

Richardson, D.H.S., and D. Dowding. 1988. *In* J.N. Cape and P. Mathy, eds. *Scientific basis of forest decline symptomatology* (Air Poll Rep Ser 15), 133–148. Comm Eur Communities, Brussels.

Richter, D.D., and S.E. Lindberg. 1988. J Environ Qual 17:619–622.

Ritter, T., I. Kottke, and F. Oberwinkler. 1986. 2. Statuskoll PEF 1986, KfK-PEF 4: 315–323, Kernforschungszentrum Karlsruhe.

Roelofs, J.G.M., A.W. Boxman, H.F.G. van Dijk, and A.L.F.M. Houdijk. 1988. *In* J. Bervaes, P. Mathy, P. Evers, eds. *Relationships between above and below ground influences of air pollutants on forest trees.* (Air Poll Rep Ser 16), 205–221. Comm Eur Communities, Brussels.

Roelofs, J.G.M., A.J. Kempers, A.L.F.M. Houdijk, and J. Jansen. 1985. Plant Soil 84:45–56.

Roloff, A. 1986. Morphologie der Kronenentwicklung von Fagus sylvatica L. (Rotbuche) unter besonderer Berücksichtigung möglicherweise neuartiger Veränderungen. Ber Forschungszentr Waldökosyst Univ Göttingen 18: 177 p.

Roloff, A. 1988. *Kronenentwicklung und Vitalitätsbeurteilung ausgewwählter Baumarten der gemässigen Breiten.* Schriften Forst. Fak. Univ. Göttingen 93. 258 pp. Sauerländer Verlag, Frankfurt.

Roloff, A. 1989. *In* J.N. Cape and P. Mathy, eds. *Scientific basis of forest decline symptomatology* (Air Poll Rep Ser 15), 73–91. Comm Eur Communities, Brussels.

Rost-Siebert, K. 1983. Rost-Siebert, K. 1985. Allgem Forstzeitschr 39:686–689. Untersuchungen zur H- und Al-Ionen-Toxizität an Keimpflanzen von Fichte (*Picea abies*) und Buche (*Fagus sylvatica*) in Lösungskultur. Diss Univ Göttingen, Ber Forschungszentrum Waldökosysteme Univ Göttingen 12. 219 p.

Rost-Siebert, K., and G. Jahn. 1988. Forst u Holz 43:75–81.

Ruck, B, and E.W. Adams 1988. *In* F. Horsch, et al., eds. 4. Statuskoll PEF. 1988. KfK-PEF 35, Vol 2:627–640. Kernforschungszentrum Karlsruhe.

Ruetze, M., U. Schmitt, W. Liese, and K. Küppers. 1988. Allgem Forst- u Jagdztng 159:195–203.

Schierl, R., A. Göttlein, E. Hohmann, D. Trubenbach, and K. Kreutzer. 1986. Forstwiss Centralbl 105:309–313.

Schlechte, G. 1986. Z Mykol 52:225–232.

Schlechte, G. 1987. *In* Czech Soc Sci and Techn, ed. *Ekologi mykorrhiz a mykorrhiznich hub: Imise a mykorrrhiza,* 82–92. Pardubice.

Schlichter, T.M., R.R. van der Ploeg, and B. Ulrich. 1983. Z Pflanzenernaehr Bodenk 146:725–735.

Schmidt, M., R. Mayer, and B. Georgi. 1985. VDI-Ber 560:423–438.

Schmitt, G. 1987. Methoden und Ergebrisse der Nebellanalyse. Ber Inst f Meteorologie und Geophysik, Univ Mainz 72. 199 p.

Schmitt, U., and W. Liese. 1987. Eur J For Path 17:202–297.

Schreiber, H., and I. Rentschler. 1988. *In* F. Horsch. et al., eds. 4. Statuskoll PEF 1988, KfK-PEF 35(2):657–622, Kernforschungszentrum Karlsruhe.

Schröder, D., J. Bauch, and R. Endeward. 1988. Trees 2:96–103.

Schröter, H., and E. Aldinger. 1985. Allgem Forstzeitschr 41:438–442.

Schulte, A., and M. Spiteller. 1987. Forst und Holz 42:150–154.

Schulte-Bisping, H., and D. Murach. 1984. Ber Forschungszentr Waldoekosys Univ Göttingen 4:207–265.

Schultz, R. 1987. Vergleichende Betrachtung des Schwermetallhaushalts verschiedener Waldökosysteme Norddeutschlands. Diss GH Kassel, Ber Forschungszentr Waldoekosys Univ Göttingen 32. 217 p.

Schultz, R., N. Lamersdorf, H. Heinrichs, R. Mayer, and B. Ulrich. 1988. Ber Forschungszentr Waldoekosys Univ Göttingen Vol B 7. 143 p.

B. Ulrich                                                                                      271

Schultz, R., M. Schmidt, J. Godt, and R. Mayer. 1987. Verh Ges f Okologie (Giessen, 1986) 16:297–303.
Schulze, E.-D.R. Oren, and R. Zimmerman. 1987. Allgem Forstzeitschr 42:725–730.
Schütt, P. 1984. Der Wald stirbt an Stress. Bertelsmann, München.
Schuurkes, J.A.A.R., M.M.J. Maenen, and J.G.M. Roelofs. 1988. Atmospheric Environment 22:1689–1698.
Senser, M., K.-A. Höpker, A. Penker, and B. Glashagen. 1987. Allgem Forstzeitschr 42:709–714.
Seufert, G., and V. Arndt. 1986. Allgem Forstzeitschr 41:545–549.
Seufert, G., and V. Arndt. 1988. GeoJournal 17:261–270.
Siefermann-Harms, D., and R. Hampp. 1987. Forschungsber KfK-PEF 10:151–175, Kernforschungszentrum Karlsruhe.
Simon, B., and G.M. Rothe. 1985. Allgem Forstzeitschr 41:931–936.
Skeffington, R.A., and T.M. Roberts. 1985. Oecologia 65:201–206.
Smith, W.H., and Th.G. Siccama. 1981. J Environ Qual 10:323–333.
Söderlund, R., L. Granat, and H. Rodhe. 1985. Dept of Meteorology, Univ Stockholm Report 1985-09-26 CM-69.
Solberg, W., and E. Steinnes. 1983. Proc Acid Rain and For Resources Conf, Quebec.
Spedding, D.I., I. Ziegler, R. Hampp, and H. Ziegler. 1980. Z Pflanzenphysiol 96:351–364.
Stienen, H. 1986. Forstarchiv 57:227–231.
Stienen, H., R. Barckhausen, H. Schaub, and J. Bauch. 1984. Forstwiss Centralbl 103:262–274.
Stienen, H., and J. Bauch. 1988. Plant Soil 106:231–238.
Stock, R. 1988. Forst u Holz 43:283–286.
Strack, S., and H. Unger. 1988. Untersuchungen zum Wassertransport und zur Wasserführung in Fichten. Forschungsber KfK-PEF 38. Kernforschungszentrum Karlsruhe. 64 p.
Süsser, P. 1987. Art, Menge und Wirkungsweise von Puffersubstanzen in Mineralbodenhorizonten forstlich genutzter Böden des Fichtelgebirges. Diss TU München-Weihenstephan.
Sutinen, S., G. Wallin, L. Skärby, and G. Sellden. 1988. *In* J. Bervaes, P. Mathy, and P. Evers, eds. *Relationships between above and below ground influences of air pollutants on forest trees* (Air Poll Rep Ser 16), 251–259. Comm Eur Communities, Brussels.
Tischner, R., U. Kaiser, and A. Hüttermann. 1983. Forstwiss Centralbbl 102:329–336.
Tyler, G. 1972. Ambio 1:52–59.
Ulrich, B. 1975. Forstwiss Centralbl 94:280–287.
Ulrich, B. 1980. Allgem Forstzeitschr 35:1198–1202.
Ulrich, B. 1981a. Z. Pflanzenernaehr Bodenk 144:289–305.
Ulrich, B. 1981b. Z. Pflanzenernaehr Bodenk 144:647–659.
Ulrich, B. 1983a. *In* B. Ulrich and J. Pankrath, eds. *Effects of accumulation of air pollutants in forest ecosystems*, 1–29. Reidel, Dordrecht.
Ulrich, B. 1983b. *In* B. Ulrich and J. Pankrath, eds. *Effects of accumulation of air pollutants in forest ecosystems*, 33–45. Reidel, Dordrecht.
Ulrich, B. 1984. Atmospheric Environment 18:621–628.
Ulrich, B. 1985. *In* C. Troyanowski, ed. *Air pollution and plants*, 151–181. VCH, Weinheim.
Ulrich, B. 1986. Z Pflanzenernaehr Bodenk 149:702–717.
Ulrich, B. 1987. Ecol Studies 61:11–49.

Ulrich, B. 1988. Z Pflanzenernaehr Bodenk 151:171–176.

Ulrich, B., and V. Malessa. 1989. Z Pflanzenernaehr Bodenk 152:81–84.

Ulrich, B., R. Mayer, and P.K. Khanna. 1979. *Deposition von Luftverunreinigungen und ihre Auswirkungen in Waldökosystemen im Solling.* Schriften Forstl Fak, Univ Göttingen 58:291 pp. Sauerländer Verlag, Frankfurt.

Ulrich, B., and H. Meyer. 1987. *Chemischer Zustand der Waldböden Deutschlands zwischen 1920 und 1960, Ursachen und Tendenzen seiner Veränderung.* Ber Forschungszentrum Waldökosysteme Univ Göttingen B6. 133 p.

Ulrich, B., D. Pirouzpanah, and D. Murach. 1984. Forstarchiv 55:127–134.

Unsworth, M.H., and A. Crossley. 1987. *In* T.C. Hutchinson and K.M. Meema, eds. *Effects of atmospheric pollutants on forests, wetlands and agricultural ecosystems,* NATO ASI Series Vol. G 16, 171–188.

Verhoeven, W., R. Herrmann, R. Eiden, and O. Klemm. 1987. Theoretical Appl Climatology 38:210–221.

Vogelei, A., and G. M. Rothe. 1988. Forstwiss Centralbl 107:348–357.

Vries, W. de, and A. Breeuwsma. 1987. Water Air Soil Pollut 35:293–210.

Werk, K.S., R. Oren, E.-D. Schulze, R. Zimmermann, and J. Meyer. 1988. Oecologia (Berlin) 76:519–524.

Wiedey, G., and M. Gerriets. 1986. Ber Forschungszentr Waldökosyst B2:26–55.

Winkler, P. 1982. Z Pflanzenernaehr Bodenk 145:576–585.

Wittig, R., and H. Neite. 1985. Vegetatio 64:113–119.

Wittig, R., and W. Werner. 1986. Düsseldorfer Geobot Kolloq 3:33–70.

Wolters, V. 1988. Pedobiologia 32:387–398.

Wolters, V. 1989. Jber. Naturwiss. Ver. Wuppertal 42 (in press).

Wolters, V., and J. Schauermann 1989. Ber. Forschungszentr. Waldökosyst. Univ. Göttingen A 49: 141–151.

Wolters, V., and S. Scheu. 1987. Statussem. Ursachenforschung Waldschäden, Jül-Spez 413: 336–340. Kernforschungsanlage Jülich.

Zech, W., and E. Popp. 1983. Forstwiss Centralbl 102:50–55.

Zech, W., Th. Suttner, and E. Popp. 1985. Water Air Soil Pollut 25:175–183.

Zezschwitz, E. von 1982. Forst u Holzwirt 37:275–276.

Zezschwitz, E. von 1985a. Geol Jb F 20:3–41.

Zezschwitz, E. von 1985b. Forstwiss Centralbl 104:205–220.

Zezschwitz, E. von 1986. Geol Jb F 21:3–61.

Zezschwitz, E. von 1987. Allgem Forst- u Jagdztng 158:136–147.

Zimmermann, R., R. Oren, E.-D. Schulze. and K.S. Werk. 1988. Oecologia (Berlin) 76:513–518.

Zöttl, H.W. 1987. *In* T.C. Hutchinson and K.M. Meema. *Effects of atmospheric pollutants on forests, wetlands and agricultural ecosystems,* 255–266. Springer Verlag NATO ASI Series G 16.

Zöttl, H., K.-H. Feger, and G. Brahmer. 1985. Naturwissenschaften 72:268–270.

Zöttl, H.W., R. Hüttl. 1985. Allgem Forstzeitschr 40:197–199.

Zöttl, H.W., and E. Mies. 1983. Mitt Dtsch Bodenkd Ges 38:429–434; Allgem Forst- u Jagdzeitung 154:111–114.

# Effects of Acidic Precipitation on the Biota of Freshwater Lakes

P.M. Stokes*, E.T. Howell[†], and G. Krantzberg[‡]

## Abstract

Acidification of freshwater systems can be natural or anthropogenic in origin. Studies of acidified waters have included field surveys of lakes ranging in pH from acidic to circumneutral (spatial comparisons), as well as temporal comparisons, and experimental manipulation of pH in large and small systems. Relevant laboratory studies have addressed the direct response of organisms to $H^+$ concentration and related chemical changes. A large body of descriptive literature exists on biotic response to acidification, including not only fish but also the lower trophic levels. In addition, there is reasonable understanding of the changes in water chemistry that accompany acidification. The challenge now is to integrate the information and derive some cause-and-effect relationships. We exclude fish from the present review.

In general, the species richness of aquatic flora and fauna decreases with increasing $H^+$ concentration. Typically this is accompanied by an increase in the abundance and biomass of species that are quantitatively less significant in circumneutral lakes. Certain functional groups of macroinvertebrates, such as scrapers, may be more susceptible to changes in pH than others, such as shredders. With the exception of zooplankton, community biomass of plants and animals is not significantly affected by acidification, presumably because of the increased contribution of acid-tolerant species.

Many potential biological effects are associated with acid-related chemical and physical changes that occur in aquatic systems—for example, decreases in dissolved inorganic carbon (DIC), changes in the speciation of inorganic carbon, mobilization and changes in speciation of metals, changes in phosphorus chemis-

---

*To whom correspondence should be addressed: Institute for Environmental Studies and Department of Botany, University of Toronto, Toronto, Ontario M5S 1A4, Canada.

†Department of Botany, University of Manitoba, Winnipeg, Manitoba R3T 2N2, Canada.

‡Ontario Ministry of the Environment, Water Resources Branch, 1 St. Clair Avenue West, Toronto, Ontario M4V 1P5, Canada.

try, and increased clarity of the water—but with the possible exception of primary producers and DIC, we are unable to find evidence that specific chemical changes alone have structured the communities in acidified systems.

We also examine the indirect effects of acidification through changes in, for example, herbivore-plant and predator-prey balances. There is some evidence that release from grazing pressure contributes to alterations in the periphyton community structure, but other causal interactions remain speculative. In common with the investigation of chemical-biological relationships, this is a fruitful area for research.

# I. Introduction

Since the recognition during the mid-1970s that acidic deposition was a widespread phenomenon occurring over several sensitive areas of North America and Europe, a great deal of descriptive work has been done on the relationship between low pH and its effect on the biota of surface waters. During the 1980s, a number of reviews have been published on effects, both chemical and biological, of acidic deposition on fresh waters (Haines, 1981; Mierle et al., 1986; National Academy Press, 1985; Altshuller and Linthurst, 1984; EPRI, 1986; Dillon et al., 1984; ORNL, 1986; Stokes, 1986; Eilers et al., 1984). In this chapter, we do not attempt to repeat these recent, often excellent, reviews of biological effects. Rather, our approach is to summarize some of the physical and chemical changes related to acidification, especially those that are expected to affect directly the aquatic biota that we are considering, and determine to what degree the observed biotic effects can be related to these chemical changes.

Changes in mineral nutrients have the potential to affect directly the primary producer communities, whereas changes in trace metals have the potential to influence all trophic levels directly through toxic action. The hydrogen ion itself can be extremely toxic to biota, but it also influences the solubility and chemical speciation of many other chemical constituents of surface waters. It is often difficult, if not impossible, to unravel cause and effect, and to differentiate between direct and indirect effects of acidification, from observations in the field. Therefore, we have included references to experimental work, mainly field manipulations but in some instances laboratory studies. With these data we attempt to determine if mechanistic explanations exist to account for phenomena observed in nature. In many instances, however, the mechanisms are still obscure, and we speak of "acid-related changes" without making a commitment as to the immediate causal factor(s).

Because of the interdependence between trophic levels and the importance of interactions between species populations, it is likely that some of the biotic changes observed in acidic systems may result from indirect effects of acidification. Indirect effects may stem from shifts in the quantity or types of herbivores and predators, changes in size structure of communities, and alterations in the supply of energy by autotrophs and detritivores. There may even be the "loss" of an entire trophic level. For example, many acidic lakes have no fish. This is expected to have a profound influence on the populations of organisms that are normally consumed by fish. The importance of fish predation on prey composition, abundance, and behavior, described from studies of nonacidic and usually nonoligotrophic systems, suggests that indirect effects of acidification via fish losses are likely to occur. Exactly what this effect would be on acidified systems is difficult to predict from models of fish predation in less stressed systems. We address several instances of possible biotic effects of acid-related changes in community structure. We generally use *acidic* to mean alkalinity less than 0 or pH below 6.0, but we also recognize that acid-stressed lakes, not yet acidic under this definition, may experience periodic fluctuations of pH as their buffering capacity is depleted. Therefore, we do not always adhere consistently to this "chemical" definition.

It seems to us that the descriptive phase of the study of lake acidification is sufficiently mature that other approaches are timely and should be intensified.

# II. Physical and Chemical Changes Related to Acidification with Potential Importance to Biota of Softwater Lakes

## A. Inorganic Carbon

The relationship between the chemical species of dissolved inorganic carbon (DIC) and pH is well known. Inorganic carbon equilibria represent a major component of the buffering system in fresh water. Additions of acid (e.g., from deposition) will result in changes in both the form and quantity of inorganic carbon in fresh water. In acid-sensitive systems, as well as in surface waters, inputs of $H^+$ are, in part, neutralized by conversion of $HCO_3^-$ to $H_2CO_3$, which decomposes to $CO_2$ (aq) and $H_2O$ (Stumm and Morgan, 1981). The equilibrium concentration of $CO_2$ (aq) in surface waters is a function of the $pCO_2$ of the atmosphere and temperature, (atmospheric $pCO_2$ can be considered a constant). The equilibrium concentration of $CO_2$ is low; at saturation, water will contain only 0.15 mg/L $CO_2$ at 25°C (Stumm and Morgan, 1981). As $HCO_3^-$ is neutralized in the course of lake acidification, DIC is likely to be lost. The $pCO_2$ of water exceeds the $pCO_2$ of the atmosphere, and $CO_2$ degases. Below pH 5.5, DIC exists predominantly as $CO_2$ (aq) (Wetzel et al., 1984). The critical pH range for quantitative and qualitative changes in DIC is the same range over which the most striking changes in the biota have been observed.

Decrease of DIC in epilimnetic waters has the potential to influence primary producers. Williams and Turpin (1987) indicate the potential importance of DIC in determining community structure of phytoplankton in acidified systems. In laboratory experiments, different species of algae showed varying affinities for DIC, and this governed the outcome of competition experiments at variable concentrations of DIC. The competitive ability of planktonic algae in acidified systems might be determined in part by their ability to utilize low concentrations of DIC.

Addition of sulfuric acid to Lake 223, Experimental Lakes Area (ELA), resulted in loss of DIC from the epilimnion as bicarbonate was converted to $CO_2$ and degasing occurred. However, in the deepest parts of the lake in summer, there was no net bicarbonate depletion because bacterial production of bicarbonate offset $CO_2$ losses (Schindler et al., 1980). Community production rates of phytoplankton in the Lake 223 acidification study did not indicate carbon limitation (Schindler et al., 1980); however, this is not incompatible with the hypothesis that species composition is affected by reduced DIC. If species with a high affinity for DIC predominate in acidic conditions, community primary production would not be expected to change from nonacidified levels. It is still feasible, however, that acid-related changes in community structure of primary producers might be influenced by decreases in DIC and also that production rates might be affected on a local scale; that is, at certain regions within a particular lake there could still be DIC-limited production. Studies of periphytic algae have indicated that the response of epilithon to acidification may be related to changes in DIC. Turner and others (1987) observed that net productivity of epilithic algae in three experimentally acidified lakes was reduced relative to nonacidic reference lakes of comparable trophic status. The decreased productivity was attributed to carbon limitation of photosynthesis resulting from the acid-related loss of DIC from the water mass.

The growth of submerged macrophytes in low-alkalinity lakes is strongly influenced by DIC, regardless of pH condition, because of the potential for carbon limitation of photosynthesis (Wetzel et al., 1985; Roelofs et al., 1984; Boston, 1986). The interstitial water of lake sediments typically has higher concentrations (1 to 2 orders of magnitude greater) of DIC than does the overlying water (Wium-Anderson and Anderson, 1972; Boston, 1987). This is a result of microbial respiration within the substrate-rich sediments. The DIC gradient between the water column and the sediment increases as $HCO_3^-$ is depleted from the water column by acidification. Many aquatic macrophyte species in different families, that is, a genetically diverse group, show anatomical and physiological adaptations for utilization of sediment DIC (Raven et al., 1988; Boston, 1986; Wetzel et al., 1984). This is best illustrated by the isoetid type of macrophyte that predominates in low-alkalinity lakes. The growth form is a rosette of short linear leaves and a large root mass, which functions as a conduit for transfer of $CO_2$ from the sediment to the leaves.

The distribution and abundance of macrophytes in acidifying lakes may be affected as a consequence of changes in DIC in the water column. Species capable of using a sediment source of DIC or having some physiological adaptation for dark fixation of carbon by crassulacean acid metabolism (CAM) will be favored. Macrophytes that are not rooted in sediment, such as *Sphagnum*, may be at a disadvantage. Wetzel and others (1984) determined that *Sphagnum*, and macroalgae in acidic lakes were likely to be acutely carbon limited. In this respect, the invasion of *Sphagnum* described for acidic waters in Sweden (Grahn, 1977) and the Netherlands (Roelofs, 1983) is perplexing. Interestingly, however, Roelofs and others (1984) reported that the *Sphagnum* invasion was restricted to acidic waters with high DIC concentrations (150–870 µmol/L, that is, above equilibrium concentration) and that the moss was not abundant in other acidic waters with DIC in the range of 50 to 60 μmol/L. These high DIC concentrations could only result from large DIC fluxes from the sediments. This may occur either because of acidification of carbonate-bearing sediments, high rates of microbial respiration, or percolation of carbon-rich groundwater. It is probable that the lack of a widespread invasion by *Sphagnum* in North American acidic lakes (Roberts et al., 1985; Wile et al., 1985; Hunter et al., 1986) with its occasional abundance in acidic lakes (Roberts et al., 1985; Hendrey and Vertucci, 1980; Catling et al., 1986) may be explained by different degrees of enrichment of the water column by DIC from the sediment.

In the Roelofs's study, the invasion of *Juncus bulbosus* in acidic waters was shown to depend upon high concentrations of DIC in the open water. This species, although rooted, fixes $CO_2$ predominantly from the water column.

The overall impression from available data is that DIC plays an important role in structuring primary producer communities in acidifying systems.

## B. Phosphorus

Phosphorus is frequently the limiting nutrient for phytoplankton in oligotrophic lakes (Lean and Nalewajko, 1976), and studies have related phytoplankton community structure to the availability of P. Although some early reports on phytoplankton production in acidified systems suggested that acidification limited the biomass of primary producers, overall there seems to be no consistent relationship between acidification and production of phytoplankton (Dillon et al., 1984). A significant relationship was found, however, between total P and primary production over a range of pH that included acidic lakes (Dillon et al., 1984). The effects of acidification may be confounded with P limitation because most acid-sensitive systems are also oligotrophic and often P limited. However, experimental acidification of Lake 223 did not result in decreased P concentration (Schindler, 1980), and there are no measurements that directly demonstrate decreases in P concentration related to acidification in field studies.

The availability rather than the total concentration of P might decrease with acidification (e.g., Dickson, 1978). In particular, aluminum (section II,C) can complex or precipitate P, rendering it unavailable to plants (Driscoll et al., 1983). Although it is possible that some of the acid-related changes in species composition in the plankton or other plant communities might be related to different species' affinities for uptake and storage of P, overall productivity of the phytoplankton does not consistently decrease with acidification, as has been pointed out above. That is to say, the process of oligotrophication, which is depletion of nutrients accompanied by decreased production (Grahn et al., 1974), does not appear to be a generally occurring phenomenon in acidic lakes, at least not in the earlier stages of acidification (Schindler, 1980).

Acid-related changes in the chemical speciation of P are not expected to pose problems in terms of P availability to organisms, as the $H_2PO_4$ form predominates over pH 2.5 to 7.0 (Nalewajko and O'Mahoney, 1988), but, as noted above for Al, binding by Al, Fe, Mn, and Ca could decrease available P concentrations, especially over the pH 5.0 to 6.0 range (Stumm and Morgan, 1970). A recent study by O'Mahoney (1985) addressed the uptake rate of P by indigenous phytoplankton from two soft water lakes and showed that the uptake was depressed more by experimental depression of pH ("downshock") in the more acidic lake (originally pH 5.6) than it was for the community from a circumneutral one (originally pH 6.8). The species composition and community structure of the phytoplankton in the more acidic lake appeared to be very similar to those described for acidic lakes elsewhere. Based on this, the author expected that the flora of the acidic lake would be adapted to the more acidic conditions, and the greater sensitivity to downshock was unexpected.

Concentrations of P in both these lakes varied seasonally and were generally low, with no clear differences between the two (O'Mahoney, 1985). However, during experimental depression of pH over a relatively short period of time, the lowest concentrations of P occurred between pH 4.8 and 5.6, whereas the highest concentrations in both lakes occurred at pH 4.0. This pattern was more marked with "whole" than with filtered water: Particles $>0.45$ $\mu$m were the major source of P release (Nalewajko and O'Mahoney, 1988). These results suggest that increasing $H^+$ concentration can have an effect on dissolved P but that the change can go in either direction, depending upon the extent of acidification. The authors also discussed evidence for direct biological effects of $H^+$ on P uptake. Because the experiments were of short duration and the pH depressions quite severe, it is premature to extrapolate from the experiment to the field. Nevertheless, the observations of this study merit further consideration.

One approach to investigating the importance of a potentially limiting nutrient is to determine experimentally the biological response(s) to increases in its concentration. Nutrient-addition experiments have been carried out for P in acidic systems by several workers. For the most part, field enclosures (limnocorrals) were used. Avalos (1981) manipulated P and pH in a P-limited system. The combination of acidification and P enrichment produced changes in phytoplankton community structure that were quite distinct from the effect of P addition alone. The rate of response was also different. When P was added to a neutral system, the response of phytoplankton was almost immediate, but in the acidified enclosures, the response to P enrichment was delayed by 4 to 8 days. Marmorek (1984) followed changes in phytoplankton (measured as chlorophyll *a*), zooplankton biomass, and community structure in enclosure experiments involving acidification, nutrient enrichment, and a combination of the two. The systems were originally low in P. Enrichment included additions of P and N (ammonium nitrate). Chlorophyll and zooplankton biomass increased initially in the fertilized treatments, whether or not they were acidified. Depression of pH resulted later in the experiment in high mortality of the major crustacean zooplankter, *Daphnia rosea*, and this was followed by a massive increase in chlorophyll in the "acidified plus fertilizer" treatments. The chlorophyll increase was attributed to release of the phytoplankton community from grazing pressure and not to increases in rates of primary production; that is, the effect on the phytoplankton was a second-order effect. In general, these types of experiments suggest that fertilization with P can influence production regardless of acidification but that the response to P addition may be influenced by $H^+$ concentration.

We conclude, tentatively, that the influence of P on primary production may be modified by the existence of elevated $H^+$.

## C. Aluminum

The solubility and chemical speciation of aluminum is pH dependent (Campbell et al., 1983a). If pH is depressed, by whatever means, the concentration of Al in surface waters increases. Furthermore, the monomeric forms of Al, thought to be the most biologically available, are favored at low pH. Below pH 4.5, $Al^{3+}$ predominates, whereas between pH 4.5 and 6.3, $Al(OH)_3$ becomes dominant. Aluminum ions can combine with organic ligands such as fulvic acids and in this way may be rendered unavailable to biota. However, the binding of Al to organic ligands is itself a pH-dependent process. If pH increases rapidly, as it character-istically does in poorly buffered waters, Al ions that are not organically complexed can form precipitates (Dalal, 1975) that are believed to interfere with fish gas exchange and ionoregulation (Overrein et al., 1980). It is entirely possible that invertebrates possessing gill-like structures would be similarly affected; that is to say, in addition to the effect of pH on absolute concentrations of Al, short-term fluctuations in pH and Al may be of importance particularly to organisms with gills or gill analogues.

Even though there is no net addition of Al to lakes and their watersheds from aerial deposition, an increase in Al concentrations is one of the most striking chemical changes accompanying acidification (Dickson, 1980). The source of Al is the parent material, that is, the soils, rocks, and sediments. These are always rich in Al, but normally the metal is insoluble at neutral pH and therefore is nonavailable. Aluminum is not even considered to be very toxic (it is a class A cation; Nieboer and Richardson, 1980). Nevertheless, in surface waters below pH 5.6, it is present in sufficiently high concentrations that it can kill fish and invertebrates and possibly plants as well.

Much of the research on the biological effects of Al and low pH has been done in the laboratory. Certain fishes are killed or adversely affected by Al concentra-tions as low as 0.1 mg/L at moderately acidic pH (NRCC, 1986). In surface waters at pH 4.5, Al concentrations can be as high as 0.5 mg/L (NRCC, 1986). Driscoll (1980) demonstrated for acidic Adirondack lakes that inorganic monomeric Al increased rapidly with decreasing pH. Dillon and others (1980) similarly reported increases of total Al, as well as Mn, in acidified lakes in Ontario. Because of the chemical dependence of Al chemistry with $H^+$ concentration, it is not possible to find decreases in pH independent of Al concentration changes in field situations, and therefore it has not been possible to attribute the biological changes in the lower trophic levels that accompany acidification to Al, $H^+$, or both.

Algae are generally rather tolerant of high Al concentrations, at least more tolerant than are most freshwater planktonic crustaceans. NRCC (1986) lists 168 species of algae that are found in acidified lakes, ponds, and other systems but admits that there is no evidence to indicate that Al is responsible for the species and community shifts in algae that occur in acidic systems. Experimental manipulations of Al and $H^+$ have been done with invertebrates, and it appears that certain crustaceans experience disturbances in ionoregulation in response to Al treatment. Havas (1985) showed that for *Daphnia magna,* mortality at pH <5.0 and Al >0.32 mg/L was associated with net loss of Na and Cl. Havas and Likens (1985) were able to differentiate the respective effects of Al and $H^+$ on *D. magna* using $^{22}$Na. Aluminum inhibited Na uptake at pH 6.5 and 5.0 but had no additional effect to that of $H^+$ at pH 4.5. Aluminum also elicited a decreased efflux of Na in daphnids at pH 4.5. It is unfortunate that no experimental data exist for pH and Al combinations that occur in the field as lakes start to exhibit biotic changes, such as losses of daphnids. Most laboratory experiments use higher concentrations of Al than are found in all but the most acidic conditions, so we are still unsure of the significance of Al in the losses of certain zooplankters from acidic waters.

A recent study of the acute toxicity of acidity related chemical factors to the distribution of the amphipod *Hyalella azteca,* which is quite sensitive to low pH and is rarely found in the field at pH below 5.6 (Stephenson and Mackie, 1986). Previous laboratory studies had shown that this organism experienced $H^+$ toxicity at a pH of 5.0 and 5.5 for adults and juveniles, respectively (France and Stokes, 1987a). The influence of Mn, Ca, and Al on $H^+$ toxicity indicated that $H^+$ was much more important in terms of evoking a toxic response than was Al at low pH (France and Stokes, 1987b). Furthermore, there was little modification of the $H^+$ effect by Ca or Mn. The concentrations of all the ions were "realistic" (i.e., concentrations and combinations of respective elements that were found in the field), but the response was a short-term one to a 96-hour test.

In Lake 223, which was acidified experimentally, Al concentrations were much lower at a given pH than would be expected based upon data from other lakes that were acidified either via aerial deposition or natural processes (Bailey and Stokes, 1984). The lower Al levels in Lake 223 were attributed to absence of watershed acidification in the ELA experiments. The ELA experimenters observed losses of crayfish, mysids, and daphnids and an increase in dinoflagellates and a few acid-tolerant diatoms, as well as blooms of filamentous green algae. Changes in the invertebrate and algal communities in Lake 223 resembled rather closely, at least in terms of species replacement, changes seen in nonexperimental acidification (Schindler et al., 1985). There were some differences, however. For example, the production rates and the standing stock of phytoplankton were remarkably high in Lake 223 as compared with normal oligotrophic lakes. At the same time, however, there was an increase in chlorophyll production in the ELA control lake, so the interpretation is confounded. It is tempting to speculate, nevertheless, that the high production in Lake 223 was related to the relatively low concentrations of Al (and other potentially toxic metals) at a given pH, whereas the species composition was affected much less by Al than by $H^+$. In another study, Yan and Stokes (1978) acidified water in enclosures in a slightly acidic lake and monitored changes in phytoplankton community structure. Although Al was not measured, the enclosures were anchored in the open lake and did not contain sediments; therefore, no net increase in total Al concentration would have been expected. After acidification, the algal community in the enclosures was very similar to those in nearby lakes at the same pH. Collectively, these indirect lines of evidence suggest for phytoplankton the effect of acidification on phytoplankton species composition may be mainly related to $H^+$ concentration, rather than a response to Al or other potentially toxic metals.

Nalewajko and Paul (1985) studied the physiological responses of phytoplankton to perturbations of Al in two lakes, one circumneutral (pH 6.1–6.9) and the other somewhat acidic (pH 5.2–6.2). They found significant decreases in both photosynthetic production and phosphate uptake at 50 μg/L Al. Both physiological processes were more affected at pH 5.2 to 6.9 than at pH 4.5, and the depressions were greater for plankton from the more acidic lake than from the neutral one. The authors provide evidence for direct toxicity of Al and an abiotic effect, namely, the precipitation of phosphate by Al (see also section II, B). They point out that the effect of Al on phytoplankton may be more important in circumneutral than in acidic lakes. Although these experiments were done at more natural concentrations of Al than some of the other work described in this chapter, and the results do correlate the relationship between Al and phosphate uptake, they still do not contribute very much to our understanding of the relative effect of Al and $H^+$ on phytoplankton in the field, especially when acidification occurs over a long period of time.

Burton and Allen (1986) investigated the interactive effects of $H^+$, Al, and organic matter on five species of benthic invertebrates. The addition of 0.5 mg/L Al to acidified treatments caused significant increases in mortality over the effect of pH depression alone. Removal of organic matter from the water, apparently as a result of precipitation by monomeric Al, also increased mortality. Although these results do provide some information on the toxicity of Al under a variety of conditions, they do not enable us to determine for the "real world" situation the importance to biota of Al versus $H^+$ in acidifying systems.

In summary, experimental evidence suggests that Al mobilization and/or precipitation in response to changes in pH may be important in structuring aquatic communities. However, at concentrations that occur in the field, the relative importance of Al and $H^+$ is still unknown.

## D. Other Trace Metals

Campbell and others (1983b) have reviewed other metals that are mobilized and/or change their speciation between pH 7 and 4. In addition to Al, they recommend that Cd, Cu, Hg, Mn, Pb, and Zn be evaluated in terms of their interaction with acidification and their potential toxicity or bioaccumulation. With the exception of Mn, each of these metals can accumulate in surface waters through long-range aerial deposition. Manganese, like Al, is mobilized from parent material, and Zn increases as a result of both types of processes. Aluminum, Cu and Hg also have their chemical speciation altered across the range of pH of concern. In terms of relating the changes in metal concentration and speciation to biological responses in the field, we are unable to conclude much at this time. There is evidence that the concentrations of Zn reported in acidified lakes are in the range that is toxic to invertebrates in laboratory tests (Bailey and Stokes, 1984). Unfortunately, most laboratory tests on metals and algae are not useful in terms of resolving the effects of metals and acidity in the field. The metal levels are frequently too high, and single-metal tests are frequently used. To complicate matters, there is also an emerging body of information that shows a decrease in toxicity and uptake of metals at pH below 6.0, independent of any changes in availability (Campbell and Stokes, 1985). Thus the metals in an acidic system, providing that the $H^+$ per se is not at a toxic level, may have their biological uptake decreased by $H^+$ and be less toxic to biota than they would be at neutral pH. From the review cited above, then, $H^+$ would be considered as a possible ameliorative or protective factor in the context of toxic trace metals. The task of sorting out the respective influence of the hydrogen ion on chemical speciation of metals and biological uptake of metals, however, remains for the future.

## E. Changes in Transparency

Acidic soft water lakes may be colored to varying degrees, or they may be extremely clear. Color, resulting from dissolved yellow or brown humic and fulvic substances as well as suspended material in the form of plankton, detritus, and suspended mineral particles, affects the transparency of a lake. Transparency is commonly measured as Secchi disk depth or light-extinction coefficient. Light penetration determines the depth below the surface at which primary production can occur in a water body, and transparency also affects the thermal regime of a lake. However, climate and mixing are of more importance to the thermal regime in most lakes. Light penetration can also have secondary effects through its influence on primary production and primary producer communities. Photoinhibition of algal photosynthesis can occur under very high light conditions, and certain algal species are better adapted to high light conditions than are others.

The transparency of lakes increases during the course of acidification (Almer et al., 1978, 1974; Schindler, 1980). This phenomenon can occur in lakes that were originally colored, as well as in lakes of low color, although it may be more evident in the latter type of lake. Originally thought to be indicative of decreased production (e.g., Kwiatkowsi and Roff, 1976), acid-related increases in transparency are now believed to be a result of chemical changes, particularly changes in dissolved organic matter. The key reactions of the organic matter probably result from a combination of increased concentrations of $H^+$ as well as increased concentrations of metals (see sections II,C and II,D), but there is some disagreement as to the exact nature of the chemical change, as discussed below.

Yan (1983) described changes in transparency of Lohi Lake, a small acidic metal (especially Ni and Cu) contaminated lake near Sudbury, Ontario. The lake had been neutralized by liming and subsequently reacidified. Secchi transparency varied from 5.6 to 9.3 m and was not correlated with total P, chlorophyll $a$, or phytoplankton biomass. Yan proposed several mechanisms for transparency changes, which include precipitation of the organic matter and change in its color without precipitation. Almer and others (1978) suggested that the increased clarity was due to precipitation of yellow organic matter by Al, and this interpretation was supported by chemical evidence. The same authors had shown a threefold decrease in dissolved organic matter during the acidification of Lake Skarsjon (Almer et al., 1974). Davis and others (1985) also hypothesized that increased clarity was due to precipitation of organic matter with increased concentration of metals. However, Schindler and Turner (1982), in the course of experimental acidification of Lake 223, found no charge in dissolved organic carbon, although they observed increased transparency, a decrease in suspended C, and a decrease in total C.

Increased transparency associated with acidification involves not only increased light penetration but also, in cases of extreme lake clarity, changes in the thermal regime. The thickness of the epilimnion of Lohi Lake was correlated with transparency, which increased with the lake's reacidification (Yan, 1983). As the hypolimnetic temperature increased, in response to increased solar heating, the density gradient across the metalimnion decreased, which would account for the positive correlation of epilimnetic thickness with transparency. Overall phytoplankton production might be expected to increase with increased depth available for photosynthesis as well as possibly with increased thermal content. Also, macrophytes and benthic algae could exploit deeper regions of the lake as clarity increased. However, these relationships have not been systematically examined. All of these changes in primary producers could alter ecological niches of the consumers. Given that increased light penetration directly affects the vertical migrations of zooplankton, predator-prey interactions involving zooplankton may be altered in acidified lakes. The implications of such an event are addressed in section D.

Another effect of hypolimnetic heating in acidified lakes, discussed by Yan (1983), includes increased respiration rates, which are temperature dependent. The implications for hypolimnetic oxygen levels are quite complex. On the one hand, enhanced rates of organic matter decomposition and decreased solubility of oxygen at higher hypolimnetic temperatures may stress the invertebrates in this stratum. On the other hand, according to Yan, "the severity of hypolimnetic oxygen depletion may be reduced because of earlier autumnal overturn and because the trophogenic zone may extend into the hypolimnion, permitting photosynthetic generation of oxygen within it." Thus, increases in water clarity concurrent with acidification are likely to have profound effects on both primary and secondary producers.

## III. Descriptive Studies of Changes in Aquatic Biota Related to Acidification

The recent literature on acidification includes a number of reviews on the subject of acid-related changes, and the material will not be repeated here. The reader is referred to the Mierle and others (1986) review of the effects of acidification on plants and invertebrate communities and Stokes's (1986) treatment of primary producers, as well as Haines's (1981) earlier review and Dillon and others (1984). Recent reports from the Electric Power Research Institute (EPRI, 1986), Oak Ridge National Laboratories (ORNL, 1986), Altshuller and Linthurst (1984), and Medeiros and others (1985) from Brookhaven National Laboratories also cover the descriptive literature. We have summarized the available information from the recent literature. These data are presented in Tables 8–1 through 8–5 for phytoplankton, benthic algae, aquatic macrophytes, zooplankton, and benthic invertebrates, respectively. In the text below, we indicate major acid-related changes where there is general agreement, as well as indicating major inconsistencies among studies.

### A. Phytoplankton

As shown in Table 8–1, there is a tendency for species composition to change with acidification, such that many diatoms and most blue-green taxa are lost. Dinoflagellates are less affected, and as a result their contribution to the biomass of the plankton increases. Total phytoplankton biomass, however, does not decrease with acidification, but species richness does. The particular species affected varies from study to study and probably from region to region, but the general patterns of change in community structure are consistent.

### B. Benthic Algae

Certain acid-related changes in periphyton are macroscopically visible and have recently attracted attention from the general public because certain forms appear to be on the increase. This is of interest because, until now, losses of fish have been the major public concern in the the context of lake acidification. Biomass, species richness, and dominant periphyton taxa have been studied in experimental as well as unmanipulated systems with respect to acidification (Table 8–2). In many of these lakes there was a conspicuous increase in biomass of green filamentous algae, both attached to substrates as well as in metaphytic clouds in the littoral zone. Although the filamentous green algae (FGA) phenomenon has been observed and described frequently, there is a paucity of quantitative information on the phenomemon. Filamentous green algae often dominate in both experimental acidification studies and in surveys that include acidic lakes, but not all acidic lakes have FGA mats or clouds, and FGA can also occur in nonacidic situations. The species involved are almost exclusively members of the family Zygnemataceae, especially species of *Mougeotia, Zygnema,* and *Zygogonium.* On the sediments of acidic lakes, blue-green algae and diatoms are also incorporated into the mats, but to a lesser extent than the FGA.

One of the mechanisms underlying the increased abundance of FGA in acidic lakes may be related to changes in inorganic carbon, but the algae must also be tolerant of the acidity and related dissolved metals. Note that FGA do not appear to be limited to acidic conditions, but they contribute more to the total community

**Table 8-1.** Examples of changes in phytoplankton related to acidification.

| Parameter | pH range | Type of study | Observation/result | Reference |
|---|---|---|---|---|
| Biomass, taxon or tax. group | 3.9–7.1 | Survey, 9 lakes, Ontario (LaCloche) | Pyrrophyta biomass increased with pH decrease; Chrysophyceae and Diatomae biomass decreased | Stokes and Yung, 1985 |
| Biomass, taxon or tax. group | 4.6–6.8 | Survey lakes in S. Sweden | Pyrrophyta and Cyanophytes increased at pH below | Dickson et al., 1975 |
| Biomass, taxon or tax. group | 5.6–6.7 | Experimental acidification of whole lake | Pyrrophyta and Cyanophyte biomass increased with pH decrease; Chrysophyceae and Diatomae decreased | Findlay, 1984; Findlay and Saesura, 1980 |
| Biomass, taxon or tax. group | 4.5–7.0 | Survey, 9 lakes, N. Florida | Loss of Cyanophyta and Diatomae between pH 6 and 7; Chlorophytes and *Peridinium* dominant pH 4.5–5 | Crisman et al., 1980 |
| Biomass, taxon or tax. group | 5.6–6.5 | Comparison two lakes in S.C. Ontario | Increased Pyrrophyta at lower pH; less Diatomae biomass. | O'Mahoney, 1985 |
| Total phytoplankton biomass | 4.3–6.7 | Survey Ontario soft water lakes, including smelter lakes | No trend for biomass and pH; more related to phosphorus level | Yan, 1979 |
| Species richness | 5.3–6.8 | Twenty N. Florida lakes | 11–22 spp. in nonacidic lakes; 8–13 spp. in acidic lakes | Brezonik et al., 1984 |
| Species richness | 3.9–7.1 | Survey, 9 lakes, Ontario (LaCloche) | Decrease in richness with decrease in pH | Stokes and Yung, 1985 |
| Species richness | 4.0–6.5 | Experimental manipulation of pH in limnocorrals | Decrease in richness with decrease in pH | Yan and Stokes, 1978 |
| Production | 4.0–6.0 | Experimental manipulation of pH and P; lab assays and field experiments | P stimulated production at all pHs in field; in lab P alone not stimulated production at low pH | DeCosta and Preston, 1980 |

**Table 8-2.** Examples of changes in benthic algae related to acidification.

| Parameter | pH range | Type of study | Observation/result | Reference |
|---|---|---|---|---|
| Biomass | 4.7–6.4 | Experimental manipulation of pH in limnocorrals | No change in biomass related to pH | Muller, 1980 |
| Biomass | 4.0–6.7 | Experimental acidification of stream channels | Increase in biomass in acidified channels after 84 days; no trends with acidification over short term | Parent et al., 1986 |
| Biomass (as Chl. *a*) | 4.0–7.3 | Experimental acidification of artificial streams | Less biomass in acidified than in control | Maurice et al., 1986 |
| Rate of biomass accrual | 5.4–6.6 | Artificial substrates in littoral zones of Ontario lakes | Higher rate at lower pH; relationship only weakly significant | Stokes, 1981 |
| Species richness | 4.5–6.5 | Experimental acidification of limnocorrals | Decreased as pH decreased | Muller, 1980 |
| Species richness | 5.4–6.6 | Artificial substrates in littoral zones of Ontario lakes | Decreased as pH decreased | Stokes, 1984 |
| Dominant taxon or biomass of one taxon | 4.5–6.5 | Experimental acidification of limnocorrals | One species of *Mougeotia* became abundant at pH 5.0; most Diatomae decreased concurrently | Muller, 1980 |

| | pH range | Treatment | Finding | Reference |
|---|---|---|---|---|
| Dominant taxon or biomass of one taxon | 4.5–6.0 | Acidification of enclosures | *Mougeotia* replaced *Oedogonium* below pH 5.0 | Brezonik et al., 1986 |
| Dominant taxon or biomass of one taxon | 5.5–6.4 | Acidification of lake basin, Wisconsin | *Mougeotia* sp replaced *Oedogonium* below pH 5.5 | Brezonik et al., 1986 |
| Dominant taxon or biomass of one taxon | 5.2–7.0 | Acidification of stream | Blooms of *Mougeotia* in acidified stream | Hall et al., 1980 |
| Dominant taxon or biomass of one taxon | 5.0–6.7 | Experimental acidification of whole lake | Blooms of *Mougeotia* in littoral at pH 5.6 | Schindler et al., 1985 |
| Dominant taxon or biomass of one taxon | 5.6–6.7 | Experimental acidification of lake basin | *Mougeotia* and *Spirogyra* shoreline cover 90% at pH 5.9 | Turner et al., 1987 |
| Dominant taxon or biomass of one taxon | 4.5–6.5 | Survey of S.C. Ontario lakes | Greatest cover *Mougeotia* and *Zygogonium* at pH 4.5 | Jackson, 1985 |
| Community composition | 4.0–7.3 | Experimental acidification of artificial stream | No filamentous green algae in acidified treatment | Maurice et al., 1986 |

**Table 8–3.** Examples of changes in aquatic macrophytes related to acidification.

| Parameter | pH range | Type of study | Observation/results | Reference |
|---|---|---|---|---|
| Biomass of *Eriocaulon septangulare* | 4.5–6.5 | Survey, 4 Maine lakes | No relationship with pH | Hunter et al., 1986 |
| Presence of *Eriocaulon septangulare* | 4.4–6.0 | Survey, 20 Nova Scotia lakes | Present in 95% of lakes; no relationship with pH | Catling et al., 1986 |
| Presence of *Eriocaulon septangulare* | 4.4–6.9 | Survey, 9 Adirondack lakes | Present in all lakes | Roberts et al., 1985 |
| Presence of *Eriocaulon septangulare* | 3.3–7.1 | Survey, 46 Ontario lakes including lakes near Sudbury smelters | Present in all lakes | Wile et al., 1985 |
| Species richness | 4.5–6.5 | Survey, 4 Maine lakes | 14–15 spp in neutral lakes; 6–10 spp in acidic lakes | Hunter et al., 1986 |
| Species richness | 4.4–6.0 | Survey, 20 Nova Scotia lakes | 15–32 spp in pH 6 lakes; 8–20 spp in acidic lakes | Catling et al., 1986 |

| | pH range | Study | Results | Reference |
|---|---|---|---|---|
| Species richness | 4.4–6.9 | Survey, 9 Adirondack lakes | Lower in lakes below pH 5.5 | Roberts et al., 1985 |
| Species richness | 3.3–7.1 | Survey, 46 Ontario lakes including lakes near Sudbury | Not consistently lower in acidic lakes except where local smelter influence | Wile et al., 1985 |
| Cover of *Sphagnum* and *Isoetes* | 4.0–5.8 | Study of 6 lakes near Gothenburg, Sweden | *Sphagnum* "invasion"; isoetids "driven out" by *Sphagnum* | Grahn, 1977 |
| Littorellean community | 3.9–6.9 | Survey, 68 lakes in lakes in Netherlands, over time and pH range | In 41 lakes, *Litorella* replaced by *Juncus bulbosus* (if low $CO_2$ in sediment) or *Sphagnum* (if high carbonate sediments); related to acidification | Roelofs, 1983; Roelofs et al., 1984 |
| Biomass and reproduction of *Litorella*, *Juncus bulbosus* and *Sphagnum cuspidatum* | 3.5–7.0 | Experimental acidification in artificial ponds, seeded with macrophytes; used $H_2SO_4$ or $NH_4SO_4$ | pH 3.5-3.8 resulted in increased *Litorella* biomass but lowered flowering, with $NH_4SO_4$ only. *J. bulbosus* flowering increased at pH 3.5–4.0 with $NH_4SO_4$ only. Cover of *Sphagnum* increased with acidification by $NH_4SO_4$. No consistent changes when pH lowered by $H_2SO_4$. | Schuurkes et al., 1987 |

Table 8-4. Examples of changes to zooplankton related to acidification.

| Parameter | pH range | Type of study | Observation/result | Reference |
|---|---|---|---|---|
| Richness of crustacean species | 4.4–7.9 | Survey, S. Sweden | Decrease as pH decreased | Almer et al., 1974 |
| Richness of crustacean species | 4.5–7.1 | Survey, Ontario lakes (LaCloche) | Decrease as pH decreased | Bleiwas et al., 1984 |
| Richness of crustacean species | 3.8–6.4 | Survey, Ontario lakes (LaCloche) | Decrease as pH decreased | Sprules, 1975 |
| Richness of crustacean species | 4.7–6.8 | Survey, 20 lakes N. Florida | Decrease as pH decreased | Brezonik et al., 1984 |
| Crustacean biomass | 4.5–7.2 | Survey, Adirondacks | Decrease as pH decreased | Confer et al., 1983 |
| Crustacean biomass | 4.3–6.0 | Survey, N. Ontario | Decrease as pH decreased | Yan and Strus, 1980 |
| Crustacean biomass | 4.5–7.1 | Survey, Ontario lakes (LaCloche) | No relationship with pH | Bleiwas et al., 1984 |
| Crustacean biomass | 4.5–7.3 | Survey, New Hampshire lakes | No relationship with pH | Singer and Boylen, 1984 |
| Biomass of *Daphnia* spp. | 4.5–7.3 | Survey, Ontario lakes (LaCloche) | Decreased as pH decreased | Bleiwas et al., 1984 |
| Biomass of *Daphnia* spp. | 4.5–6.5 | Experimental acidification, B.C. lake | Decreased as pH decreased | Marmorek, 1983 |
| Biomass of *Daphnia* spp. | 4.5–7.0 | Survey, Norwegian lakes | Decreased as pH decreased | Nilsson, 1980 |
| Biomass of *Daphnia* spp. | 4.5–7.3 | Survey, Adirondack lakes | Decreased as pH decreased | Sutherland et al., 1984 |
| Individual species density or biomass | 5.6–5.9 | Experimental acidification of whole lake | *Mysis relicta* lost | Nero and Schindler, 1983 |

| | | | | Reference |
| --- | --- | --- | --- | --- |
| Individual species density or biomass | 4.7–7.2 | Experimental acidification | Increase in *Bosmina longirostra* | Havens and DeCosta, 1986 |
| Individual species density or biomass | 4.5–6.5 | Experimental acidification, B.C. lake | Increase in *Bosmina longirostra* | Marmorek, 1983 |
| Individual species density or biomass | 4.0–6.9 | Survey, N. Ontario | Increase in *Bosmina longirostra* | Keller and Pitblado, 1984 |
| Individual species density or biomass | 4.5–7.3 | Survey, New Hampshire lakes | Increase in *Diaptomus minutus* | Singer and Boylen, 1984 |
| Individual species density or biomass | 4.5–7.1 | Survey, Ontario lakes | Increase in *Diaptomus minutus* | Bleiwas et al., 1984 |
| Individual species density or biomass | 4.9–8.4 | Survey, Newfoundland lakes | Increase in *Diaptomus minutus* | Chengalath et al., 1984 |
| Individual species density or biomass | 4.7–8.2 | Survey, Atlantic Canada | Increase in *Keratella taurocephala* | Carter et al., 1986 |
| Individual species density or biomass | 3.8–5.6 | Survey, Adirondacks | Increase in *Keratella taurocephala* | Siegfried et al., 1984 |
| Individual species density or biomass | 4.5–6.5 | Experimental acidification, B.C. lake | Increase in *Keratella taurocephala* | Marmorek, 1983 |

**Table 8-5.** Examples of changes in benthic invertebrates related to acidification.

| Parameter | pH range | Type of study | Observation/result | Reference |
|---|---|---|---|---|
| Abundance of functional groups | 4.0–6.4 | Experimental stream acidification | Collectors greatly decreased; scrapers some decrease; shredders little effect | Hall et al., 1980 |
| Abundance of functional groups | 4.3–5.9 | Comparison streams of differing pH | In acidic streams most abundant, then predators and deposit feeders | Friberg et al., 1980 |
| Abundance of functional groups | 5.0–6.0 | Survey, 100 Norwegian lakes | Grazing and scraping gastropods decline at pH 6.0, completely gone at pH 5.2 | Okland, 1980a; Okland, 1980b |
| Abundance of functional groups | <5.7 | Lake survey | Reduction in scraper abundance | Sutcliffe and Carrick, 1973 |
| Abundance of functional group | Not given | Review with survey, lakes in Norway and experimental work | Predator (*Chaoborus* and corixids) dominate at low pH | Henrikson et al., 1980 |
| Biomass (all groups) | 4.6–7.2 | Survey, lakes in Ontario | No changes related to pH | Collins et al., 1981 |
| Biomass (all groups) | 4.0–6.4 | Experimental acidification of a stream | Decreased total density in acidified stream | Hall et al., 1980 |
| Individual taxa or taxonomic groupings: biomass or numbers | 4.9–6.8 | Comparison of some sites between 1937 and 1985, Algonquin Park, Ontario | Decrease in many mayfly and stone fly taxa; a few acid-tolerant mayfly spp. replaced other in 1985 | Hall and Ide, 1987 |
| Individual taxa or taxonomic groupings: biomass or numbers | 5.0–6.6 | Survey, 1,000 Norwegian lakes | *Gammarus* lost at pH 6.0. Fingernail clams eliminated between pH 5.0–5.7 | Okland, 1980b |

| Individual taxa or taxonomic groupings: biomass or numbers | 4.0–6.4 | Experimental stream acidification | Emergence of stone flies increased at low pH | Hall and Likens, 1980 |
|---|---|---|---|---|
| Individual taxa or taxonomic groupings: biomass or numbers | 4.5–8.5 | Survey, Ontario lakes laboratory bioassays | *Hyalella azteca* lost at pH 5.6 | Stephenson and Mackie, 1986 |
| Individual taxa or taxonomic groupings: biomass or numbers | 5.0–8.0 | Experimental acidification | *Hyalella azteca* most sensitive to low pH | Zischke et al., 1983 |
| Individual taxa or taxonomic groupings: biomass or numbers | 5.2–6.8 | Experimental acidification of whole lake | *Orconectes virilis* lost at pH 5.7 | France, 1983 |
| Individual taxa or taxonomic groupings: biomass or numbers | 4.5–6.3 | Survey, Norwegian lakes | Fewer spp of Chironomids in acidic lakes | Raddum and Saether, 1981 |
| Individual taxa or taxonomic groupings: biomass or numbers | 4.5–6.3 | Survey, Norwegian lakes | Reproduction of *Amnicola* failed at pH below 5.0 | Raddum and Saether, 1981 |
| Individual taxa or taxonomic groupings: biomass or numbers | 4.5–6.3 | Survey, Norwegian lakes | *Procladius* declines | Raddum and Saether, 1981 |

**Table 8-5.** (*Continued*)

| Parameter | pH range | Type of study | Observation/result | Reference |
|---|---|---|---|---|
| Individual taxa or taxonomic groupings: biomass or numbers | 4.5–6.5 | Experimental acidification | 4 spp. of stone flies tolerant to pH 4.5 | Raddum, 1979 |
| Individual taxa or taxonomic groupings: biomass or numbers | 4.8–6.5 | Lake survey | Bivalues disappeared at low pH | Raddum, 1980 |
| Individual taxa or taxonomic groupings: biomass or numbers | 4.3–5.9 | Comparison of streams with differing pH | Ephemeropterans and helminthid beetles absent | Friberg et al., 1980 |
| Individual taxa or taxonomic groupings: biomass or numbers | <4.0–>6.0 | Field observations and experimental acidification | The mayfly *Baetis* disappeared at pH <6.0, but the mayfly *Leptophlebia* found at <4.0 | Englom and Lingdell, 1984 |
| Individual taxa or taxonomic groupings: biomass or numbers | 3.6–4.8 | Field observations | Sialids, coleopterans, zygopteran odonates, notonectide, corixids, chaoborids, oligochaetes, tolerant at pH 4.0 | Kerekes et al., 1984 |

biomass under acidic conditions. This community change will also be evaluated in the context of changes in grazer communities (section IV).

## C. Aquatic Macrophytes

Table 8–3 indicates that the common isoetid macrophyte *Eriocaulon septangulare* (pipewort) is present in most lakes over a wide range of pH. The probable reason for the success of macrophytes with this type of morphology and physiology in acidic conditions has been discussed in section II,A. Other common soft water species tend to be less acid tolerant than the pipewort, and there are certain geographic differences as well. Studies in Sweden in the early 1970s described the invasion of the bog moss *Sphagnum* in acidic lakes, and until recently North American workers have been puzzled by the absence of this phenomenon on this continent. Roelofs (1983) and Roelofs and others (1984) suggested that the *Sphagnum* invasion follows acidification only when accompanied by high DIC, a relatively unusual occurrence. In the Netherlands lakes studied by Roelofs's group, *Juncus bulbosus* replaced *Littorella* (another isoetid macrophyte) at low pH. There is consistency in the reporting of decreased species richness of macrophytes as pH declines, as shown in the table. Not shown here, but discussed in Stokes (1986), is the fact that total macrophyte biomass does not decrease with acidification and in a few instances even increases (Wile et al., 1985).

## D. Zooplankton

Rather clear and consistent results have been reported on the effects of acidification to zooplankton (Table 8–4). A general decrease in species richness and loss of *Daphnia* spp. is reported repeatedly. The crustaceans that become important as lakes acidify include *Diaptomus minutus* and *Bosmina longirostris*, both small-bodied organisms. Rotifer biomass increases in acidic lakes, and certain species (e.g., *Keratella taurocephala*) may show a dramatic increase.

## E. Benthic Invertebrates

An overview of responses by benthic invertebrates, including insects and other groups, is presented in Table 8–5. In general, crustacean taxa are lost as pH declines. Although occasionally reported in lakes of pH <5.0, *Hyallela azteca*, *Gammarus*, and *Orconectes* are rarely found at pH below 5.6. Nero and Schindler (1983) demonstrated that *Mysis relicta* was the first macroinvertebrate to disappear due to the acidification of Lake 223. Gastropod richness and biomass also decline with decreasing pH, and gastropods are virtually absent as lake pH falls below 5.0. Some insects, such as mayflies and stone flies, are also lost with acidification. As systems become increasingly acidic, community structure shifts towards taxa dominated by predators, shredders, and deposit feeders, while scrapers and collectors disappear. However, the net impact on standing stock, whether expressed as biomass or density, is variable. It is not clear whether acidification depresses benthic macroinvertebrate productivity, although diversity may decrease as taxa are eliminated. It appears that invertebrates that possess large external gills are more sensitive to acidic conditions than are those with less permeable exoskeletons.

## IV. Effects on Community and Ecosystem Function from Acid-Related Biotic Changes and Indirect Effects of Acidification

There are few universally accepted criteria for assessing the state or condition of an ecosystem (Stokes and Piekarz, 1987). Nevertheless, potentially useful indicators at the ecosystem level include rates of primary production, secondary production, nutrient cycling, and decomposition. In order to interpret changes in these and other parameters, however, it is necessary to have a basis for comparison in the form of reference, control, or pretreatment values. These are often lacking in studies of acidification. It is also essential to have a measure of the natural variability of the parameter of interest. Because of the general lack of information of this type, much of the following is speculative.

Phytoplankton primary productivity has been measured in a number of experimentally and "naturally" acidified lakes. Although some early work suggested that acidification resulted in decreased production (cited in Dillon et al., 1984), more recent studies have shown no such effect. In general, it appears that when acid-tolerant species replace acid-sensitive forms, primary production at the community level is not diminished. Therefore, changes in the phytoplankton community structure are regarded as more sensitive indicators of acidification than is the functional measure of primary production.

The distribution of the plankton, particularly its vertical stratification, may be greatly altered as lakes acidify, due in part to changes in clarity that affect light penetration and thermal structure (section II,E). Changes in the light regime can influence the distribution and productivity of phytoplankton, benthic algae, and aquatic macrophytes. The distribution of plants might in turn dictate the distribution of zooplankton and benthos, but to date this has not been systemicatally studied in the context of acidification.

The communities grazing upon the algal plankton are unlikely to be food limited in terms of the amount of algal biomass present, but changes in the size, palatability, or digestibility of algae as the community structure shifts in response to acidification are a real possibility. Vanni (1987) has evidence that zooplankton density is strongly influenced by phytoplankton availability. For example, the large-bodied dinoflagellates that dominate the biomass of many acidic lakes are probably too large to be grazed by the small zooplankters that survive acidification. Even though there is evidence that these dinoflagellates are not grazed in acidic systems (Yan and Strus, 1980; Havens and DeCosta, 1985), this does not demonstrate that the hervibores in these systems are food limited. In fact, the Havens and DeCosta study, which involved experimental manipulation in enclosures, suggested that they were not. Therefore, we conclude that if secondary production is affected by acidification, it is more likely to be from direct chemical effects than from indirect means.

Major changes in the species composition and biomass of periphyton communities have been observed to coincide with the onset of acidification (section III, B). Periphyton may influence nutrient cycling in a lake, at least in the littoral zone. These taxa also have the potential to compete for nutrients with planktonic algae and possibly macrophytes and to limit light for macrophytes. None of these aspects has so far been examined satisfactorily in the field. The proliferation of metaphytic and attached FGA may cause physical changes in the littoral zone that modify the habitat for fish spawning or for the reproduction of other animals.

The mechanism behind the acid-related proliferation of FGA is complex. The algae whose biomass increases must be acid tolerant and in many cases adapted to low DIC. There is probably an indirect cause as well. The invertebrates that normally graze upon periphyton are themselves affected by acidification. Many invertebrates, including crayfish, snails, amphipods, and smaller crustaceans, are absent from acidic lakes (Mierle et al., 1986). The evidence to date implicates release from grazer influence as a partial explanation for the increased biomass of FGA in acidified lakes. Increases in algal biomass due to shifts in grazer community composition have been induced in whole-lake manipulation studies (Carpenter et al., 1987). Alterations in the phytoplankton size spectra, due to zooplankton community changes, were supported by empirically derived regression analysis based on data from enclosure experiments (McQueen et al., 1986) and a phytoplankton size-distribution model developed by Carpenter and Kitchell (1984). The plankton ecology group (PEG) model of phytoplankton dynamics also supports the zooplankton-mediated changes in phytoplankton composition and abundance (Sommer et al., 1986).

Further implications of acidification can be seen in the selective loss of certain feeding groups of macroinvertebrates in the benthos. As shown in Table 8–5, shredders seem to be the least affected by acidification, whereas collectors and scrapers are apparently more sensitive to low pH. The processing of detrital matter by invertebrates is of major importance in the nutrient cycling of aquatic systems. Prior to bacterial or fungal decomposition of organic matter, invertebrates prepare the detritus physically and chemically. The buildup of coarse organic debris in acidic systems has been described in a number of studies. It is possible that the delayed decomposition is due to a combination of the loss of functional groups of invertebrates and possibly (not discussed here) decreased rates of microbial decomposition, with a strong interaction between the two. Loss of invertebrate taxa from the benthos may also have an effect on the predators of these organisms, both in aquatic systems and terrestrial habitats, where biota feed on insects after they emerge.

No discussion of indirect effects associated with acidification would be complete without considering the impacts on fish and the functioning and structure of food chains. In lakes where fish populations are lost or severely reduced, changes in invertebrate populations may be anticipated as a consequence of altered predator-prey interactions. Eriksson and others (1980) suggested that shifts in community structure in acidified lakes were linked with loss of fish and the concurrent changes in predator-prey regimes. Indeed, the hypothesis that fish predation is of significant importance in structuring invertebrate communities has received considerable support (Zaret, 1980). The introduction of yellow perch in a small Precambrian lake resulted in a 60% reduction in benthic biomass of the littoral zone (Post and Cucin, 1984). Fish predation accelerated the decline of herbivorous plankton and was accompanied by a shift from larger- to smaller-sized species (Sommer et al., 1986). When planktivores are reduced or absent, invertebrate predators influence the structure of the biotic assemblages (McQueen et al., 1986). The strong functional link between certain invertebrate predators and fish has been well established for some systems (Kerfoot and DeMott, 1984). However, few studies have examined shifts in invertebrate community structure in acidified systems and their relationship to the presence or absence of fish.

In studies of benthos of acidic lakes in Ontario, no significant changes in benthic community structure could be attributed solely to pH (Collins et al., 1981; Kelso et al., 1981). Bendell and McNicol (1987) examined the relationship between fish presence or absence and aquatic insects, for lakes of different pH. They found that regardless of pH lakes without fish had aquatic insect communities that were similar to each other. Most notable in their study was an abundance of notonectids and corixids, families that have been shown to be strongly influenced by fish predation (Weir, *in* Bendell and McNicol, 1987). A shift in the species distribution of *Chaoborus* was also reported, with *C. americanus* being abundant only in fishless lakes. This is in agreement with the earlier study of Yan and others (1985). The pattern could be explained by the fact that unlike other species of *Chaoborus*, which are diurnally benthic, *C. americanus* is planktonic during the day and would be readily eliminated by predatory fish. The loss of fish, then, can be a driving force for community shifts. This, of course, does not contradict observations that direct toxicity can result in elimination of some invertebrate species in response to acidification. A change from a predatory system dominated by fish to one dominated by invertebrates may have profound implications on other biotic assemblages. For example, if invertebrate body size and abundance is substantially altered by a release from fish predation, one can anticipate shifts in invertebrate grazing pressure on algal communities, as discussed above.

Therefore, although first-order effects of acidification can be demonstrated in many instances, higher-order effects on ecosystem structure can be expected but are extremely difficult to demonstrate.

## V. Conclusions

1. In general, the flora and fauna of acidic lakes are depauperate in species compared with circumneutral lakes.
2. Community composition in terms of species biomass and numbers of species for all the groups described is altered with acidification.
3. With the exception of the zooplankton, there is no significant relationship between community biomass and pH, but for zooplankton biomass decreases at low pH.

4. Production rates are not decreased by acidification, so acidifying lakes can be as productive as circumneutral ones. In general, structural biological properties of lakes are more susceptible to change from acidification than are functional properties.

5. In addition to increases in $H^+$ concentration, other acid-related chemical changes can directly affect aquatic biota. Of these, there is evidence that decreases in the form and concentration of dissolved inorganic carbon do have effects on the primary producer communities, especially the benthic algae and macrophytes, as lakes acidify. For other mineral nutrients, such as phosphorus, there is less direct evidence of causal relationships between chemical change and biological response in acid waters.

6. Of the metals, both increased concentrations and increased bioavailability of aluminum have the potential for biological effects on producer and consumer organisms, but the evidence for this element being responsible for observed acid-related changes is still speculative. The change in chemical form of metals other than aluminum also has the potential to increase metal toxicity, but this factor has not been clearly demonstrated in the field.

7. Increased $H^+$ concentration may protect organisms from metal uptake and toxicity.

8. The clarity of acidic lakes is increased compared with circumneutral ones, and this affects the thermal and light regimes, with potential for direct and indirect biological effects.

9. Of the species that are intolerant of acidity and are therefore rare or absent from acidifying waters, some important functional groups should be noted. These include fish, especially predatory fish, whose absence is expected to influence the invertebrates and other fish upon which they normally feed. Also absent are whole groups of insect larvae and other invertebrates that normally graze on attached algae and other plant communities. The proliferation of filamentous green algae is believed to be due in part to the absence of certain grazers.

10. The potential for other indirect effects or interactions resulting from acidification on aquatic biota is considerable, but clear demonstration that changes are due to indirect rather than direct effects of acidification is rarely available.

## Acknowledgments

The authors wish to thank Judy Eichmanis and Christine Radewych for assistance in searching the literature and preparing the manuscript and Robert France for advice on the invertebrate literature.

## References

Almer, B., W. Dickson, C. Ekstrom, and E. Hornstrom. 1974. Ambio 3:30–36.
Almer, B., W. Dickson, C. Ekstrom, and E. Hornstrom. 1978. *In* J. O. Nriagu, ed. *Sulfur in the environment, Part II. Ecological impacts*, 271–311. J Wiley & Sons, New York.

Altshuller, A.P., and Linthurst, R.A. 1984. *The acidic deposition phenomenon and its effects: Critical assessment review papers.* EPA-600/8-83-016aF. 2 vol. U.S. EPA, Washington, D.C.

Avalos, E.A. 1981. The effect of phosphorus addition and pH manipulation in bag experiments on phytoplankton communities of two small lakes in West Virginia. M.Sc. thesis. West Virginia University.

Bailey, R.C., and P.M. Stokes. 1984. *In* R.D. Cardwell, R. Purdy, and R.C. Bahner, eds. *Aquatic toxicology and hazard assessment: Seventh symposium,* ASTM STP 854, 5–26. American Society for Testing and Materials, Philadelphia.

Bendell, B.E., and D.L. McNicol. 1987. Hydrobiol 150:193–202.

Bleiwas, A.S.H., P.M. Stokes, and M.M. Olaveson. 1984. Verh internat Verein Limnol 22:332–337.

Boston, H.L. 1986. Aquat Bot 26:259–270.

Boston, H.L. 1987. Ann Bot 65(5):485–494.

Brezonik, P.L., L.A. Baker, J.R. Eaton, T.M. Frost, P. Garrison, T.K. Kratz, J.J. Magnusson, J.E. Perry, W.J. Rose, B.K. Shepard, W.A. Swenson, C.J. Watros, and K.E. Webster. 1986. Water Air Soil Pollut 31:115–121.

Brezonik, P.L., T.L. Crisman, and R.L. Schulze. 1984. Can J Fish Aquat Sci 41:46–56.

Burton, T.M., and J.W. Allen. 1986. Can J Fish Aquat Sci 43:1285–1287.

Campbell, P.G.C., M. Bisson, R. Bougie, A. Tessier, and J.P. Villeneuve. 1983a. Anal Chem 55:2246–2252.

Campbell, P.G.C. and P.M. Stokes. 1985. Can J Fish Aquat Sci 42(12) 2034–2049.

Campbell, P.G.C., P.M. Stokes, and J. Galloway. 1983b. *Effects of atmospheric deposition on the geochemical cycling and biological availability of metals.* Proc. Int. Conf. on Heavy Metals in the Environment.

Carpenter, S.R., and J.F. Kitchell. 1984. Am Nat 124:159–172.

Carpenter, S.R., J.F. Kitchell, J.R. Hodgson, P.A. Cochran, J.J. Elser, M.M. Elser, D.H. Lodge, D. Kretchmer, X. He, and C.N. von Ende. 1987. Ecology 68:1863–1876.

Carter, J.H.C., W.D. Taylor, R. Chengalath, and D.A. Scruton. 1986. Can J Fish Aquat Sci 43:444–456.

Catling, P., B. Freedman, C. Stewart, J.J. Kerekes, and C.P. Lefkovitch. 1986. Can J Bot 64:724–729.

Chengalath, R.W., W.J. Bruce, and D.A. Scruton. 1984. Verh Internat Verin Limnol 22:419–430.

Collins, N.C., A.P. Zimmerman, and R. Knoechel. 1981. *In* R. Singer, ed. *Effects of acidic precipitation on benthos.* 35–48. North American Benthological Soc, Hamilton, N.Y.

Confer, J.L., T. Kaaret, and G.E. Likens. 1983. Can J Fish Aquat Sci 40:36–42.

Crisman, T.L., R.L. Schulze, P.L. Brezonik, and S.A. Bloom. 1980. *In* D. Drablos and A. Tollan, eds. *Proc int conf Ecological Impact of Acid Precipitation,* 296–291. SNSF project. Norway.

Dalal, R.C. 1975. Soil Sci 119:127–131.

Davis, R.B., D.S. Anderson, and F. Berge. 1985. Nature 316:436–438.

Dickson, W. 1978. Verh Internat Verein Limnol 20:851–856.

Dickson, W. 1980. *In* D. Drablos and A. Tollan, eds. *Proc. int. conf. Ecological Impact of Acid Precipitation,* 851–856. SNSF project. Norway.

Dickson, W., E. Hornstrom, C. Ekstrom, and B. Almer. 1975. *Rodingsjoar seder OmDalalben.* Information from Drottingholm 7. 140 p.

Dillon, P.J., D.S. Jefferies, W.A. Scheider, and N.D. Yan. 1980. *In* D. Drablos and A.

Tollan, eds. *Proc. int. conf. Ecological Impact of Acid Precipitation*, 212–213. SNSF project. Norway.

Dillon, P.J., N.D. Yan, and H.H. Harvey. 1984. Control 13:176–194.

Driscoll, C.T. 1980. *In* D. Drablos and A. Tollan, eds. *Proc. int. conf. Ecological Effects of Acid Precipitation*, 214–215. SNSF project. Norway.

Driscoll, C.T., J.P. Baker, J.J. Bisogni, and C.L. Scofield. 1983. *In* O.R. Bricker, ed. *Acid precipitation: Geological aspects*. Ann Arbor Sci., Ann Arbor, Mich.

EPRI. 1986. *An evaluation and compilation of the reported effects of acidification on aquatic biota*. EPRI EA-4825 Project.

Eilers, J.M., G.L. Lien, and R.G. Berg. 1984. *Aquatic organisms in acidic environments: A literature review*. Wisc Dept Nat Resour Tech Bull 150. 18 p.

Englom, E., and P.E. Lingdell. 1984. Rep Inst Freshw Res Drottinghom 61:60–68.

Eriksson, M.O.G., L. Henrikson, B.I. Nilsson, G. Nyman, H.G. Oscarson, A.E. Stenson, and K. Larsson. 1980. Ambio 9:248–249.

Findlay, D.L. 1984. *Effects on phytoplankton biomass, succession and composition in Lake 223 as a result of lowering pH levels from 5.6 to 5.2. Data from 1980–1982*. Can MS report of Fish Aquat Sci No 1761. 10 p.

Findlay, D.L., and G. Saesura. 1980. *Effects on phytoplankton biomass, succession and composition in Lake 223 as a result of lowering pH levels from 7.0 to 5.6. Data from 1974–1979*. Can MS Report of Fish Aquat Sci No 1585. iv + 16 p.

France, R.L. 1983. *In* C.R. Goldman, ed. *Freshwater crayfish V: Papers from 5th int symp freshwater crayfish* (Davis, Ca.) 98–111. AVI Publishing Co., Westport Conn.

France, R.L., and P.M. Stokes. 1987a. Can J Fish Aquat Sci 65:3071–3078.

France, R.L., and P.M. Stokes. 1987b. Can J Fish Aquat Sci 44:1102–1111.

Friberg, F., C. Otto, and B.S. Svenson. 1980. *In* D. Drablos and A. Tollan, eds. *Proc int. conf. Ecological Impact of Acidification*, 304–305. SNSF project, Norway.

Grahn, O. 1977. Water Air Soil Pollut 7:295–305.

Grahn, O., H. Hultberg, and L. Lander. 1974. Ambio 3:93–94.

Haines, T.A. 1981. Trans Amer Fish Soc 110:669–707.

Hall, R.J. and F.P. Ide. 1987. Can J Fish Aquat Sci 44:1652–1657.

Hall, R.J., and G.E. Likens. 1980. *In* D. Drablos and A. Tollan, eds. *Proc. Int. Conf. Ecological Impact of Acidification*, 375–576. SNSF project. Norway.

Hall, R.J., G.E. Likens, S.B. Fiance, and G.R. Hendrey. 1980. Ecology 61:976–989.

Havas, M. 1985. Can J Fish Aquat Sci 42:1741–1748.

Havas, M., and G.E. Likens. 1985. Proc Natl Acad Sci 82:7345–7349.

Havens, K.E., and J. DeCosta. 1985. J Plankton Res 7:207–222.

Havens, K.E., and J. DeCosta. 1986. Water Air Soil Pollut 33:277–293.

Hendrey, G.R., and F. Vertucci. 1980. *In* D. Drablos and A. Tollan, eds. *Proc. int. conf. Ecological Effects of Acid Precipitation*, 314–315. SNSF project. Norway.

Henrikson, L., H.G. Oscarson, and J.A.E. Stenson. 1980. *In* D. Drablos and A. Tollan, eds. *Proc. int. conf. Ecological Impact of Acidification*, 316 p. SNSF project. Norway.

Howell, E.T. (1988) Ecology of periphyton associated with isoetid plants in low-alkalinity lakes. Thesis in preparation, University of Toronto.

Howell, E.T., R.L. France, and P. Turner. 1988. Personal communication.

Hunter, M.L., J.J. Jones, and J.W. Whitham. 1986. Aquat Bot 24:91–95.

Jackson, M.B. 1985. Filamentous algae in Ontario softwater lakes. Abstract Muskoka Conference '85, an inter. symp. on acidic precipitation, Muskoka, Canada.

Keller, W., and J.R. Pitblado. 1984. Water Air Soil Pollut 23:271–291.

Kelso, J.R.M., R.J. Lowe, J.H. Lipsit, and T.R. Dermott. 1981. *In Proc. effects of acid precipitation on ecological systems,* Inst. of Water Research, Michigan State Univ., East Lansing, Mich.

Kerekes, J., B. Freedman, G. Howell, and P. Clifford. 1984. Water Poll Res Canada 19:1–10.

Kerfoot, W.C., and W.R. DeMott. 1984. *In* D.G. Meyers and J. R. Strickler, eds. *Trophic interactions within aquatic ecosystems,* 347–382.

Kwiatkowski, R.E., and J.C. Roff. 1976. Can J Bot 54:2546–2561.

Lean, D.R.S., and C. Nalewajko. 1976. J Fish Res Board Can 33:1312–1323.

McQueen, D.J., J.R. Post, and E.L. Mills. 1986. Can J Fish Aquat Sci 43:1571–1581.

Marmorek, D.R. 1983. Effects of lake acidification on zooplankton community structure and phytoplankton-zooplankton interactions: An experimental approach. M.Sc. thesis, University of British Columbia. 397 p.

Marmorek, D.R. 1984. *In* G.R. Hendrey, ed. *Early biotic responses to advancing lake acidification,* 23–41. Butterworth Publishers, Stoneham, Mass.

Maurice, C.G., R.L. Lowe, T.M. Burton, and R.M. Stanford. 1986. Water Air Soil Pollut 33:165–177.

Medeiros, W.H., A.F. Meinhold, and P.D. Moskowitz. 1985. *Effects of acidification on microorganisms, plankton, macrophytes, periphytic algae, macroinvertebrates and amphibians.* Interagency Agreement D.W. 89-932006-01-0 BNL 52056. 71 p.

Mierle, G., K. Clarke, and R. France. 1986. Water Air Soil Pollut 31:593–604.

Muller, P. 1980. Can J Fish Aquat Sci 37:355–363.

NRCC. 1986. *Aluminum in the Canadian environment.* National Research Council of Canada report no. 24759. 331 p.

Nalewajko, C., and M.A. O'Mahoney. 1988. Can J Fish Aquat Sci 45:254–260.

Nalewajko, C., and B. Paul. 1985. Can J Fish Aquat Sci 42:1946–1953.

National Academy Press. 1985. *Acid deposition. Effects on geochemical cycling and biological availability of trace elements.* Report of sub group on metal, Tri-Academy Committee on Acid Deposition. National Academy Press, Washington, D. C. 83 p.

Nero, R.W., and D.W. Schindler. 1983. Can J Fish Aquat Sci 40:1905–1911.

Nieboer, E.D., and H.S. Richardson. 1980. Environ Pollut (B) 1:3–36.

Nilsson, J.P. 1980. Hydrobiol 94:217–221.

ORNL. 1986. *Assessment of acidic deposition effects on aquatic systems.* (Environment Sciences Division, Oak Ridge National Laboratory). Contract DE-AC05-840R21400. US Environmental Protection Agency Publication 2769. 176 p.

Okland, J. 1980a. *In* D. Drablos and A. Tollan, eds. *Proc int conf Ecological Impact of Acidification,* 322–323. SNSF project. Norway.

Okland, J. 1980b. *In* D. Drablos and A. Tollan, eds. *Proc int conf Ecological Impact of Acidification,* 324–325. SNSF project. Norway.

O'Mahoney, M.A. 1985. A seasonal study of an acidifying and circumneutral shield lake: Effect of pH shock on phosphorus concentration and kinetics, and photosynthesis. M.Sc. thesis, University of Toronto.

Overrein, L.N., H.M. Seip, and N. Tollan. 1980. *Acid precipitation: Effects in forest and fish.* SNSF research report. 175 p.

Parent, L., M. Allard, D. Planas, and G. Moreau. 1986. *In* B.G. Isom, S.D. Dennis, and J.M. Bates, eds. *Impact of acid rain and deposition on aquatic biological systems,* 28–41. ASTM STP 928. American Society for Testing and Materials, Philadelphia.

Post, J.R., and D. Cucin. 1984. Can J Fish Aquat Sci 41:1496–1501.

Raddum, G.G. 1979. *VVirkinger av low pH pa insektlarner.* SNSF IR 45/79. 58 p.

Raddum, G.G. 1980. *In* D. Drablos and A. Tollan, eds. *Proc int conf Ecological Impact of Acidification,* 330–331. SNSF project. Norway

Raddum, G.G., and O.A. Saether. 1981. Verh Internat Verein Limnol 21:367–373.

Raven, J.A., L.L. Handley, J.J. Macfarlane, S. McInroy, L. McKenzie, J.H. Richards, and G. Samuelsson. 1988. New Phytol 108(2):125–148.

Roberts, D.A., R. Singer, and C.W. Boylen. 1985. Aquatic Bot 21:219–235.

Roelofs, J.G.M. 1983. Aquatic Bot 17:139–155.

Roelofs, J.G.M., J.A.A.R. Schuurkes, and A.J.M. Smits. 1984. Aquatic Bot 18: 389–411.

Schindler, D.A. 1980. *In* D. Drablos and A. Tollan, eds. *Proc int. conf. Ecological Impact of Acidification,* 370–374. SNSF project. Norway.

Schindler, D.W., R.H. Hesslein, R. Wagemann, and W.S. Broecker. 1980. Can J Fish Aquat Sci 37:373–377.

Schindler, D.W., K.H. Mills, D.F. Malley, D.L. Findlay, J.A. Shearer, I.J. Davis, M.A. Turner, G.A. Lindsay, and D.R. Cruikshank. 1985. Science 228:1395–1401.

Schindler, D.W., and M.A. Turner. 1982. Water Air Soil Pollut 18:259–271.

Schuurkes, J.A.A.R., M.A. Elbers, J.J.F. Gudden, and J.G.M. Roelofs. 1987. Aquatic Bot 28:199–226.

Siegfried, C.A., J.W. Sutherland, S.D. Quinn, and J.A. Bloomfield. 1984. Verh Internat Verein Limnol 22:549–558.

Singer, R., and J. Boylen. 1984. *Biological field survey of northeastern acidified lakes.* Final report, EPA/NCSU Acid Deposition Program, North Carolina State University, Raleigh, N.C.

Sommer, V., Z.M. Gliwicz, W. Lampert, and N. Duncan. 1986. Arch Hydrobiol 106:433–471.

Sprules, W.G. 1975. J Fish Res Board Can 32:389–395.

Stephenson, M., and G.L. Mackie. 1986. Can J Fish Aquat Sci 43:288–292.

Stokes, P.M. 1981. *In* R. Singer, ed. *Effects of acid precipitation on benthos,* 119–138. North American Benthological Society, Hamilton, N.Y.

Stokes, P.M. 1984. *In* G.R. Hendrey, ed. Early biotic responses to advancing lake acidification, 43–61. Butterworth Publishers, Stoneham, Mass.

Stokes, P.M. 1986. Water Air Soil Pollut 30:421–438.

Stokes, P.M., and D. Piekarz. 1987. *Ecological indicators of the state of the environment.* Report of a workshop held at the University of Toronto. 152 p.

Stokes, P.M., and Y.K. Yung. 1985. *In* J.P. Smol, R.W. Battarbee, R.B. Davis, and J. Merilainen, eds. *Silicous algae as indicators of lake acidification,* 57–72. Developments in Hydrobiology (Hydrobiologia) Junk, The Hague.

Stumm, W., and J. Morgan. 1970. *Aquatic chemistry.* John Wiley and Sons, New York. 780 p.

Stumm, W., and J. Morgan. 1981. *Aquatic Chemistry,* 2d ed. John Wiley and Sons, New York. 780 p.

Sutcliffe, D., and T. Carrick. 1973. Freshwater Biol 3:437–462.

Sutherland, J.W., S.O. Quinn, J.A. Bloomfield, and C.A. Siegfried, 1984. *In Lake and Reservoir Management* (Proc. 3d Ann. Symp of the North Amer. Lake Manag. Soc., Knoxville, Tenn.), 380–384. EPA 440/5-84-801.

Turner, M.A., M.B. Jackson, D.A. Findlay, R.W. Graham, E.R. DeBruyn, and E. Vendermeer. 1987. Can J Fish Aquat Sci 44(Supp 1):135–149.

Vanni, M.J. 1987. Ecological Monographs 57(1):61–88.

Wetzel, R.G., E.S. Brammer, and C. Forsberg. 1984. Aquatic Bot 19:329–342.

Wetzel, R.G., E.S. Brammer, K. Lindstrom, and C. Forsberg. 1985. Aquatic Bot 22:107–120.

Wile, I., G. Miller, G.G. Hitchin, and N.D. Yan. 1985. Can Field Natural 99:308–312.

Williams, T.C., and D.H. Turpin. 1987. Oecologia 73(2):307–311.

Wium-Anderson, S., and J.M. Anderson. 1972. Limnol Oceanogr 17(6):943–951.

Yan, N.D. 1979. Water Air Soil Pollut 11:43–55.

Yan, N.D. 1983. Can J Fish Aquat Sci 40:621–626.

Yan, N.D., R.W. Nero, W. Keller, and D.C. Lazenby. 1985. Holarct Ecol 8:93–99.

Yan, N.D., and P.M. Stokes. 1978. Environ Conserv 5:93–100.

Yan, N.D., and R. Strus. 1980. Can J Fish Aquat Sci 37:2282–2293.

Zaret, T.M. 1980. *Predation and freshwater communities*. Yale University Press, New Haven, Conn. 187 p.

Zischke, J.A., J.W. Arthur, K.S. Nordlie, R.O. Hermanutz, D.A. Standen, and T.P. Henry. 1983. Water Res 17:47–63.

# Effects of Acidic Precipitation on Soil Microorganisms

## A.J. Francis*

## Abstract

Organic matter decomposition, soil respiration, and several steps in the nitrogen cycle are affected to varying degrees by acidic rain and soil acidification. Variations in microbial activities reported in the literature are due largely to the complex properties of the different soil types studied, the diversity of the microbial community involved, the interdependence of various microbial processes, the experimental methodology used, and the duration of the experiments. Therefore, no generalizations can be made that are applicable to all ecosystems. Nevertheless, convincing evidence exists in the literature that acidic rain does indeed affect certain key microbial processes, particularly nitrogen transformation in poorly buffered and unmanaged soils, and thus affects the overall nutrient recycling and soil fertility. These effects could be long lasting and, in some instances, irreversible, and the responsible organisms eliminated altogether from the ecosystem.

## I. Introduction

Soil microorganisms play an important role in nature and are critical to ecosystem functioning and to the well-being of plants, animals, and humans. They are responsible for converting many organic forms of essential nutrient elements to the inorganic forms and are known to solubilize several major and minor trace metals that are available for higher plants. They occupy a significant place in carbon, nitrogen, phosphorous, and sulfur cycles and play a key role in the degradation of toxic organic contaminants to innocuous products. Soil is considered the ultimate sink for at least certain atmospheric pollutants, such as carbon monoxide, ethylene, sulfur dioxide, nitrogen dioxide, and hydrocarbons. These gaseous

*Department of Applied Science, Brookhaven National Laboratory, Upton, NY 11973, USA.

pollutants are removed from the atmosphere when reacted with soil by microbial and/or chemical action. Any alteration in the soil microbial activity due to physical or chemical action may affect the overall productivity of the ecosystem. If acidic precipitation has a significant impact on soil microbial processes such as organic matter decomposition, nitrogen transformation, and nutrient recycling, then the effects will be more pronounced in unmanaged range and forest soils where soil fertility is reduced, resulting in greater economic loss. In agricultural soils the effect may be negligible, if any, because of the buffering capacity of the soils as well as liming practices necessary to keep the soil pH near neutral. Examination of both laboratory and field experimental data in the literature indicates that acid precipitation affects soil and aquatic microbial processes (Alexander, 1980; Francis, 1986). In this chapter, the effects of soil acidity and acid precipitation on soil microorganisms, with particular emphasis on soil microbial processes, are reviewed.

## II. Effects on Soil Microorganisms

The type, abundance, and activities of microorganisms in soils are influenced not only by the nature and the availability of carbon and nitrogen but also by environmental factors. Microorganisms in general are sensitive to acidity. A change in soil pH due to acidic rain would be expected to affect microbial numbers and activities and thus alter the biogeochemical processes brought about by them. Microbial growth, activity, and numbers are reduced by soil acidification. Microbial activity is inhibited by a direct $H^+$ effect, an indirect pH-induced effect such as an increase in dissolution of and bioavailability of toxic metals, or both. The effect on soil microorganisms is evident only when the soil becomes more acidic due to acidic precipitation. Wang and others (1980) studied the effect of acidic deposition on microbial populations in Adirondack forest soils. They examined the microbial community in soil and litter under a beech and maple stand and in red pine plantations. In the hardwoods the populations of bacteria, actinomycetes, and fungi closely paralleled changes in soil acidity. Microbial populations in softwood sites were not correlated with characteristics of the site. However, different soil types respond to acidic rain differently, and one may or may not see any change in soil pH. For example, poorly buffered soil will respond to acidic rain more quickly and rapidly both chemically and biologically than medium and highly buffered soils. The effects of acidic rain on soils have been reviewed (Tabatabai, 1985).

### A. Bacteria

Bacteria in general are sensitive to acidity, except for chemoautotrophic thiobacilli, which can survive under extreme acidic environments (Alexander, 1980). Bacterial numbers were significantly reduced in acidified soils, and total soil microbial activity was severely affected in an acid soil of pH 3.0 (Bryant et al.,

1979). Fewer heterotrophic bacteria were isolated from soils exposed to acidic rain and heavy atmospheric pollution than from similar but unexposed soils (Wainwright, 1979). Similarly, Baath and others (1979, 1980a, 1980b) reported that artificial acidification of field plots in a pine forest spodosol decreased bacterial numbers and cell size. They also found significant changes in the functional characteristics of the bacterial population due to acidic treatment, as determined by factor analyses. They noted a shift towards spore-forming bacteria in soils receiving $H_2SO_4$ inputs for 6 years, compared with control soils.

Total numbers of bacteria and actinomycetes generally declined in soil acidified from pH 4.6 to 3.0 with the addition of $H_2SO_4$ (Francis et al., 1981). Nevertheless, Wainwright (1980) reported that in soil receiving rain of pH 3.0 and dry deposition, the bacterial numbers did not change significantly over a 1-year period, even though the soil pH changed from 4.2 to 3.7. Over the long term, it is conceivable that only those organisms that can tolerate these conditions are able to grow and survive, and hence they become dominant because of less competition. Studies by Francis (1982) indicated that bacteria involved in N transformations were sensitive to soil acidity. The abundance of N-fixing free-living bacteria was low in acidic and acidified forest soils.

## B. Fungi

Studies have shown the dominance of fungi in acidic soils compared with other groups of microorganisms. Wainwright (1979) isolated more fungi from soils exposed to acidic rain and atmospheric pollutants than from the control unexposed soils, whereas Baath and others (1980a) observed a decrease in fungal biomass in an artificially acidified forest spodosol soil. Baath and others (1980a) also observed that fluorescein diacetate (FDA) active fungal biomass decreased significantly. The microbial diversity of poorly buffered Adirondack soils decreased, with fungi *Phythium* and *Phytophthora* becoming dominant organisms (Hileman, 1982). In soil microcosm studies, Wang and others (1980) also found that acidic treatments increased fungi and actinomycetes populations, which they suggested may be caused by reduced competition from bacterial populations. Alexander (1980) also suggested decreased competition from other heterotrophs as a possible explanation for the relative abundance of fungi with decreased pH.

The growth of *Aspergillus niger*, *A. flavipes*, *Trichoderma viridae*, and *Penicillium brefeeldianum* was reduced or completely inhibited in soils acidified below pH 3.5. The addition of montmorillonite enhanced fungal growth under these acidic conditions, but kaolinite had no effect (Bewley and Stotzky, 1983a). The fungal species composition in the humus layer of a coniferous forest in Sweden was found to have been altered by treatments of sulfuric acid (100 and 150 kg/ha each year over 6 years) (Baath et al., 1984). However, very few individual species were significantly affected by the experimental acidification; *Penicillium spinulosum* and *Oidiodendron* cf. *echinulatum* II increased with increasing acidic application, but only small changes were found for other isolated fungal taxa (Baath et al., 1984).

## III. Organic Matter Decomposition

Microbial communities are the major participants in the decomposition of plant residues, pesticides, and organic pollutants. Microbial decomposition of organic matter is important to plant growth because plants mineralize organic carbon and release carbon dioxide into the atmosphere for photosynthesis. During the decomposition process, nutrients (e.g., nitrogen contained in the plant material) are transferred back to the inorganic state and so made readily available for plant growth. Any alterations in the environment that affect the decomposition of organic matter will have a significant impact on nutrient cycling. A diverse group of microorganisms participate in the decomposition of natural organic materials in soil, and many of these organisms are sensitive to acidity. A major concern regarding the effect of acidic deposition is the reduction in the rate of organic matter decomposition and release of essential nutrients. Decomposition of simple organic compounds and naturally occurring complex organic materials subjected to acidic precipitation or acidification of soils, either in the laboratory or in the field, has been studied by several investigators.

### A. Simple Organic Compounds

Mineralization of $^{14}C$-glucose was studied in soils exposed to pH 4.1 and 3.2 acidic rain treatments by Strayer and Alexander (1981). The pH 4.1 treatment had no apparent effect on glucose mineralization, but the pH 3.2 treatment decreased the glucose mineralization rate by 30% to 66%. Bewley and Stotzky (1984) observed progressively decreasing amounts of mineralization of vanillin in montmorillonite-amended soil with increasing acidification with $H_2SO_4$ (pH 3.4, 2.8, 2.2, and 1.6) and complete inhibition of mineralization at a soil pH of 1.6.

### B. Naturally Occurring Organic Materials

Acidification effects on leaf litter decomposition in forest soils vary with the type of material studied and the experimental conditions. Exposure to simulated acidic rainfalls increased the rate of decomposition of pine (*Pinus contorta, P. sylvestris* L., *P. nigra*) needles (Abrahamsen et al., 1980; Baath et al., 1980a, 1980b; Roberts et al., 1980) but had no detectable effect on that of spruce (*Picea abies*) needles or aspen (*Populus tremula*) sticks. Abrahamsen and others (1980) studied the decomposition of lodgepole pine (*Pinus contorta*) needles, Norway spruce (*Picea abies*) needles, and raw coniferous humus with various acidic rain treatments. They found that lodgepole pine needles incubated in the field for 70 to 90 days at pH 5.6 and 3.0 showed an increase in decomposition (29%) at pH 3.0 over that at pH 5.6. Norway spruce needles given twice-weekly waterings with pH 5.6, 3, or 2 at a rate of 100 mm/month or 200 mm/month for up to 9 months showed relatively small effects from the acidic treatments. At 100 mm/month, no significant effect was noted, and at 200 mm/month the pH 3 and 2 treatments decreased decomposition by <5%. In litterbag experiments, raw coniferous humus was treated with pH 5.3, 4.3, and 3.5 treatments. The pH 4.3 treatment decreased the rate of decomposition by 8% and the pH 3.5 treatment decreased the rate of decomposition by 10%. Baath and others (1980a) studied the decomposition of Scots pine (*Pinus sylvestris* L.) needle and root litter in litterbags placed in field plots exposed to $H_2SO_4$ treatments at the rate of 50 and 150 kg/ha. Acidic treatments lowered the decomposition rate of both needle and root litter.

Small effects on decomposition of Norway spruce needles in lysimeters exposed to pH 5.6, 3.0, and 2.0 solution at 100 and 200 mm/month were observed by Hovland and others (1980). The pH 3 and 2 treatments at 100 mm/month initially increased the decomposition rate. However, with pH 3 and 2 treatments at 200 mm/month, decomposition had decreased after 38 weeks relative to that of the controls. The effect of acidic treatments on monosaccharide content was not consistent, but with pH 3 and 2 treatment at 200 mm/month, some indication of reduced lignin decomposition was observed. Roberts and others (1980) incubated litterbags for 5 months in field plots subjected to biweekly 5 mm applications of pH 3.1 and 2.7 acidic rain. They observed no significant effect of acidic treatments on respiration, but they found a significant increase (15%) in weight loss of the litterbags with increased acidity. This is probably due to the acidic effect on organic matter, leading to weight loss by leaching and/or making the material susceptible to enhanced biodegradation. Abrahamsen and others (1980) suggested that decomposition of organic matter in acidic coniferous forest soils is apparently only slightly affected by acidification and that the decomposition of fresh litter and cellulose is influenced only at pH <3.

Acidic rain application has significantly reduced organic matter decomposition in forest soils where increases in soil acidity have been observed. Long-term acidification of a field soil severely retards its biological potential for degrading protein and complex polysaccharides (Bryant et al., 1979). Tamm and others (1976) found decreased $CO_2$ respiration with increased $H_2SO_4$ in coniferous samples from field plots given 0, 50, and 100 kg/ha/yr applications of $H_2SO_4$. Lohm (1980) exposed litterbags for 2 years in plots given 0, 50, and 150 kg/ha $H_2SO_4$ per year and found that acidic treatments lowered the decomposition rate by 5% to 7%. With a slight decrease in soil pH after application of pH 2.5 rainfalls, rates of cellulose decomposition decreased; with increased acidification, the rate of humus decomposition decreased significantly; and changes in pH of the humus samples were observed (Abrahamsen et al., 1980). Effects of acidity on microbial decomposition of oak (*Quercus alba*) leaves in naturally acidic and acidified soils were studied by Francis (1982). Acid soil with pH adjusted to 3.5 showed a 52% decrease in total $CO_2$ production relative to that in natural control soil at pH 4.6. Highly significant differences ($p < 0.01$) in the rates of $CO_2$ production were observed among soils amended with organic material. A 37% reduction of total $CO_2$ evolution was observed in acidified (pH 3.5) soils. Acid rain treatments lowered soil pH values, increased total and exchangeable acidity, increased exchangeable Al, and decreased the heterotrophic microbial activity (Strayer and Alexander, 1981). The rate of organic matter decomposition in soils decreases as exchangeable $H^+$ increases. Among the soils tested (control and amended), there was a highly significant ($p$ <0.01) correlation ($y$ = 0.8160) between the relative amount of $CO_2$ produced and the exchangeable hydrogen ion content of the soil (Francis, 1982).

Simulated sulfuric acidic rain (pH 3.0, 3.5, or 4.0) and control (pH 5.6) were applied to decomposing leaf packs of ten hardwood species. Changes in weight and chemical element concentrations were followed for 408 days. Leaf-pack weight decreased most rapidly under the intermediate acidic treatments, especially at pH 3.5. The ratio of the decomposition rate of leaves at pH 3.5 to that of the control leaves varied from 1.10 for Garry oak (*Quercus garryana*) to 12.57 for pin oak (*Quercus palustris*) (Lee and Weber, 1983). Moloney and others (1983) reported that in vitro studies on degradation of balsam fir (*Abies balsamea* (L) Mill.) and red spruce (*Picea rubens*) mixed needle litter showed reduction of microbial $CO_2$ evolution from litter under acidic conditions (pH 3 or 4) and further reduction in the presence of Pb and Zn but not in the presence of Al or Cu. They found cellulose breakdown to be unaffected in acidic, metal-containing soils treated with water acidified to pH 3.8. They attributed the reduction in litter decomposition to repression of the metabolism of litter-degrading microflora by precipitation acidity and metals present in soils and polluted rain. Kelly and Strickland (1984) used both field and laboratory measurements of $CO_2$ evolution as an index of decomposer activity. Forest microcosms were used to evaluate the impact of simulated acidic precipitation on decomposition. Treatments with annual average pH 5.7, 4.5, 4.0, and 3.5 were applied for a 30-month period. No statistically significant effect of treatment on decomposition could be found in the field measurements. When the microcosm was partitioned into 01 and 02 litter, mineral soil (A and B horizons), and roots within the mineral soil horizons for laboratory determination of $CO_2$ efflux, only the 02 litter exhibited a statistically significant decrease as a function of treatment. Efflux of $CO_2$ from the 02 layer was small compared with that from the other layers, and this may account for failure to detect a significant response in field measurements. The inhibition effect observed in the 02 layers may be ecologically important because many plants derive a major portion of their nutritional requirements directly from the 02 litter layer.

## C. Degradation of Pesticide

Information on the effects of acidity or acidic rain on biodegradation of pesticides and synthetic organic compounds is scanty. Biodegradation of pesticides would be affected by acidic precipitation via changes in microbial populations, or changes in pH levels in water and soil solutions, which affect adsorption characteristics of pesticides by soil particulate matter and in some cases makes them unavailable for biodegradation. The pesticides captan, Dicamba, amitrole, Vernolate, Chloramben, Crotoxyphos (Hamaker, 1972), Metribuzin (Ladlie et al., 1976), 2,4-D and MCPA (Torstensson, 1975), and Prometryne (Best and Weber, 1974) were reported to persist longer under acidic than under neutral conditions. Conversely, Diazinon and Diazoxan (Hamaker, 1972) were degraded more readily at lower pH levels. The effects of the pesticides 2,4-D, cacodylic acid, Dylox, methoxychlor, Sevin, and paraquat on soil microbial activity and the fate of $^{14}$C-labeled 2,4-D, Sevin, and paraquat in naturally acid (pH 4.7) and pH-adjusted acidic (pH 3.5) and neutral soils (pH 6.8) were studied (Francis, unpublished results). The data suggest that the addition of these compounds to soils did not have any effect on soil respiration. However, variations in the rate and degradation of 2,4-D, Sevin, and paraquat were observed. Paraquat was not degraded in acidic soils. Persistence of pesticides due to reduction in the rates of degradation may indeed increase the residue levels in soils as well as contaminate surface and groundwater. Nevertheless, further studies are needed to evaluate fully the impact of acidic rain on biodegradation of pesticides and other organic contaminants.

## IV. Soil Respiration and Soil Enzymatic Activities

Evolution of $CO_2$ from degradation of native or amended organic carbon (glucose, cellulose, leaf litter) and the activities of certain enzymes such as dehydrogenase, cellulose, and urease in the soil are good indicators of the overall microbial activity. Reductions in soil respiration and in many soil enzymatic activities have been reported in acidified soils or soils treated with acidic rain.

Soil acidification studies (Ruschmeyer and Schmidt, 1958; Schmidt and Ruschmeyer, 1958; Smith and Whitehead, 1940; White et al., 1949) have in some cases shown decreased cellulolytic activity with decreasing pH. Bewley and Stotzky (1983a) studied microbial respiration in soil mixtures containing 9% kaolinite, 9% montmorillonite, or no clay supplements amended with 1% glucose and treated with $H_2SO_4$ to give a soil pH range from 5.4 to 0.8. Acidification had little effect on soil respiration ($CO_2$, evolution) until the pH was lowered below 3. Glucose was not degraded at approximately pH 2 but was degraded once the soil pH was raised to noninhibitory levels, that is, pH 4.1 to 4.3. When the soil pH was reduced to 1.0 or below, it was necessary to reinoculate the soil and raise the pH to a noninhibitory level to obtain $CO_2$ evolution. The addition of clay minerals, particularly montmorillonite, mitigated the toxic effect of $H_2SO_4$, especially at pH values below 3 (Bewley and Stotzky, 1983a).

Natural acidic soils (pH 4.7), acidified soils (pH 3.5), and pH-adjusted near-neutral soils (pH 6.8) that had been preincubated under moist conditions for 60 and 270 days were assayed for urease and dehydrogenase activities (Francis et al., 1981). These activities showed no apparent differences between the natural and pH-adjusted neutral soils, but they were much lower in the acidified soils. Surface samples of a brown earth soil exposed to heavy atmospheric pollution for 1 year resulting in a decrease in soil pH from 4.2 to 3.7 showed no significant change in the activities of the soil enzymes arylsulfatase, cellulase, dehydrogenase, phosphatase, rhodanese, and urease (Walker and Wickramasinghe, 1979).

Norway spruce needle litter collected from field plots exposed to pH 6.1, 4.0, 3.0, and 2.5 rains over a 5-year period showed little effect on biological activity as measured by respiration and cellulase activity (Hovland, 1981). The lack of effect is probably due to a shift in the microbial community in the litter, as evidenced by enhanced growth of litter-decomposing fungi, basidiomycetes, mainly of the genera *Mycena* and *Marasmius*. Killham and others (1983) exposed a Sierran forest soil (pH 6.4) planted with ponderosa pine seedlings to simulated rain (pH 3.0, 3.0, 4.0, and 5.6) with 15 cm of precipitation over a 12-week period. Changes in microbial activity were most significant in surface soils. Only the pH 2.0 rain caused inhibition both of respiration and of enzyme activities of urease, phosphatase, dehydrogenase, and arylsulfatase.

Following short-term (92 days) exposure of soil cores to acidic rain at pH 3.7 and 3.0, no alteration in enzyme activities was observed (Bitton and Boylan, 1985). Nitrogen fixation was reduced only in one of the soils by the rain at pH 3.0. Following long-term (690 days) exposure to acidic rain, alterations in enzyme activities and $CO_2$ evolution were dependent on the soil type. No reduction in adenosine triphosphate (ATP) content or $N_2$ fixation was observed.

The effects of acidic precipitation on selected microbiological processes were investigated under field conditions. The study site, near Lake McCloud, Florida, was composed of three transects, each including three plots. Each transect was spray-irrigated with lake water at various levels of acidity. Acidic rain did not significantly reduce dehydrogenase, phosphatase, or urease activity; however, protease activity was decreased. Similarly, soil respiration was not altered by acidic rain of pH 3.7 or 3.0 (Bitton et al., 1985).

The effects of acidic precipitation on microbial and enzyme activities were determined in a field site near Melrose, Florida, by Will and others (1986). Three transects of two plots each were irrigated with lake water acidified with 7:3 (v/v) $H_2SO_4/HNO_3$ to pH 3.0 or 3.6, or with unacidified lake water (pH 4.6, control) at a rate of 10 cm/week for 20 weeks. The experimental plots were divided into covered and uncovered subplots to determine the effect of excluding natural rainfall (pH 4.6). The application of 200 cm of simulated acidic precipitation did not significantly influence soil pH, respiration, or arylsulfatase activity. Phosphatase activity decreased under the pH 3.0 treatment, and urease activity was stimulated by the pH 3.6 treatment. The amount of N mineralized over a 12-week incubation period was lower in soil from the pH 3.0 treatment than in soil from the pH 3.6 and 4.6 treatments. After the application of 200 cm of simulated acidic rain, soil pH, respiration, and phosphatase activity levels were lower in the plots protected from natural rainfall. Twenty-four weeks after the last application there were no differences in any of the measured parameters between the pH treatments (Will et al., 1986).

Soils that had previously been subjected to long-term acidification were tested for microbial activity after 3 years. Total soil respiration was decreased by acidification; mineral N accumulation was not influenced, but the rate of microbial N turnover was markedly decreased (Lohm et al., 1984). Klein and others (1984) observed significantly reduced rates of carbon and nitrogen mineralization in soil that had received simulated precipitation at pH 3.5. The soils of both a limed plot and a control plot in an acidic beech stand were compared microbiologically by

Lang and Beese (1985). Samples were collected in intervals of 1 to 2 months during one vegetation period. Liming caused marked changes in the humic layers, but there was little effect in the mineral soil, even in the upper 5 cm. Bacterial counts were elevated in the organic horizons on the limed plot. Oxygen consumption after addition of glucose showed that the microbial biomass had increased. The mineralization of organic matter was accelerated, but there was no increase of net mineralization of nitrogen.

# V. Nitrogen Cycle

Nitrogen transformation has been studied very extensively in soil, water, and sewage treatment systems. Several steps in the nitrogen cycle are sensitive to acidity. Nitrogen is the major nutrient limiting plant growth. The various parts of the nitrogen cycle, for example, ammonification, nitrification, denitrification, and nitrogen fixation, are affected by acidic rain and soil acidity to varying degrees. The pH optima have been reported for pure cultures and cultures grown in sewage and in some soils. There is some information available on nitrogen transformation in soils subjected to short-term and prolonged exposure to acidic precipitation.

## A. Ammonification

The conversion of organic nitrogenous compounds to ammonium is called *ammonification* and is brought about by a large group of heterotrophic bacteria, fungi, and actinomycetes. Because of the participation of a diverse group of microflora, ammonification may be less sensitive to environmental changes. Several studies have shown that acidification can slightly enhance mineralization of organic nitrogen (Strayer et al., 1981; Tamm et al., 1977; Wainwright, 1980). However, ammonification in a pH-adjusted acidic forest soil (pH 4.6) and highest rates of ammonification were observed in the pH-adjusted neutral soil (Francis, 1982).

## B. Nitrification

*Nitrification* is the sequential oxidation of ammonia to nitrite and then to nitrate by autotrophic and heterotrophic microbial communities. Nitrification in acidic soils is inhibited, and the responsible organisms are probably eliminated at high acidities. Nitrification results in production of $H^+$ ions; it is both a source of acidity and a process regulated by acidity. Nitrification is the rate-controlling step not only in the oxidation of ammonium but also in the generation of the product for its loss to the atmosphere through denitrification.

### 1. Autotrophic Nitrification

Autotrophic nitrification is principally accomplished by *Nitrosomonas* sp. and *Nitrobacter* sp. Members of the genus *Nitrosomonas* oxidize ammonia to nitrite only, and the genus *Nitrobacter* is limited to the oxidation of nitrite to nitrate. The nitrifying bacteria are very sensitive to acidity. Nitrification occurs optimally at neutral to slightly alkaline pH. In acidic environments, nitrification proceeds slowly even in the presence of an adequate supply of substrate, and the responsible species are rare or totally absent at high acidities. Typically, nitrification decreases markedly below pH 6.0 and becomes negligible below pH 5.0 (Dancer et al., 1972), yet nitrate may occasionally be present in field soils of pH 4.0 or lower (Alexander, 1977). Some soils nitrify at pH 4.5; others do not. The difference is possibly attributable to acid-adapted strains or to chemical differences in the two habitats. Neutral to alkaline soils have the largest nitrifier populations. Accumulation of nitrate has been observed in acidic soils of pH as low as 3.9 (Ishaque and Cornfield, 1976; Ishaque et al., 1971; Weber and Gainey, 1962). Of particular concern are the rates of nitrification in forest soils impacted with acidic pollutants because autotrophic nitrification is far more sensitive to acidity than are other steps in the nitrogen cycles. Pure cultures of ammonium-oxidizing autotrophic bacteria were isolated from acidic soils (pH 4.0–4.5) by Walker and Wickramasinghe (1979), whereas nitrite-oxidizing bacteria were detected in several of the acidic soils but were not isolated in pure cultures. Application of acidic rain (pH 4) reduced nitrification rates in a soil with pH 4.4 (Hovland and Ishac, 1975). In a beech forest soil (pH 3), very small numbers of nitrite- and nitrate-forming organisms were present, but substantial amounts of nitrate were detected (Niese, 1971; Runge, 1971).

Wainwright (1980) reported that N mineralization was enhanced by exposing soil to atmospheric pollutants. Acidification of soil by addition of either powdered sulfur or sulfuric acid decreased $CO_2$ evolution (decomposition) by soils, whereas it increased the amount of mineral N (ammonification) in the sample but lowered the amount of nitrate (Tamm et al., 1976). In naturally acidic soils in which the pH was adjusted with $H_2SO_4$, Francis (1982) reported that ammonium and nitrate formation decreased as soil acidity increased. Strayer and others (1981) found that nitrate formation from added ammonium, but not from native organic N, was inhibited following continuous exposure of several forest soils to simulated acidic rain. Bitton and others (1985) reported that acidic rain of pH 3.7 and 3.0 significantly reduced nitrification.

In studies of nitrate formation in three forest soils from the Adirondack Mountains of New York, Klein and others (1983) found that nitrate was formed when the soils were treated with artificial rain at pH 3.5, 4.1, or 5.6. Compared with simulated rain at pH 5.6, simulated rain at pH 3.5 enhanced nitrate formation in one soil and inhibited it in two others. The rates of N mineralization in acidic forest soil exposed to the simulated precipitation were less for rain at pH 3.5 than at pH 5.6 (Novick et al., 1984). Soil columns from "hardwood" (550 to 790 m), "transition" (790 to 1050 m), and "conifer" (1050 to 1160 m) zones of Camels Hump Mountain, Vermont, were treated with simulated acidic rain (pH 4.0) or nonacidic rain (pH 5.6) (Like and Klein, 1985). The percolates were examined for ammonium and nitrate ions. Nitrification in soils from all three ecosystems was unaffected by acidic treatments, but mineralization was stimulated by acidic treatment of soil from the transition zone. Irrespective of treatment, conifer zone soils released less nitrate than did either transition or hardwood zone soils.

The formation of $NO_3^-$ and $NH_4^+$ was measured in columns containing samples from the surface horizons of 12 forest soils both during and after exposure to simulated rain applied at three times the ambient deposition rates for 116 days (Stroo and Alexander, 1986). The relative responses to increased acidity were correlated with organic matter and N levels of the soils. The average inhibition for 12 soils was linearly related to the amount of acidity added. The quantity of N mineralized was less in some soils after their exposure to simulated rain at pH 3.5 than at pH 5.6 and greater in other soils, but the average amount mineralized after exposure of the 12 soils was not significantly affected by the pH of the simulated rain during the treatment period. The mean percentage of the inorganic N produced in the 12 soils that was in the $NO_3^-$ form was lower during but not after the exposure to simulated rain at pH 3.5 than at 5.6. Sahrawat and others (1985) evaluated the nitrogen transformations (mineralization, nitrification, and nitrous oxide production) in acidic forest floor soils collected from six climax forest sites on Blackhawk Island, Wisconsin. Soil acidity ranged from pH 3.9 to 5.1, and organic matter concentrations varied from 2.4% to 59.0%. Nitrification of mineralized N ranged from complete in the alfisol to almost nonexistent in the histosol. Addition of P had little effect on ammonification or nitrification. Liming, however, greatly enhanced ammonification on nonnitrifying or slowly nitrifying soils and both ammonification and nitrification in nitrifying soils. Lang and Beese (1985) found that in the humic layers of the unlimed plot of a beech stand, nitrification accounted for 50% to 70% of the net N mineralization. Liming resulted in a nitrification of 100%. Chemoautotrophic nitrifiers were found only in the limed soils. Laboratory experiments showed that additions of powdered lime increased the bacterial counts and oxygen consumption during 2 to 8 weeks of incubation (Lang and Beese, 1985).

## 2. Heterotrophic Nitrification

Although the role of autotrophic nitrifiers in the natural ecosystem is well recognized, little is known about the activities of heterotrophic nitrifiers. Many heterotrophic microorganisms have been reported to oxidize various nitrogen-containing compounds in culture medium. Few studies have addressed the relative importance of autotrophic versus heterotrophic nitrification in forest ecosystems. Heterotrophic nitrification has been hyopthesized to be of importance in forest soils, but the actual extent to which it occurs in soil and its ecological significance are not clearly understood. Nitrification occurs in many forest soils under conditions more acid than pH 5.0, but which organisms are nitrifying in these soils is not known (Federer, 1983; Vitousek et al., 1982; Weber and Gainey, 1962). Some have found autotrophic nitrifiers (Bhuiya and Walker, 1977; Martikainen and Nurmiaho-Lassila, 1985; Walker and Wickramasinghe, 1979; Josserand and Cleyet-Marel, 1979; Hankinson and Schmidt, 1984, 1988) in acidic forest soils and some have not, and some identified heterotrophic nitrification (Adams, 1986;

Schimel et al., 1984; Strayer et al., 1981; Stroo et al., 1986) in acidic forest soil. These observed differences by many investigators may be due to (1) differences in pH values of the microsites that are conducive to supporting autotrophic nitrifiers, (2) presence of acid-tolerant or adapted autotrophic nitrifier strains, or (3) activities of heterotrophic nitrifiers. Nitrification by heterotrophs might be of major importance in acidic soils and in highly alkaline, nitrogen-rich aqueous environments, where autotrophic nitrification is not detected (Focht and Verstraete, 1977; Remacle, 1977). Organic heterotrophic nitrification may result in other significant environmental implications because many of the intermediary products such as hydroxylamines, amine oxides, nitroso- and nitro-compounds are toxic and even mutagenic or carcinogenic. Nevertheless, the formation and accumulation of the heterotrophic nitrification intermediates or metabolites in acidic soils, such as forest soils, are yet to be documented.

Formation of nitrate by heterotrophic nitrification in acidic soils has been suggested by Abrahamsen and others (1980) and Ishaque and Cornfield (1976). Lang and Jagnow (1986) found no autotrophic nitrifying organisms in a podzolic brown earth (pH 3.5) forming nitrate. They isolated 350 fungi and aerobic heterotrophic bacteria and found that about 25% of the isolates produced 0.05 to 0.90 mg N/L nitrite or nitrate in peptone solution, soil extract mixture, or sterilized soil. The nitrification rate of the most active fungus, *Verticillium lecanii*, was highest at pH 3.5 in defined media. Nitrate formation is rapid in the organic horizons of forest soils in the Adirondacks with pH values ranging from 3.6 to 4.0 (Klein et al., 1983). Nitrate production in samples of these soils is not affected by the addition of either nitrapyrin, an inhibitor of autotrophic nitrifiers, or of ammonium, whereas additions of peptone stimulated nitrate formation (Kreitinger et al., 1985). These data suggest that the active organisms in these soils are heterotrophs, but the responsible organisms have not been characterized. Recently, Stroo and others (1986) isolated a fungus from acidic soil capable of nitrification under acidic conditions. They also presented evidence that heterotrophic organisms bring about nitrate formation in an acidic forest soil (pH 3.8) and that the nitrifiers are adapted to acidic conditions.

Schimel and others (1984) used short-term assay with slurries of freshly sampled soils to quantify potential fluxes through the $NO_2^-$ and the $NO_3^-$ pools in a forest soil and examined the effects of clear-cutting. The authors used two inhibitors (chlorate and acetylene) to block the oxidation of $NO_2$ and $NH_4$ by chemically isolating the process of interest. By separately measuring the production rates of $NO_2^-$ and $NO_3^-$, the pathways responsible for autotrophic nitrification of $NH_4^+$ and heterotrophic nitrification of $NO_3^-$ from organic matter were measured. The authors concluded that the potential for heterotrophic nitrification in the forest soil studied was greater than that for autotrophic nitrification. Similarly, Tate (1977) found that heterotrophic microorganisms were responsible for some of the nitrate produced in organic-rich muck soils.

In aquatic systems, heterotrophic nitrification was identified as the sole bacterial process responsible for the production of $NO_2^-$ and $NO_3^-$ (Tam, 1985). The presence of the autotrophic inhibitor allythiourea did not inhibit $NO_2^-$ and

$NO_3^-$ production in water samples. The addition of ammonium sulfate, $(NH_4)_2SO_4$, did not stimulate, and in some cases depressed, the production of $NO_2^-$ and $NO_3^-$. The addition of sodium acetate to give a final C:N ratio of 3:1 stimulated $NO_2^-$ production. Autotrophic nitrifying bacteria were not found in the waters, whereas several heterotrophic nitrifying bacteria were isolated from the lakes.

Inhibition of nitrification in acidic forest soils and soils undergoing acidification due to acidic precipitation would indeed result in the accumulation of $NH_4^+$ in soils, that is, nitrogen conservation as opposed to nitrogen mineralization (nitrification) and loss of $NO_3^-$ due to leaching and denitrification. Therefore, it would appear that the inhibition of nitrification in acidic soils would result in conservation of N in forest soils and benefit forest growth in the long term. It has been suggested that increased nitrogen compounds in acidic rain could be an important factor in forest dieback (Nihlgard, 1985). Whether such accumulation and increased availability of ammonium in forest soils can be attributed to forest dieback observed across vast areas of northern Europe and North America is not clearly understood. Nevertheless, the significance of nitrification in forest soils needs to be critically examined. Furthermore, the extent of natural processes, particularly nitrification in the watershed soils, contributing to surface water acidification needs further examination.

## C. Denitrification

The reduction of $NO_3^-$-N to $N_2$ gas by soil microorganisms generally under anoxic conditions in the presence of available carbon is a process called *denitrification*. Biological denitrification has received much attention because of its importance in the removal of nitrates from waste waters. Of particular interest to soil and atmospheric scientists is the biogenic evolution of $N_2O$ and its subsequent effect in depleting atmospheric ozone. Denitrification is partially inhibited in acidic soils releasing $N_2O$ instead of $N_2$ as the end product. Thus, microbial nitrogen metabolism in acidic soils may increase the quantity of $N_2O$ produced. Nitrous oxide is a product of denitrification, but it is also produced during nitrification (Bremner and Blackmer, 1978) and during reduction of $NO_3^-$ to $NH_4^+$ (Yoshida and Alexander, 1970). Denitrification in soil produces both $N_2O$ and $N_2$; hence this process may serve either as a major source of $N_2O$ or as a sink for $N_2O$, through the reduction of $N_2O$ to $N_2$.

Soil acidity is known to affect the evolution rate and the composition of the gaseous end products of denitrification. Chemical $NO_2^-$ decomposition is predominant at low pH in sterile and nonsterile soils incubated under aerobic and anaerobic conditions. Little chemical decomposition of $NO_3^-$ is observed in neutral or alkaline environments. Major gases evolved during chemical $NO_2^-$ volatilization are NO and $NO_2$. Nitrous oxide and molecular nitrogen are the major gaseous end products of biological denitrification. Denitrification is favored by relatively high pH values, and at pH values below 6 the reduction of $N_2O$ is often strongly inhibited, resulting in an increase in the ratio $N_2O$ to $N_2$ (Firestone et al.,

1980; Bollag et al., 1973; Bremner and Shaw, 1958; Bryant et al., 1979; Koskinen and Kenney, 1982; Nommik, 1956; Waring and Gilliam, 1983; Wijler and Delwiche, 1954). Francis (1982) reported that the rate of denitrification was rapid at soil pH 6.5, with little $N_2O$ accumulation, indicating further reduction to $N_2$, whereas in acidic (pH 4.6) and acidified (pH 3.5) soils, the rate of denitrification was slow, with $N_2O$ as the major end product. It has been observed that soils with an acidic pH produce proportionately more $N_2O$ than soils with a neutral or alkaline pH. The production of $N_2O$ relative to $N_2$ during denitrification in soils is strongly influenced by the physical environment and by the physiological characteristics of the microbial community (Munch, 1987).

Nitrous oxide emission rates in acidic forest soils were related to nitrification (Sahrawat et al., 1985). From 0.03% to 0.3% of the $NH_4^+$-N nitrified was released as $N_2O$. Phosphorus addition had little effect, but liming increased $N_2O$ emission rates in soil where nitrification was also enhanced. Nitrous oxide production under aerobic conditions was studied in nitrifying humus from a urea-fertilized pine forest soil. Acetylene and nitrapyrin inhibited both $NH_4^+$ oxidation and $N_2O$ production, indicating that $N_2O$ production was closely associated with autotrophic $NH_4^+$ oxidation. Low soil pH enhanced $N_2O$ production; it was negligible above pH 4.7. When soil pH decreased from 4.7 to 4.1, the relative amount of $N_2O$-N produced from $NH_4^+$-N oxidized increased exponentially to 20%. The results suggest that enhanced autotrophic $NH_4^+$ oxidation is a potential source of $N_2O$ in fertilized acidic forest soil (Martikainen, 1985). Poth and Focht (1985) demonstrated with a series of $^{15}N$ kinetic tracer experiments that under conditions of oxygen stress, *Nitrosomonas ueuropea* used nitrite as a terminal electron acceptor and produces $N_2O$.

## D. Nitrogen Fixation

One of the most important microbial processes in the nitrogen cycle is nitrogen fixation, the conversion of $N_2$ gas into combined nitrogen. This process is principally carried out by microorganisms, and higher plants benefit by using the nitrogen that is fixed. There are two types of nitrogen-fixing microorganisms: free-living and symbiotic. The former include the cyanobacteria and the soil bacteria of the genera *Azotobacter* and *Clostridium*. Nitrogen is released as a result of excretion by these organisms and is produced as a result of decomposition of the cellular biomass by other microorganisms. The symbiotic nitrogen fixation is carried out by bacteria and actinomycetes in association with higher plants, the most important of which are the bacteria of the genus *Rhizobium*, which live in association with legumes. Members of the actinomycete genus *Frankia* also fix nitrogen in the nodules of nonleguminous plants, such as Alder (*Alnus* spp.). The nitrogen fixed in the symbiotic associations is incorporated into plant tissues. Acidic precipitation affects N fixation by free-living organisms and by symbiotic associations. Toxicity resulting from iron or aluminum in acidic soils also has a profound effect upon nitrogen fixation.

## 1. Asymbiotic N Fixation

Nitrogen-fixing microorganisms differ in tolerance to acidity. Among free-living bacteria, *Azotobacter, Beijerinckia,* and *Clostridium* have been extensively studied. Environments more acidic than pH 6.0 contain few or no nitrogen-fixing bacteria. Nitrogen fixation by free-living bacteria in freshly collected forest soil (pH 4.6) was not detectable. Only soil samples of pH 5.7 amended with glucose and preincubated under aerobic or anaerobic conditions exhibited slight activity (Francis, 1982). Blue-green algae, which also fix nitrogen, grow poorly and are found to be sparse in acidic environments. Chang and Alexander (1983a) reported that the rates of nitrogen fixation in three forest soils exposed to acidic precipitation were significantly lower if the pH of the simulated rain was 3.5 than if it was 5.6 and that $CO_2$ fixation also was significantly less in soils exposed to acidic rain of pH 3.5. They suggest that algae in terrestrial ecosystems may be especially susceptible to acidic precipitation.

## 2. Symbiotic N Fixation

Soil acidification affects symbiotic nitrogen fixation in legumes. Effects on symbiotic nitrogen fixation by legumes in unmanaged range soils could be significant. Soil acidity affects plant growth, survival of rhizobia, and the symbiotic relationship. Successful infection, formation of nodules, and fixation of nitrogen will all be affected. Reductions in nodulation and plant growth due to reduced N fixation have been reported (Chang and Alexander, 1983b; Evans et al., 1980; Munns et al., 1981; Shriner, 1977; Shriner and Johnston, 1981). Simulated acidic rain generally decreased plant growth, nodulation, and nitrogenase activity (Harris and Paul, 1984). In some cases, the bacteria are sensitive to acidity (Bromfield and Jones, 1980; Lowendorf et al., 1981).

Little is known of the effects of acidic rain on N fixation by actinomycete association with nonleguminous plants in forest soils. *Alnus* seedlings with *Frankia* were grown in soils of pH 3.6 to 7.6 to study the influence of acidity on nodulation of black alder used to revegetate acidic mine soils (Griffiths and McCormick, 1984). In some of the soil types tested, nodulation was reduced at pH below 5.5, and there was evidence of decreased viability of the endophyte at pH below 4.5; root growth and root hair development were inhibited at pH below 5.0.

# VI. Mycorrhizae Associations

A unique fungus association with higher plants is found in the structure known as the *mycorrhiza* or *fungus root,* a two-membered relationship consisting of root tissue and a specialized mycorrhizal fungus. The formation of mycorrhizae is particularly pronounced in land low in phosphorus and nitrogen. Mycorrhizae are important in forestry, and poor stands of seedlings are encountered in the absence of the fungus. Acidic rain may have an impact on the symbiotic association between roots and the soil fungi. Mycorrhizae are essential for the survival and growth of most forest tree species in the natural environment. Therefore, the sensitivity of the mycorrhizal association to acidic rain may be detrimental to forest trees and might contribute to the decline of forests. Little is known of the response of mycorrhizal associations to acidic rain. Because most fungi are able to proliferate under acidic conditions, one can expect minimal effects on survival or growth of mycorrhizal fungi and consequently their associations under acidic conditions. A reduction in the fungi mantle of spruce mycorrhizae receiving heavy atmospheric pollution was noted by Sobotka (1974). Haines and Best (1975) found no visible damage to endomycorrhizae of sweet gum exposed to pH 3.0 treatments.

Application of pH 3.5 rain at ambient rates had no detectable influence on mycorrhizal infection in six of nine soils tested (Stroo and Alexander, 1985). However, stimulation of mycorrhizal infection was observed in three soils if the simulated acidic rain contained only sulfate and in one soil if it had both sulfate and nitrate (Stroo and Alexander, 1985). Shafer and others (1985) exposed ectomycorrhizal seedlings to varying acidities over 16 weeks and found that rain of pH 4.0 and 3.2 inhibited mycorrhizae formation and that increased soil acidity or other factors at pH 2.4 enhanced formation. Furthermore, mycorrhizal associations are known to alter the response of seedlings to high concentrations of metals such as Cu, Ni, and Al but vary with fungal species (Jones et al., 1986). Near ambient deposition rates and concentrations of acidic rain and ozone caused significant changes in mycorrhizal infection of red oak (*Quercus rubra* L.) and white pine (*Pinus strobus*) seedlings regardless of whether plant growth was affected (Reich et al., 1986).

## VII. Heavy Metal Effects

Heavy metals that are introduced along with acidic rain (Campbell and Stokes, 1985) may have significant effects on microorganisms and microbial activities, particularly in forest soils (Spalding, 1979; Ruhling and Tyler, 1973; Tyler, 1974). It is often difficult to separate effects of some metals from $H^+$ acidity. There are numerous studies on the effects of toxic metals on axenic cultures of microorganisms. Many studies on the effects of metals on soil microorganisms and microbial populations are inconclusive because the bioavailability of the added metal was not clearly established.

Firestone and others (1983) examined the effects of Al or Mn mobilized by acidic precipitation on soil microbial activity using growth of *Aspergillus flavus* spores as bioassay for toxicity. They concluded that (1) there is little evidence indicating that acidic rain–induced Al or Mn inhibits soil microbial activity under field conditions; (2) under severe conditions, the potential for inhibition may exist; and (3) toxicity is a function not only of the quantity of Al mobilized but also of the anions present in the soil solution. Litter decomposition and soil enzymatic activity in forests surrounding metal-processing industries are influenced by high concentrations of deposited heavy metals (Ruhling and Tyler, 1973; Tyler, 1974). Decomposition of hardwood leaves amended with Pb (0–1,000 $\mu g/g^{-1}$) and

$H_2SO_4$ was not affected (Crist et al., 1985). The possible explanation for the lack of an effect is that the added Pb may have resulted in the formation of insoluble $PbSO_4$ in the experimental system. Bewley and Stotzky (1983b) reported little effect on the rate of $CO_2$ evolution from glucose-supplemented soil adjusted to pH 3.2, and concomitant additions of 100 or 1,000 ppm of Cd or of 1,000 or 10,000 ppm of Zn added as sulfates. In soil adjusted to pH 2.8, the lag in $CO_2$ evolution was increased by 1 day and was extended further by the concomitant addition of 10,000 ppm but not 1,000 ppm of Zn or of 1,000 ppm but not 100 ppm of Cd. The growth of *Aspergillus niger* in soil acidified to pH levels of 3.6 to 4.2 was further reduced by the addition of either 100 or 250 ppm of Cd or 1,000 ppm Zn. In vitro studies on litter decomposition showed that acidic conditions reduced microbial $CO_2$ evolution from litter (Moloney et al., 1983). The presence of Pb and Zn but not Al or Cu further repressed $CO_2$ evolution. However, cellulose breakdown was unaffected in acidic, metal-containing soils treated with water acidified to pH 3.8. Reduction in litter decomposition, which could adversely alter the rates of mineral cycling in natural ecosystems, may be due to repression of the metabolism of litter-degrading microflora by precipitation acidity and metals present in soils and introduced by polluted rain.

Reports on nitrogen mineralization and nitrification rates in mineral soils treated with heavy metals are inconsistent. However, Tyler (1975) determined the relationship between the nitrogen mineralization rate and the degree of Cu and Zn pollution in conifer forest soils and found that three times the background-level concentration of Cu and Zn impaired nitrogen mineralization rates significantly. He also found that moderate amounts of metal pollutants are sufficient to reduce the supply of nitrogen available to plants. Mineralization and nitrification of glycine and ammonium sulfate in acidified soils amended with several concentration levels of Cd were studied by Bewley and Stotzky (1983c). Ammonification was relatively insensitive to both Cd and acidity, occurring even in soil exposed to pH 2.0 or 1,000 ppm of Cd. Nitrification was more sensitive. It was slow in glycine-supplemented soils exposed to pH 2.5 and inhibited in soil exposed to pH 2.0. In $NH_4$-N supplemented soils, nitrification was slow at pH 3.0 and inhibited at pH 2.5. Nitrification was slowed down at pH 500 ppm Cd and to a greater extent at 1,000 ppm Cd, where nitrate accumulated. The combined effect of exposure to Cd and $H_2SO_4$ at pH 3.0 was additive rather than synergistic, and even with 1,000 ppm Cd at pH 3.0, some nitrification eventually occurred.

A broad spectrum of microorganisms, including eubacteria (nonmarine and marine), actinomycetes, yeasts, and filamentous fungi, were evaluated for their sensitivities to nickel (Babich and Stotzky, 1982). Wide extremes in sensitivity to Ni were noted among the filamentous fungi, whereas the range of tolerance to Ni of the yeasts, eubacteria, and actinomycetes was narrower. With all microorganisms, the toxicity of Ni was potentiated as the pH was decreased to acidic levels. The mechanism(s) of how pH affects the toxicity of Ni has not been defined, although the formation of hydroxylated Ni species with differing toxicities was not involved. The enhanced toxicity of Ni at acidic levels may have implications for the toxicity of Ni in environments stressed by acidic precipitation. The toxicity of a combination of nickel (Ni) and copper (Cu) toward growth of heterotrophic microorganisms was greater than the sum of the toxicity of each metal individually. This synergism between Ni and Cu was potentiated by acidic pH levels. The potentiation of the Ni-Cu synergistic interaction at acidic pH levels has relevance to the deposition of acidic precipitation into environments contaminated with these metals (Babich and Stotzky, 1983).

## VIII. Future Research Needs

Although a large body of information on the effects of acidic rain and soil acidification on soil microorganisms and microbial processes exists, we still are unable to determine the rate and extent to which acidic precipitation is affecting soil microbial processes and nutrient cycling. A critical evaluation of the existing literature on microbial ecology, microbial processes, and nutrient cycling in acidic soils and soils undergoing acidification should be made so that such information, in conjunction with ongoing research on acidic precipitation, can be used to determine the long-term effects of acidic rain on soil microbial processes and soil fertility. Long-term studies must consider the shifts in population dynamics of soil microorganisms, reductions in species diversity, and development of acid-resistant or -tolerant organisms and their functions in the ecosystem. In particular, the organisms responsible in the transformations of organic carbon and nitrogen (nitrification, denitrification, nitrogen fixation) should be emphasized. The role of autotrophic and heterotrophic nitrification in acidic soils and in soils undergoing acidification due to acidic precipitation, as well as acidification of surface waters by nitrification processes in watershed soils, needs further study. Also, acidic rain effects on microbial degradation of organic pollutants and microbial transformation of toxic metals in wastes should be investigated. Comprehensive information on overall microbial processes, in conjunction with soil and atmospheric studies, should be developed over a long-term period.

## IX. Conclusion

Effects of acidic rain on soil acidity on soil microorganisms and soil microbial processes have been extensively studied. These studies include laboratory investigations with soil samples or soil columns and field studies to which simulated acidic rain has been applied at an accelerated rate. Because of the complexity of the soil environment, the diversity of the microbial community, and variation in the organic matter composition and the $H^+$ loading, it is difficult to make any generalizations. Nevertheless, there are some general observations that clearly suggest that acidic rain (1) affects soil microbial population and diversity, (2) retards organic matter transformation and possibly degradation of pesticides and other synthetic organic pollutants, and (3) affects certain steps in the nitrogen cycle, particularly nitrification. Many of the processes are interlinked and dependent on each other so that conclusions drawn about one specific process in a ecosystem may be misleading.

## Acknowledgments

This research was performed under the auspices of the United States Department of Energy under Contract No. DE-AC02-76CH00016.

By acceptance of this article, the publisher and/or recipient acknowledges the U.S. Government's right to retain a nonexclusive, royalty-free license in and to any copyright covering this paper.

## References

Abrahamsen, G., J. Hovland, and S. Hagvar. 1980. *In* T.C. Hutchinson and M. Havas, eds. *Effects of acid precipitation on terrestrial ecosystems*, 341–362. Plenum Press, New York.

Adams, J.A. 1986. Soil Biol Biochem 18:339–341.

Alexander, M. 1977. *Introduction to soil microbiology*. 2d ed. John Wiley and Sons, New York.

Alexander, M. 1980. *In* T.C. Hutchinson and M. Havas, eds. *Effects of acid precipitation on terrestrial ecosystems*, 363–374. Plenum Press, New York.

Baath, E., B. Berg, U. Lohm, B. Lundgren, H. Lundkvist, T. Rosswall, B. Soderstrom, and A. Wiren. 1980a. Pedobiologia 20:85–100.

Baath, E., B. Berg, U. Lohm, B. Lundgren, H. Lundkvist, T. Rosswall, B. Soderstrom, and A. Wiren. 1980b. *In* T.C. Hutchinson and M. Havas, eds. *Effects of acid precipitation on terrestrial ecosystems*, 375–380. Plenum Press, New York.

Baath, E., B. Lundgren, and B. Soderstrom. 1979. Bull Environ Contam Toxicol 23:737–740.

Baath, E., B. Lundgren, and B. Soderstrom. 1984. Microb Ecol 10(3):197–203.

Babich, H., and G. Stotzky. 1982. Environmental Research 29:335–350.

Babich, H., and G. Stotzky. 1983. Ecotoxicology and Environmental Safety 7:576–587.

Best, J.A., and J.B. Weber. 1974. Weed Sci 22:364–373.

Bewley, R.J.F., and G. Stotzky. 1983a. Soil Biol Biochem 15:425–429.

Bewley, R.J.F., and G. Stotzky. 1983b. Environmental Research 31:332–339.

Bewley, R.J.F., and G. Stotzky. 1983c. Archives Environmen Contam Toxicol 12:285–291.

Bewley, R.J.F., and G. Stotzky. 1984. Soil Sci 137:415–418.

Bhuiya, Z.H., and N. Walker. 1977. Appl Bacteriol 42:253–257.

Bitton, G., and R.A. Boylan, 1985. J Environ Qual 14:66–69.

Bitton, G., B.G. Volk, D.A. Graetz, J.M. Bossart, R.A. Boylan, and G.E. Byers. 1985. J Environ Qual 14:69–71.

Blackmer, A.M., and J.M. Bremner. 1978. Soil Biol Biochem 10:187–191.

Bollag, J.M., S. Drzymala, and L.T. Kardos. 1973. Soil Sci 116:44–50.

Bremner, J.M., and A.M. Blackmer. 1978. Science 199:295–296.

Bremner, J.M., and K. Shaw. 1958. J Agric Sci 51:40–52.

Bromfield, E.S.P., and D.G. Jones. 1980. J Applied Bacteriol 48:253–264.

Bryant, R.D., E.A. Gordy, and E.J. Laishley. 1979. Water Air Soil Pollut 11:437–445.

Campbell, P.G.C., and P.M. Stokes. 1985a. Can J Fish Aquat Sci 42:2034–2049.

Chang, F.-H., and M. Alexander. 1983a. Environ Sci and Technol 17:11–13.

Chang, F.-H., and M. Alexander. 1983b. Bull Environ Contam Toxicol 30:379–387.

Crist, T.O., N.R. Williams, J.S. Amthor, and T.G. Siccama. 1985. Environ Pollut 38:295–303.

Dancer, W.S., L.A. Peterson, and G. Chesters. 1972. Soil Sci Soc Amer Proc 37:67–69.

Evans, L.S., K.F. Lewin, and P.A. Vella. 1980. Plant and Soil 56:71–80.

Federer, C.A. 1983. Soil Sci Soc Am J 47:1008–1014.

Firestone, M.K., R.B. Firestone, and J.M. Tiedje. 1980. Science 208:749–751.

Firestone, M.K., K. Killham, and J.G. McColl. 1983. Appl Microbiol 46:758–761.

Focht, D.D., and M. Verstraete. 1977. *In* M. Alexander, ed. *Advances in microbial ecology*, 1:135–214. Plenum Press, New York.

Francis, A.J. 1982. Water Air Soil Pollut 18:375–394.

Francis, A.J. 1986. Experentia 42:455–465.

Francis, A.J., D. Olson, and R. Bernatsky. 1980. *In* D. Drablos and A. Tollan, eds. *Proceedings of the International Conference on Ecological Impact of Acid Precipitation*, 166–167. SNSF project, Oslo, Norway.

Francis, A.J., D. Olson, and R. Bernatsky. 1981. *Microbial activity in acid and acidified forest soils*. BNL 51379, Brookhaven National Laboratory, Upton, N.Y.

Griffiths, A.P., and L.H. McCormick. 1984. Plant and Soil 79:429–434.

Haines, B., and G.R. Best. 1975. *In* L.S. Dochinger and T.A. Seliga, eds. *Proceedings of the First international symposium on acid precipitation and the forest ecosystem*, 951–961. USDA For Gen Tech Rep NE-23. Upper Darby, Pa.

Hamaker, J.W. 1972. *In* C.A. Goring, and J.W. Hamaker, eds. *Organic chemicals in the soil environment*, 253–340. Marcel Dekker, New York.

Hankinson, T.R., and E.L. Schmidt. 1984. Can J Microbiol 30:1125–1132.

Hankinson, T.R., and E.L. Schmidt. 1988. Appl Environ Microbiol 54:1536–1540.

Harris, D., and E.A. Paul. 1984. *In Effects of acid rain on plant microbial associations in California*. NTIS, PCA04/MFA01. 69 p.

Hileman, B. 1982. Environ Sci Technol 16:323A–327A.

Hovland, J. 1981. Soil Biol Biochem 13:23–26.

Hovland, J., G. Abrahamsen, and G. Ogner. 1980. Plant Soil 56:365–378.

Hovland, J., and Y.Z. Ishac. 1975. *Effects of simulated acid precipitation and liming on nitrification in forest soil*. SNSF project. Oslo, Norway IR/14. 15 p.

Ishaque, M., and A.H. Cornfield. 1976. Trop Agric (Trinidad) 53:157–160.

Ishaque, M., A.H. Cornfield, and P.A. Cawse. 1971. Plant and Soil 34:201–204.

Jones, M.D., M.H.R. Browning, and T.C. Hutchinson. 1986. Water Air Soil Pollut 30:441–448.

Josserand, A., and R. Bardin. 1981. Rev Ecol Biol Sol 18:435–446.

Josserand, A., and J.C. Cleyet-Marel. 1979. Microb Ecol 5:197–205.

Kelly, J.M., and R.C. Strickland. 1984. Water Air and Soil Pollut 23:431–440.

Killham, K., M.K. Firestone, and J.G. McColl. 1983. J Environ Qual 12:133–137.

Klein, T.M., J.P. Kreitinger, and M. Alexander. 1983. Soil Sci Soc Amer J 47:506–508.

Klein, T.M., N.J. Novick, J.P. Kreitinger, and M. Alexander. 1984. Bull Environ Contam Toxicol 32:698–703.

Koskinen, W.C., and D.R. Kenney. 1982. Soil Sci Soc Amer J 46:1165–1167.

Kreitinger, J.P., T.M. Klein, N.J. Novick, and M. Alexander. 1985. Soil Sci Soc Am J 49:1407–1410.

Ladlie, J.S., W.F. Meggitt, and D. Penner. 1976. Weed Sci 24:477–481.

Lang, E., and F. Beese. 1985. Allgemeine Forst Zeitschrift 41:1166–1169.

Lang, E., and G. Jagnow. 1986. FEMS Microbiology Ecology 38:257–265.

Lee, J.J., and D.E Weber. 1983. Can J Bot 61:872–879.

Like, D.E., and R.M. Klein. 1985. Soil Sci 140:352–355.

Lohm, U. 1980. *In* D. Drablos and A. Tollan, eds. *Ecological impact of acid precipitation,* 178–179. SNSF Project, Oslo, Norway.

Lohm, U., K. Larsson, and H. Nommik. 1984. Soil Biol Biochem 16:343–346.

Lowendorf, H.S., A.M. Baya, and M. Alexander. 1981. Appl Environ Microbiol 42:951–957.

Martikainen, P.J. 1985. Appl Environ Microbiol 50:1519–1525.

Martikainen, P.J., and E.-L. Nurmiaho-Lassila. 1985. Can J Microbiol 31:190–197.

Moloney, K.K., L.J. Stratton, and R.M. Klein. 1983. Can J Bot 61:3337–3342.

Munch, J.C. 1987. 8th International Symposium on Environmental Biogeochemistry, Sept. Nancy, France. Abstract 166 p.

Munns, D.N., J.S. Holtenberg, T.L. Tighetti, and D.J. Lauter. 1981. Agron J 73:407–410.

Niese, G. 1971. Ecol Stud 2:119–122.

Nihlgard, B. 1985. Ambio 14:2–8.

Nommik, H. 1956. Acta Agr Scand 6:195–288.

Novick, N.J., T.M. Klein, and M. Alexander. 1984. Water Air and Soil Pollut 23:317–330.

Poth, M., and D.D. Focht. 1985. Appl Environ Microbiol 49:1134–1141.

Reich, P.R., A.W. Schoettle, H.F. Stroo, and R.G. Amundson. 1986. J Air Pollut Cont Assoc 36:724–726.

Remacle, J. 1977. Ecol Bull (Stockholm) 25:560–561.

Roberts, T.M., T.A. Clark, P. Ineson, and T.R. Gray. 1980. *In* T.C. Hutchinson and M. Havas, eds. *Effects of acid precipitation on terrestrial ecosystems,* 381–393. Plenum Press, New York.

Ruhling, A., and G. Tyler. 1973. Oikos 24:402–416.

Runge, M. 1971. Ecol Stud 2:191–202.

Ruschmeyer, O.R., and E.L. Schmidt. 1958. Appl Microbiol 6:115–120.

Sahrawat, K.L., D.R. Kenney, and S.S. Adams. 1985. Forest Sci 31:680–684.

Schimel, J.P., M.K. Firestone, and K.S. Killham. 1984. Appl Environ Microbiol 48:802–806.

Schmidt, E.L., and O. R. Ruschmeyer. 1958. Applied Microbiology 6:108–114.

Shafer, S.R., L.F. Grand, R.I. Bruck, and A.S. Heagle. 1985. Can J For Res 15:66–71.

Shriner, D.S. 1977. Water Air and Soil Pollut 8:9–14.

Shriner, D.S., and J.W. Johnston. 1981. Environ Exp Bot 21:199–209.

Smith, F.B., and T. Whitehead, Jr. 1940. Proc Soil Sci Soc Amer 5:248–253.

Sobotka, A. 1974. *In International Tagung uber die Luftverunreinigung und Fortwirtschaft.* Marianske Lazne, Tschechoslowakei, Tagungsbericht.

Spalding, B.P. 1979. J Environ Qual 8:105–109.

Strayer, R.F., and M. Alexander. 1981. J Environ Qual 10:460–465.

Strayer, R.F., C.J. Lin, and M. Alexander. 1981. J Environ Qual 10:547–551.

Stroo, H.F., and M. Alexander. 1985. Water Air Soil Pollut 25:107–114.

Stroo, H.F., and M. Alexander. 1986. Soil Sci Soc Am J 50:110–114.

Stroo, H.F., T.M. Klein, and M. Alexander. 1986. Appl Environ Microbiol 52:1107–1111.

Tabatabai, M.A. 1985. CRC Critical Reviews in Environmental Control 15:65–110.

Tam, T.Y. 1985. *Effect of acid precipitation on heterotrophic nitrification in Canadian Precambrian Shield Lakes*. Ph.D. Thesis, University of Manitoba, Canada.

Tamm, C.O. 1976. Ambio 5:235–238.

Tamm, C.O., G.W. Wiklander, and B. Popovic. 1976. *Effects of application of sulfuric acid to poor pine forests*. Upper Darby, Pa. USDA Forest Service General Tech Rep NE-23.

Tamm, C.O., G. Wiklander, and B. Popovic. 1977. Water Air Soil Pollut 8:75–87.

Tate, R.L. 1977. Appl Environ Microbiol 33:911–914.

Torstensson, N.T.L. 1975. *In* F. Coulston and F. Korte, eds. *Environmental quality and safety*, 262–265. Georg Thieme, Stuttgart, FRG.

Tyler, G. 1974. Pl Soil 41:303–311.

Tyler, G. 1975. Nature 255:701–702.

Vitousek, P.M., J.R. Gosz, C.C. Grier, J.M. Melillo, and W.A. Reiners. 1982. Ecol Monogr 52:155–177.

Wainwright, M. 1979. Soil Biol Biochem 11:95–98.

Wainwright, M. 1980. Plant and Soil 55:199–204.

Walker, N., and K.N. Wickramasinghe. 1979. Soil Biol Biochem 11:231–236.

Wang, C.J.K., R.J. Ziobro, and D.L. Setliff. 1980. *In Actual and potential effects of acid precipitation on a forest ecosystem in the Adirondack Mountains*. ERDA-80, Section 5:1–60. New York State Energy Research and Development Authority, Albany, N.Y.

Waring, S.A., and J.W. Gilliam. 1983. Soil Sci Soc Amer 47:246–251.

Weber, D.F., and P.L. Gainey. 1962. Soil Sci 94:138–145.

White, J.W., F.J. Holbech, and C.D. Jeffries. 1949. Soil Sci 68:229–235.

Wijler, J., and C.C. Delwiche. 1954. Plant and Soil 5:155–169.

Will, M.E., D.A. Graetz, and B.S. Roof. 1986. J Environ Qual 15:399–403.

Yoshida, T., and M. Alexander. 1970. Soil Sci Soc Am Proc 34:880–882.

# Aluminum Toxicity and Tolerance in Plants

Gregory J. Taylor*

## Abstract

Acidic precipitation may affect vegetation as a result of increased mobility and phytotoxicity of Al in acidified soils. Soil acidity is a well-known edaphic stress and a serious agricultural problem throughout large regions of the world. Not surprisingly, this has led to an intense research effort addressing the biology of Al stress. This research base provides a valuable resource for scientists investigating the impact of acidic precipitation on natural ecosystems.

Review of literature dealing with the physiology and biochemistry of Al toxicity indicates that Al may have several direct toxic effects on plants. At the root-soil interface, potential effects include alterations in membrane structure and function and inhibition of membrane-bound enzymes. These effects may be the basis of Al-induced mineral deficiencies. Cell elongation may be reduced because of effects of Al on cell wall components or cell wall assembly. In the cytosol, Al appears to bind to phosphate in DNA, repressing template activity and inhibiting DNA synthesis. Toxic effects on enzyme-mediated reactions, especially in relation to phosphate metabolism and calcium-calmodulin are also likely.

Aluminum tolerance appears to be a result of both exclusion and internal tolerance mechanisms. The plasma membrane and the cell wall likely play a role in exclusion of Al from the symplasm. Formation of a plant-induced pH barrier in the rhizosphere and exudation of chelate ligands may also contribute to this exclusion. In the symplasm, chelation in the cytosol by carboxylic acids or Al-binding proteins, compartmentation in the vacuole, and evolution of Al-tolerant enzymes are all of potential significance.

## I. Introduction

Acidic precipitation can affect plants as a direct result of acidic deposition on aboveground tissues or indirectly as a result of soil acidification. Soil acidity is a well-known edaphic stress that reduces plant growth as a result of a number of

---

*Department of Botany, University of Alberta, Edmonton, Alberta T6G 2E9, Canada.

detrimental factors, including nutrient deficiencies, drought intolerance, and Mn toxicity. It is Al toxicity, however, that has been identified as one of the most important growth-limiting factors on acidic soils (Foy et al., 1978). Recently, interest has focused on the possibility that soil acidification could lead to mobilization of phytotoxic concentrations of Al into the soil solution. Although the literature addressing this hypothesis is not extensive, literature from other fields of study provide support for the concept that Al toxicity limits growth of plant communities exposed to acidic deposition.

Acidic soils are not restricted to ecosystems receiving considerable acidic deposition. Soil acidity can arise from the equilibrium of water with carbon dioxide to produce carbonic acid, decomposition of organic material, oxidation of ammonium ($NH_4^+$) to nitrate ($NO_3^-$), and oxidation of pyrite (Bache, 1980). In agricultural systems, oxidation of $NH_4^+$ to $NO_3^-$ is especially important because $NH_4^+$ is used as an N fertilizer. In many cases this practice has exacerbated existing acidity problems (Pierre et al., 1971; Jarvis and Robson, 1983a, 1983b, 1983c), and in some cases has created acidity problems in areas where no problem previously existed (Unruh and Whitney, 1986).

Regardless of the source, soil acidity is a serious agricultural problem throughout large geographic regions (Van Wambeke, 1976), possibly affecting as much as 40% of the world's arable soils (Haug, 1984). Thus, Al phytotoxicity is a phenomenon of considerable agronomic and economic importance. Not surprisingly, this has led to an intense research effort addressing both basic and applied aspects of the issue. This research is a valuable source of information for scientists investigating the potential impact of acidic deposition on plants and plant communities. The volume of this literature makes it impossible to review the entire subject here. Furthermore, such a review would duplicate other recent reviews on the subject (Foy, 1974, 1983a, 1983b, 1984; Foy et al., 1978; Rhue, 1979; Haug, 1984; Haug and Caldwell, 1985; Martin, 1986; Rorison, 1986). For these reasons, I have chosen to give an overview of the literature dealing specifically with the physiological and biochemical mechanisms of Al phytotoxicity and tolerance. Readers who wish to pursue this subject in greater detail may wish to refer to recent reviews on Al phytotoxicity (Taylor, 1988a) and tolerance (Taylor, 1988b).

## II. Expression of Aluminum Phytotoxicity

The most dramatic effects of Al toxicity are reduced root and shoot growth and symptoms resembling phosphate deficiency, Ca deficiency, and Fe deficiency. Roots are generally affected more than shoots. Elongation is inhibited, and roots become thickened, stubby, brown, brittle, and occasionally necrotic. Root systems are dramatically reduced, and coralloid in appearance, lack fine branching, and are generally inefficient in absorbing nutrients and water (Clarkson, 1965; Fleming and Foy, 1968; Keser et al., 1975, 1977; Kirkpatrick et al., 1975; Foy, 1984). Cells of the root periphery lose turgor, become vacuolated, and show induction of callose formation, disorganization of the plasma membrane, disruption of golgi function and plastid development, alteration of nuclear structure, loss of cytoplasm, and disintegration. Meristematic regions become disorganized to the extent that the root cap and vascular elements cannot be distinguished (Clarkson, 1965; Fleming and Foy, 1968; Keser et al., 1975, 1977; Kirkpatrick et al., 1975; Hecht-Buchholz and Foy, 1981; Hecht-Buchholz, 1983; Bennet et al., 1985a, 1985b, 1985c, 1985d, 1987b; Wagatsuma et al., 1987; Wissemeier et al., 1987).

Not all species respond equally to Al stress; indeed, differential tolerance to Al among cultivars and ecotypes of at least 39 species has been demonstrated (see Taylor, 1988a). The phenomenon of differential tolerance can be dramatic. In *Triticum aestivum,* root growth of Al-tolerant and Al-sensitive cultivars varied ninefold (13–116% of control) with exposure to 74 μM Al (Taylor and Foy, 1985a; Figure 10–1). Such differences are genetically based in *Glycine max* (Hanson and Kamprath, 1979), *Hordeum vulgare* (Stolen and Andersen, 1978), *Medicago sativa* (Buss et al., 1975; Devine et al., 1976), *Nicotiana plumbaginifolia* (Conner and Meredith, 1985a), *Phalaris aquatica* (Culvenor et al., 1986), *Secale cereale* (Aniol and Gustafson, 1984), *Sorghum bicolor* (Boye-Goni and Marcarian, 1985), *Triticum aestivum* (Kerridge and Kronstad, 1968; Rhue, 1979;

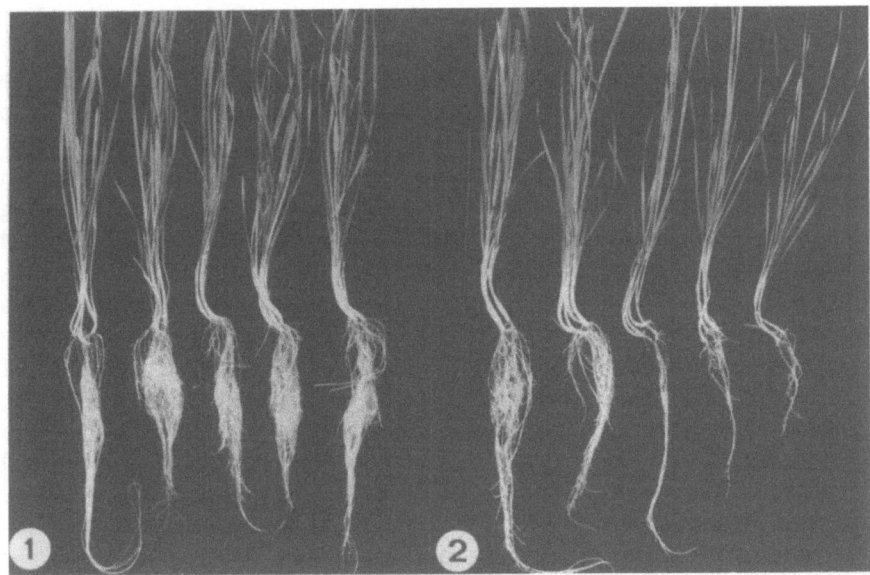

**Figure 10–1.** Growth of five cultivars of *Triticum aestivum* in nutrient solutions without (*1*) or with (*2*) 74 μM Al. Root growth under conditions of Al stress ranged from 13% to 116% of control; shoot growth ranged from 64% to 97% of control. The occurrence of differential tolerance to Al among cultivars or ecotypes of a number of plant species provides a powerful research tool for the investigation of Al phytotoxicity and tolerance. (Reproduced by permission from G.J. Taylor and C.D. Foy, 1985a.)

Aniol, 1984a; Aniol and Gustafson, 1984), *Zea mays* (Rhue et al., 1978; Rhue, 1979), and presumably a host of other species.

Even within a single cultivar or ecotype, the appearance of Al toxicity symptoms does not correlate well with a "threshold" concentration of Al in solutions or soils. This may reflect the importance of factors such as pH, formation of insoluble precipitates, protective effects of certain ions, ionic strength of the growth medium, and the presence of chelate ligands. Each of these factors may affect plant performance through effects on Al speciation. Unfortunately, there is little agreement concerning the species of Al that are phytotoxic, despite much recent research activity (Pavan and Bingham, 1982; Blamey et al., 1983; Kinraide et al., 1985; Wagatsuma and Ezoe, 1985; Alva et al., 1986a, 1986c, 1986d; Hue et al., 1986; Kinraide and Parker, 1987a; Parker et al., 1988).

A number of factors have contributed to our lack of knowledge concerning the toxic species of Al. In addition to the factors mentioned above, it is now clear that the speciation of Al in experimental solutions is affected by conditions such as initial Al concentration, OH/Al ratio, neutralization rate, and mixing conditions (Bertsch et al., 1986; Bertsch, 1987). Until recently, techniques suitable for differentiation between mononuclear and polynuclear Al species have not been available (see Bertsch et al., 1986; Jardine and Zelazny, 1986, 1987a, 1987b; Hodges, 1987; Wright et al., 1987). Although mononuclear species have been computed using equilibrium speciation models, the reliability of thermochemical constants used in these calculations has been questioned (Parker et al., 1987, 1988). Even if reliable thermodynamic data were available, the problem of collinearity still remains (Blamey et al., 1983; Parker et al., 1988); because the activities of monomeric Al species are closely interrelated, it will be difficult to identify the actual toxic species.

With the data available, it would appear that products of hydrolysis and polymerization reactions are potentially phytotoxic. In *Coffea arabica,* seedling growth and leaf Al concentrations were most closely related to $Al^{3+} \cdot 6H_2O$ activity (Pavan and Bingham, 1982; Pavan et al., 1982). Aluminum$^{3+} \cdot 6H_2O$ was also the best indicator of Al stress in *Hordeum vulgare* (Tanaka et al., 1987) and *Triticum aestivum* (Parker et al., 1988). In *Glycine max,* root growth was closely correlated with the sum of monomeric Al species, followed by the activities of $Al(OH)^{2+} \cdot 5H_2O$, $AlSO_4^+$, $Al(OH)_2^+ \cdot 4H_2O$, and $Al^{3+} \cdot 6H_2O$ (Alva et al., 1986d). In *Glycine max* (Blamey et al., 1983; Alva et al., 1986b, 1986c, 1986d), *Gossypium hirsutum* (Hue et al., 1986), *Helianthus annuus, Medicago sativa,* and *Trifolium subterraneum* (Alva et al., 1986b, 1986c), growth was also closely related to the sum of monomeric Al species (Figure 10–2). Recently it has been suggested that hydroxy-Al polymers may also be phytotoxic (Wagatsuma and Ezoe, 1985; Wagatsuma and Kaneko, 1987; Parker et al., 1988), whereas the toxicity of $AlSO_4^+$ would appear to be minimal (Kinraide and Parker, 1987a).

In discussions concerning potential phytotoxic species of Al, it might be useful to distinguish between membrane-mobile species and toxic species. The large pH gradient at the plasma membrane will ensure that the membrane-mobile species is present in the rhizosphere as an acidic species, whereas the species toxic in the

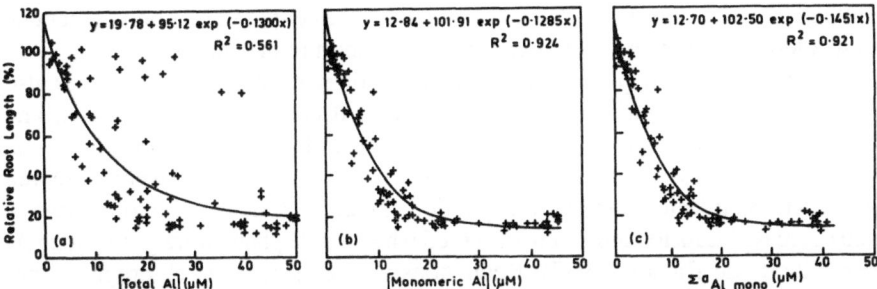

**Figure 10–2.** The relationship between relative root length of *Glycine max* and the concentration of total Al, the concentration of monomeric Al species, and the sum of activities of monomeric Al species in nutrient solutions. Use of the sum of the activities of monomeric Al species as the putative toxic species may be warranted. (Reproduced by permission from A.K. Alva, F.P.C. Blamey, D.G. Edwards, and C. J. Asher, 1986. © Marcel Dekker, Inc., N.Y.)

cytosol must be present at near neutral pH. The toxic effects of Al, however, are not necessarily restricted to the cytosol. If Al exerts a toxic effect on the membrane itself, then the toxic species for membrane injury might be acidic. Thus, we probably have a system with multiple toxic species. For this reason, use of the sum of monomeric Al species as the putative toxic species seems warranted, at least until more definitive techniques for speciation have been developed or more reliable thermodynamic data for equilibrium speciation models are available.

## III. Mechanisms of Aluminum Phytotoxicity

Several lines of evidence indicate that the toxic action of Al is primarily a root-related phenomenon: (1) reductions in root elongation are typically the first observable symptom of Al toxicity (Osborne et al., 1981; Sarkunan and Biddapa, 1982; Schier, 1985; Jarvis and Hatch, 1986), (2) root biomass production is normally more sensitive to Al than shoot biomass production (Brenes and Pearson, 1973; Buss et al., 1975; Sarkunan and Biddapa, 1982; Taylor and Foy, 1985a, 1985b; Zhang and Taylor, 1988), (3) a correlation between the supply of Al and its accumulation in shoots is not a universal feature of Al-stressed plants (Foy et al., 1972), and (4) reciprocal grafting experiments have shown that the tolerance of a grafted plant closely approximates the tolerance of the cultivar from which the rootstock was derived (Klimashevskii et al., 1970; Klimashevsky, 1970).

This is not to suggest that intact roots are required for the expression of Al toxicity and tolerance. Callus or suspension cultures of *Daucus carota*, *Lycopersicon esculentum*, *Nicotiana plumbaginifolia*, and *Sorghum bicolor* derived from different cultivars or cell lines responded differently to Al (Meredith, 1978a, 1978b; Ojima and Ohira, 1982, 1983; Smith et al., 1983; Conner and Meredith,

1985a). Thus, sensitivity to Al was expressed at the cellular level and was not due to whole-plant parameters such as reduced transport to the shoots.

Data demonstrating phytotoxic effects of Al on metabolism in leaf tissues also exist. In *Triticum aestivum,* growth, chlorophyll content, photosynthesis, and transpiration were negatively correlated with Al in leaf blades (Ohki, 1986). In *Phaseolus vulgaris,* reduced photosynthesis and translocation were observed when Al was injected into the transpiration stream, but increased photosynthesis and decreased translocation were observed when whole plants were grown with Al (Hoddinott and Richter, 1987). Reductions in photosynthesis, chlorophyll content, chlorophyll *a/b* ratio, soluble carbohydrates, respiration, and RNA content in leaves were also observed in plants of *Oryza sativa* grown with Al (Sarkunan et al., 1984). Finally, in isolated *Spinacia oleracea* chloroplasts, Al depressed photosynthetic $CO_2$ fixation without a major influence on the photosynthetic enzymes ribulose-1,5,-diphosphate carboxylase and ribulose-5-phosphate kinase (Hampp and Schnabl, 1975).

Regardless of whether Al toxicity is viewed as a root- or a shoot-related phenomenon, a number of possible mechanisms by which Al might disrupt cellular function have been proposed. These include (1) disruption of membrane structure and function, (2) inhibition of DNA synthesis and mitosis, (3) inhibition of cell elongation, (4) disruption of mineral nutrition, and (5) disruption of phosphate and Ca metabolism. Each of these potential mechanisms is reviewed below.

## A. Effects on Membrane Structure and Function

The plasma membrane represents the ultimate barrier between the cytosol and its external environment. As such, it is potentially the first metabolically active target for Al-induced injury. Despite the possible importance of Al injury to the plasma membrane, little experimental attention has been directed towards this issue.

Aluminum has no intrinsic disruptive effect on artificial bilayers, but $Al(OH)_4^- \cdot 2H_2O$ forms strong bonds with carboxylic groups of cationic surfactants used in membrane generation (Tracey and Boivin, 1983). In plasma membrane–enriched microsomes isolated from *Zea mays,* membrane fluidity was reduced by Al stress (Suhayda and Haug, 1986). Membrane assembly in vivo may also be disrupted as evidenced by Al-induced inhibition of golgi development and secretory function (Bennet et al., 1985b, 1987b). Aluminum also appears to affect membrane permeability. In *Quercus rubra,* Al increased membrane permeability to the nonelectrolytes urea, methyl urea, and ethyl urea and decreased permeability to water (Zhao et al., 1987). Such results suggest that Al alters the chemical environment of membrane lipids by binding to polar regions of phospholipids or to membrane proteins (Zhao et al., 1987). Enhanced K efflux has also been reported in Al-stressed plants of *Trifolium repens* (Lee et al., 1984), *Agrostis capillaris* (McCain and Davies, 1984), and *Agrostis tenuis* (Woolhouse, 1970), although enhanced K efflux was not observed with an ecotype of *Agrostis tenuis* adapted to growth on acidic soils (Woolhouse, 1970).

Aluminum also inhibits membrane carriers as exemplified by competitive inhibition of a magnesium $(Mg)^{2+}$-dependent, $K^+$-stimulated ATPase in plasma membrane preparations from *Pisum sativum* (Figure 10–3) and *Zea mays* (Matsumoto and Yamaya, 1986; Suhayda and Haug, 1986). Several mechanisms might be involved in the inactivation of membrane-bound enzymes. Aluminum ions may bind to hydrophilic regions of membrane phospholipids and alter the nature of interactions between lipids and integral proteins. This could affect the conformation and activity of these proteins. Aluminum might also act directly on membrane proteins through formation of an Al-enzyme complex or indirectly by alteration of surface charge (Suhayda and Haug, 1986). Finally, Al could bind to the enzyme substrate, preventing formation of an enzyme-substrate complex. Formation of an Al-ATP complex has been suggested as a possible mechanism for the Al-induced inhibition of membrane-bound ATPases (Matsumoto and Yamaya, 1986).

## B. Effects on DNA Synthesis and Mitosis

Aluminum clearly disrupts normal function of the nucleus, and it has been suggested that Al acts on metabolic processes associated with cell division or DNA replication during interphase (Clarkson, 1965, 1966a). In Al-stressed roots of

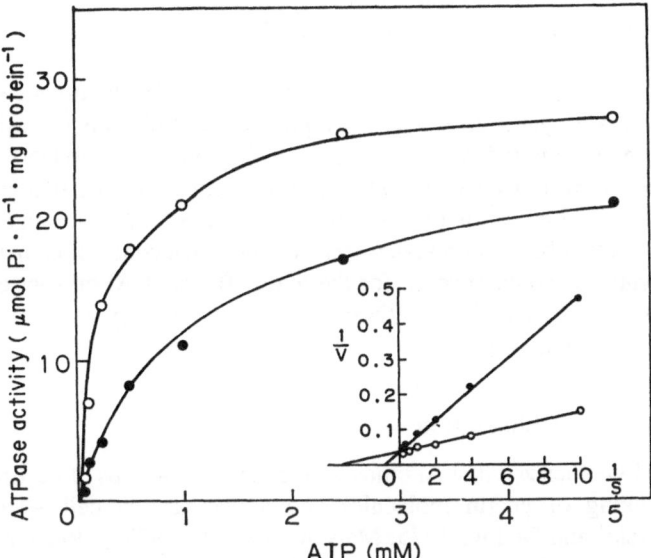

**Figure 10–3.** Effect of ATP concentrations on $Mg^{2+}$-ATPase activity associated with plasma membrane-enriched fractions of roots of *Pisum sativum*. Open circles represent ATPase activity in the absence of Al; closed circles represent ATPase activity in the presence of 500 μM Al. Double reciprocal rate-concentration plots are presented as an inset. One of the primary toxic lesions of Al may be inhibition of membrane-bound enzymes, especially transport enzymes. (Reproduced by permission from H. Matsumoto and T. Yamaya, 1986.)

*Allium cepa,* disintegration and extrusion of the nucleolus from the nucleus (Fiskesjo, 1983) and inhibition of root elongation were associated with the disappearance of mitotic figures in the root tip (Clarkson, 1965, 1969; Morimura et al., 1978). Inhibition of cell elongation and decreased mitotic activity were characteristic of Al stress in *Vigna unguiculata* (Horst et al., 1983) and *Hordeum vulgare* (Sampson et al., 1965). Rapid inhibition of root elongation has also been reported in *Allium cepa* (Clarkson, 1965, 1969; Morimura et al., 1978), *Glycine max* (Wissemeier et al., 1987), *Pisum sativum* (Matsumoto et al., 1976), *Triticum aestivum* (Wallace et al., 1982; Wallace and Anderson, 1984), and *Vigna unguiculata* (Horst et al., 1983).

Localization of Al in the nucleus has been reported in *Allium cepa* (Morimura et al., 1978), *Gossypium hirsutum* (Naidoo et al., 1978), *Phaseolus vulgaris,* and *Pisum sativum* (Matsumoto et al., 1976, 1977b). In *Pisum sativum,* Al was bound specifically to DNA, and binding was inhibited by masking of DNA by histones (Matsumoto et al., 1976). Binding to RNA (Matsumoto et al., 1976) or histones (Matsumoto et al., 1976, 1977b; Matsumoto and Morimura, 1980) was not observed. Aluminum may bind to phosphate in DNA, increasing the rigidity of the double helix, promoting aggregation of chromatin (Matsumoto, 1988), and repressing template activity (Matsumoto et al., 1976, 1977b; Morimura and Matsumoto, 1978; Matsumoto and Morimura, 1980). Template activity of chromatin from *Pisum sativum* was decreased by 40% in both in vivo (Figure 10–4) and in vitro Al-treated preparations (Morimura and Matsumoto, 1978; Matsumoto and Morimura, 1980).

In *Hordeum vulgare,* DNA synthesis was not inhibited by treatment with Al, but the type of DNA synthesized was not typical of control roots (Sampson et al., 1965). Aluminum did reduce incorporation of $^3$H-thymidine into DNA in roots of *Triticum aestivum;* however, inhibition of root elongation preceded any measurable effects on incorporation of $^3$H-thymidine into DNA (Wallace and Anderson, 1984). This latter observation places some doubt on the conclusion that inhibition of DNA synthesis could account for the toxic effects of Al on root elongation. Nonetheless, effects on DNA synthesis and mitosis may represent several of the primary toxic lesions of Al.

## C. Effects on Cell Elongation

It has been hypothesized that Al binds to free carboxyl groups of pectin, resulting in cross-linking of pectin molecules and a decrease in cell wall elasticity (Klimashevskii and Dedov, 1975; Matsumoto et al., 1977a). Reduced cell wall elasticity could inhibit root elongation and cause the formation of brittle roots, both of which are characteristic symptoms of Al toxicity. Aluminum was reported to be associated with pectins in cell wall preparations from *Pisum sativum,* where root plasticity, elasticity, and growth were decreased by treatment with Al (Klimashevskii and Dedov, 1975). In another study, nearly 70% of the total Al in roots of *Pisum sativum* was recovered in the pectin fraction; however, upon ethanol precipitation of polysaccharides and treatment with pectinase, Al did not

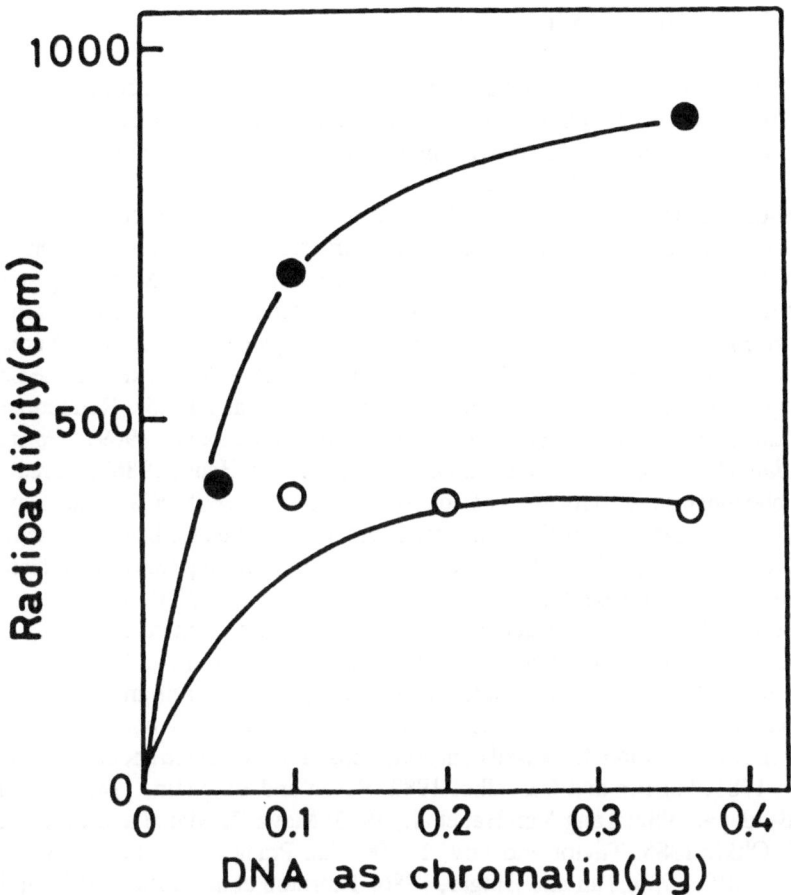

**Figure 10–4.** Effect of in vivo Al treatment of roots of *Pisum sativum* on template activity of chromatin for RNA synthesis. Synthesis of RNA was measured by incorporation of [3]H-UTP into RNA supported by *E. coli* RNA polymerase. Reduced template activity and inhibition of DNA synthesis may be a primary toxic lesion of Al. (Reproduced by permission from H. Matsumoto and S. Morimura, 1980.)

appear to be directly associated with pectins (Matsumoto et al., 1977a). These data suggest that inhibition of cell elongation in *Pisum sativum* did not occur by cross-linking of pectins by Al (Matsumoto et al., 1977a).

Cell elongation could also be reduced by an Al-induced reduction in the rate of cell wall synthesis. A toxic effect of Al on golgi secretory function in *Zea mays* has been demonstrated (Bennet et al., 1985b, 1987b), and treatment of *Gossypium hirsutum* with 1.0 $\mu g\ g^{-1}$ Al impaired the incorporation of [14]C labeled sucrose into cell wall materials. Interestingly, this effect was accompanied by the occurrence of polynucleate cells (Huck, 1972), which might be expected if synthesis of cell wall materials limited growth.

## D. Induced Mineral Deficiencies

Many authors have suggested that Al may induce mineral nutrient deficiencies that form the basis of Al phytotoxicity. Broad disruptions in patterns of mineral accumulation occur in Al-stressed plants. Such disruptions could arise from reductions in mycorrhizal associations (Entry et al., 1987) or N-fixation symbioses (addressed later in this section). Induced mineral deficiencies could also be the result of toxic effects of Al on membrane structure or membrane carriers. For example, Al competitively inhibited (with respect to ATP) a $Mg^{2+}$-dependent, $K^+$-stimulated ATPase in plasma membranes isolated from roots of *Pisum sativum* (Figure 3) and inhibited K uptake by intact seedlings (Matsumoto and Yamaya, 1986). In intact roots of *Picea rubens*, the effect of Al on K uptake was mediated through changes in both $V_{max}$ and $K_m$ (Cumming et al., 1985). Reductions in ATPase activity from roots of *Agrostis tenuis* (Woolhouse, 1969), *Hordeum distichon* (Veltrup, 1983), and *Zea mays* (Suhayda and Haug, 1986) and reduced acid phosphatase activity in roots of *Agrostis capillaris* (McCain and Davies, 1984) and *Agrostis tenuis* (Woolhouse, 1969, 1970) have also been reported.

Several lines of evidence have been used to support the hypothesis that the toxic effects of Al could be the result of an Al-induced phosphate deficiency: (1) symptoms of Al toxicity frequently resemble those of phosphate deficiency (Foy and Brown, 1963, 1964; Foy et al., 1973; Clark, 1977, Alam and Adams, 1979; Osborne et al., 1981; Furlani and Clark, 1981; Foy, 1984; Jarvis and Hatch, 1986; Foy et al., 1987b), (2) Al-stressed plants show increased concentrations of phosphate in roots and decreased concentrations of phosphate in shoots (Malavolta et al., 1981; Fageria and Carvalho, 1982; Pavan and Bingham, 1982; Sarkunan and Biddapa, 1982; Tang Van Hai et al., 1983; Mayz De Manzi and Cartwright, 1984; Ohki, 1985; Taylor and Foy, 1985e; Van Praag et al., 1985; Jarvis and Hatch, 1985, 1986; Tan and Binger, 1986; Thornton et al., 1986, 1987; Baligar et al., 1987; Bennet et al., 1987a), and (3) supply of phosphate to growth substrates has a protective effect against Al injury (Bache and Crooke, 1981; Furlani and Clark, 1981; Blamey et al., 1983; Sheppard and Floate, 1984; Conner and Meredith, 1985b; Kinraide et al., 1985, Alva et al., 1986c).

Clarkson (1966b) hypothesized that inorganic phosphate was precipitated with an Al hydroxide at the root surface or in the apoplasm, reducing entry of phosphate into the root symplasm and subsequent translocation to the shoot. Although Clarkson (1966b) was unable to find evidence of direct inhibitory effects of Al on phosphate uptake into the symplasm, or its subsequent translocation to the leaves, recent studies with short-term uptake of $^{32}P$ have been consistent with this model. For example, increased $^{32}P$ uptake by roots of *Picea rubens* treated with Al was accompanied by decreased translocation of $^{32}P$ to foliage (Cumming et al., 1986). Also, short-term uptake of $^{32}P$ (per unit root weight) into excised and intact roots of several legumes increased with pretreatment with Al, whereas the efficiency of translocation to leaves decreased (Andrew and Vanden Berg, 1973). However, in whole plants of *Oryza sativa*, uptake of phosphate was depressed by pretreatment with Al (Tang Van Hai et al., 1983).

The similarity of some Al toxicity symptoms and Ca deficiency symptoms (Armiger et al., 1968; Foy et al., 1969, 1987b; Soileau et al., 1969; Long and Foy, 1970; Edwards et al., 1976) has also led to the hypothesis that Al may interfere with the absorption and transport of Ca in plants. Inhibition of Ca uptake by Al has been reported in *Gossypium hirsutum* (Lance and Pearson, 1969), *Hordeum vulgare* (Clarkson and Sanderson, 1971), *Pennisetum clandestinium* (Awad et al., 1976), and *Triticum aestivum* (Johnson and Jackson, 1964). In *Triticum aestivum*, Al stress reduced the transport of Ca to shoots in a manner apparently independent of effects on Ca uptake (Johnson and Jackson, 1964).

Concentrations of Ca in roots and/or shoots of numerous species have been shown to be reduced by Al stress (Sarkunan and Biddapa, 1982; Hohenburg and Munns, 1984; Horst, 1985; Jarvis and Hatch, 1985; Konishi et al., 1985; Ohki, 1985; Ownby and Dees, 1985; Schier, 1985; Taylor and Foy, 1985e; Van Praag et al., 1985; Wagatsuma and Yamasaku, 1985; Thornton et al., 1986, 1987; Baligar et al., 1987; Bennet et al., 1987a). Furthermore, increased supply of Ca to Al-toxic growth solutions decreased the extent of Al injury and/or the accumulation of Al by roots of a number of species (Furlani and Clark, 1981; Aniol, 1983; Wagatsuma, 1983b; Kinraide et al., 1985; Alva et al., 1986a, 1986c; Rechcigl et al., 1986; Kinraide and Parker, 1987b; Zhao et al., 1987). The protective effect of Ca is not unique and can be achieved by other ions (Rhue and Grogan, 1977; Rhue, 1979; Kinraide and Parker, 1987b), perhaps suggesting that this effect may be due in part to effects of ionic strength on the speciation of Al (Alva et al., 1986c). However, several reports have indicated that amelioration of Al toxicity by Ca cannot solely be attributed to increases in ionic strength (Kinraide et al., 1985; Alva et al., 1986a, 1986c; Kinraide and Parker, 1987b).

Aluminum has also been reported to interfere with the uptake and translocation of Mg. As with phosphate and Ca, reduced concentrations of Mg are a common characteristic of Al stress (Fageria and Carvalho, 1982; Pavan and Bingham, 1982; Sarkunan and Biddapa, 1982; Grimme, 1983; Horst, 1985; Konishi et al., 1985; Ohki, 1985; Schier, 1985; Taylor and Foy, 1985e; Wagatsuma and Yamasaku, 1985; Thornton et al., 1986, 1987; Baligar et al., 1987; Bennet et al., 1987a), and Mg protects roots from Al injury (Rhue and Grogan, 1977; Aniol, 1983; Kinraide et al., 1985; Kinraide and Parker, 1987b). Lower concentrations of Mg in leaves and roots of *Avena sativa* grown under Al stress have been reported even when the size of the root mass was controlled, suggesting a specific inhibition of Mg uptake (Grimme, 1983). Furthermore, Mg concentrations in leaf and root tissues of *Zea mays* decreased more than any other element under conditions of Al stress (Clark, 1977). Interestingly, a Mg-efficient cultivar of *Zea mays* was more tolerant of Al than a Mg-inefficient cultivar (Clark, 1977), and an Al-tolerant cultivar of *Solanum tuberosum* was able to absorb and translocate Mg more efficiently than an Al-sensitive cultivar under conditions of Al stress (Lee, 1971).

Aluminum stress also disrupts N assimilation. For example, Al reduced $NO_3^-$ assimilation in *Sorghum bicolor* (Keltjens and Van Ulden, 1987a, 1987b), *Triticum aestivum* (Fleming, 1983; Taylor and Foy, 1985d), and *Trifolium repens* (Jarvis and Hatch, 1986), whereas in *Oryza sativa*, both $NO_3^-$ and $NH_4^+$ uptake were reduced (Tang Van Hai et al., 1983). In *Sorghum bicolor*, Al reduced uptake and translocation of N and altered the amino acid composition of tissues (Gomes et al., 1985), whereas in *Camellia sinesis* (Konishi et al., 1985) and *Zea mays* (Santaro et al., 1984), Al increased foliar N. Reduced nitrate reductase activity could be partially responsible for reduced $NO_3^-$ assimilation under conditions of Al stress (Santaro et al., 1984), but $NO_3^-$ did not accumulate in Al-stressed *Sorghum bicolor*, suggesting some other mechanism must be involved (Keltjens and Van Ulden, 1987a). In *Triticum aestivum*, an association between tolerance to Al and high nitrate reductase activity has been reported (Foy and Fleming, 1982), as well an association between Al tolerance and high protein content of grain (Mesdag et al., 1970)

Reduced N fixation on Al-toxic substrates has been reported in a number of species, reflecting toxic effects of Al on growth of the host and free-living bacteria, as well as infection, nodule initiation, nodule growth, nitrogenase activity, and N fixation (Andrew, 1976; Mengel and Kamprath, 1978; Keyser and Munns, 1979a, 1979b, Munns et al., 1979; Carvalho et al., 1981a, 1981b; Franco and Munns, 1982; Hohenburg and Munns, 1984; Mayz De Manzi and Cartwright, 1984; Wood et al., 1984a; Jarvis and Hatch, 1985, 1986; Whelan and Alexander, 1986; Alva et al., 1987). Although the exact cause of reduced nodule initiation in acid, Al-toxic soils is not yet resolved, nodulation improves when these soils are treated with lime (Sartain and Kamprath, 1975; Rice et al., 1977; Mengel and Kamprath, 1978; Carvalho et al., 1980; Lowther, 1980; Haynes and Ludecke, 1981).

Wood and others (1984a, 1984b) reported a complex interaction between Al, Ca, P, and pH on growth and nodulation of *Trifolium repens*. Generally, low pH and high Al decreased growth and nodulation, whereas both Ca and phosphate had a protective effect on Al and pH toxicity. Not surprisingly, considerable variation exists in the acidic soil and Al tolerance of different strains of *Rhizobium* (Keyser et al., 1979; Munns et al., 1979; Lowendorf et al., 1981; Hartel and Alexander, 1983; Hartel et al., 1983; Cunningham and Munns, 1984a, 1984b), as well as in the ability of plant species to nodulate in the presence of Al (Carvalho et al., 1981b; Hohenburg and Munns, 1984).

## E. Effects on Phosphate and Calcium Metabolism

The hypothesis that the toxic effects of Al are a result of induced mineral deficiencies has been criticized because the toxic effects of Al are ameliorated in a nonspecific fashion by phosphate, Ca, Mg, K, or simply by ionic strength (Rhue, 1979), and because such deficiencies cannot explain the rapid toxic effects of Al. Clarkson (1965, 1969) observed complete and irreversible inhibition of root elongation in *Allium cepa* after 6 to 8 hours of treatment with Al, whereas Morimura and others (1978) reported reductions in root elongation after 1 hour and near complete inhibition after 10 hours. Rapid inhibition of root elongation has also been reported in *Pisum sativum* (Matsumoto et al., 1976), *Triticum aestivum*

(Wallace et al., 1982; Wallace and Anderson, 1984), and *Vigna unguiculata* (Horst et al., 1983). Clearly, induced mineral deficiencies would require a longer period of time to be manifest. This type of argument can lead one to the conclusion that Al-induced mineral deficiencies are not direct stress injuries; rather, they are the secondary result of another stress injury, perhaps to the plasma membrane. Closer examination of the literature, however, reveals that rapid toxic effects could result from effects on mineral ion metabolism that are not directly related to induced mineral deficiencies.

It is well established that Al has a rapid effect on cytoplasmic phosphate metabolism, although results are inconsistent. In roots of *Hordeum vulgare,* phosphorylation of hexose sugars was inhibited by Al, possibly through inhibition of hexokinase. Increases in the amount of ATP and other nucleotide phosphates were detected, and the fact that the specific activities of $^{32}$P in ATP were similar in Al-treated and control roots suggested that the rates of ATP synthesis were unaffected by Al stress (Clarkson, 1966b, 1969). In roots of *Glycine max,* increased ATP and pyruvate levels were observed under conditions of Al stress, but this was interpreted to be the result of stimulation of ATP synthesis (Hanson and Kamprath, 1979). A decrease in the incorporation of $^{32}$P into both nucleotides and hexose phosphates was observed in roots of *Onobrychris sativa* (Rorison, 1965), whereas a decline in free nucleotides and an increase in hexose phosphates in Al-treated roots of *Pisum sativum* have been reported (Klimashevskii et al., 1970).

$^{31}$Phosphate nuclear magnetic resonance (NMR) spectroscopy has permitted direct observation of phosphate compounds in vivo. Under acidic conditions (pH 4.0), an increase in soluble, inorganic phosphate was observed in *Zea mays,* possibly reflecting hydrolysis of phosphate from phospholipids of the plasma membrane. Under conditions of Al stress, this phosphate pool decreased concurrently with a decline in surface acid phosphatase activity (Pfeffer et al., 1987). Reduced concentrations of several nucleotides (ATP, UDPG) and respiratory rates compared to controls were also observed after 20 hours treatment (Figure 10–5; Pfeffer et al., 1986). Addition of glucose to treatment solutions ameliorated the toxic effects of Al, suggesting that detoxification of Al in the cytosol may be an energy-dependent process (Pfeffer et al., 1986). Taken as a whole, these results suggest that Al may affect phosphate metabolism via effects on ATP hydrolysis, phosphate esterification, or the hydrolysis of phosphoric acid monoesters. The direction and magnitude of these effects are not well established.

Toxic effects on Ca metabolism are also possible through disruption of calcium-calmodulin function. Calmodulin exerts regulatory control over a wide variety of cellular functions in which Ca is the second messenger (Haug and Caldwell, 1985). Because Al affects the structural integrity of calmodulin (Siegel et al., 1983; Siegel and Haug, 1983a, 1983b; Haug, 1984; Suhayda and Haug, 1986; Weis and Haug, 1987), potential exists for a large and rapid impact on plant performance (Haug, 1984; Haug and Caldwell, 1985).

Aluminum was bound stochiometrically and cooperatively to calmodulin isolated from bovine brain powder (a protein chemically similar to plant calmod-

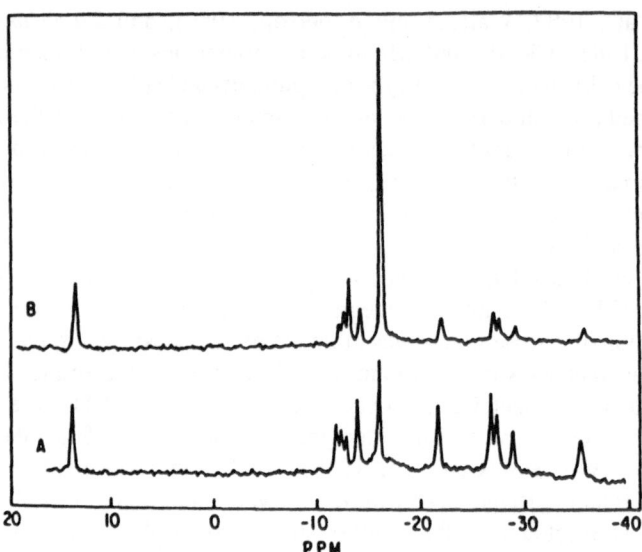

**Figure 10–5.** $^{31}$Phosphorus nuclear magnetic resonance spectra of roots of *Zea mays* perfused for 4 hours with an oxygenated solution of 0.1 mM CaSO$_4$ at pH 4.0 (*A*), or perfused for 20 hours in an oxygenated solution of 0.1 mM CaSO$_4$ and 50 μM Al$_2$(SO$_4$)$_3$ at pH 4.0. Peaks from left to right are believed to represent (1) an internal phosphate standard, (2) glucose-6-phosphate, (3) fructose-6-phosphate, (4) AMP, (5) cytoplasmic inorganic phosphate, (6) vacuolar inorganic phosphate, (7) gamma phosphate of ATP, (8) alpha phosphate of ATP, (9, 10) UDPG and NAD nucleotides, and (11) beta phosphate of ATP. Treatment with Al for 20 hours induced a 65% decline in levels of ATP and UDPG. Levels of glucose-6-phosphate decreased by 20% whereas other nucleotides remained unchanged. Effects of Al on intracellular phosphate metabolism may represent a primary toxic lesion of Al. (Reproduced by permission from P.E. Pfeffer, S.-I. Tu, W.V. Gerasimowicz, and J. R. Cavanaugh, 1986.)

ulins) (Siegel et al., 1982; Siegel and Haug, 1983b). Binding of Al to calmodulin at a molar ratio of 2:1 resulted in structural changes to the protein, including loss of alpha helical content and enhancement of the protein's hydrophobic surface domains (Siegel et al., 1982; Siegel and Haug, 1983b; Suhayda and Haug, 1984). As a result, flexibility of the protein may be decreased, and its ability to interact with other proteins impaired (Siegel and Haug, 1983b; Weis and Haug, 1987). In support of this hypothesis, an Al:calmodulin molar ratio of 3:1 resulted in 95% inhibition of calmodulin-stimulated, Mg$^{2+}$-dependent, membrane-bound ATPase activity in membrane vesicles from roots of *Hordeum vulgare* (Siegel and Haug, 1983a). An Al:calmodulin molar ratio of 4:1 was required to block completely the activity of a Ca-calmodulin-dependent phosphodiesterase (Siegel and Haug, 1983b). Siegel and Haug (1983b) estimated that the binding affinity of the first Al to calmodulin was about one order of magnitude stronger than that of Ca to its comparable site, but these equilibrium constants may have been overestimated because they were calculated at pH 6.5 without taking into account the limited solubility of $Al(OH)_3 \cdot 3H_2O$ (Martin, 1986).

Aluminum may also affect Ca metabolism through effects on growth regulators. Cytokinins have been shown to regulate Ca uptake and binding to cell membranes. Furthermore, both liming and application of benzylaminopurine to lateral shoot meristems ameliorated inhibition of lateral branch initiation in plants of *Glycine max* grown in an acidic, Al-toxic soil (Pan et al., 1988). The common site of cytokinin synthesis and Al toxicity in the root meristem suggests that impaired cytokinin biosynthesis or supply may have been important (Pan et al., 1988). Further research into the effects of Al on Ca metabolism is warranted.

# IV. Mechanisms of Aluminum Tolerance

In light of the uncertainty that exists concerning the physiological and biochemical basis of Al phytotoxicity, it is not surprising to find that our understanding of the mechanisms of Al tolerance is still in its infancy. The lack of progress in this field is discouraging because research is aided by a powerful research tool, the phenomenon of differential tolerance to Al. Because individual cultivars or ecotypes are similar genetically, factors responsible for determining tolerance or sensitivity to Al should be easy to identify. In practice, an intense research effort has failed to produce a unified view of the physiological and biochemical basis of Al tolerance. Perhaps this is because plants have evolved a number of mechanisms to reduce the impact of this powerful selection force.

In several reviews of metal tolerance in plants (Antonovics et al., 1971; Ernst, 1975, 1976; Foy et al., 1978), a distinction was made between two types of tolerance mechanisms; internal tolerance mechanisms where metals enter the organism and tolerance is achieved by some means of detoxification, and external tolerance mechanisms, or exclusion mechanisms, where metals are prevented from entering the organism and reaching sensitive metabolic sites. The role of exclusion mechanisms in metal tolerance has been downplayed in the literature, possibly because of the lack of a clear definition of *exclusion*; a definition based upon metal uptake or exclusion from an "organism" is imprecise. In this chapter I have based the distinction between internal and exclusion mechanisms on the site of metal detoxification or immobilization, either in the symplasm (internal) or apoplasm (exclusion) (Taylor, 1987).

## A. Exclusion Mechanisms

It is clear that the plasma membrane and/or the root apoplasm constitute a considerable barrier to the uptake and accumulation of Al. Exclusion at the root-soil interface occurs to some extent in the apoplasm due to precipitation of Al-phosphate compounds or the formation of insoluble Al monomers $(Al(OH)_3 \cdot 3H_2O)$ and polymers (McCormick and Borden, 1972, 1974; Naidoo et al., 1978; Wagatsuma and Kaneko, 1987). Aluminum has been found localized in intercellular spaces of the epidermis and cortex, with little penetration beyond the endodermis (Wagatsuma, 1984; Huett and Menary, 1980), although in some cases penetration into the cortex was also restricted (Rasmussen, 1968; Bennet et al., 1985c).

The extent to which exclusion of Al from the symplasm plays a role in differential tolerance to Al, however, is still a matter of debate. Historically, gross tissue analyses have been used to make inferences about the role of exclusion in Al tolerance. Lower concentrations of Al were found in aboveground tissues of Al-tolerant compared with Al-sensitive cultivars of *Medicago sativa* (Ouellette and Dessureaux, 1958), whereas similar leaf Al concentrations were required for production of toxicity symptoms. This type of pattern implies the operation of an exclusion mechanism; however, the pattern is not universal (Wallace et al., 1982). Cultivars of *Glycine max* (Foy et al., 1969), *Hordeum vulgare* (MacLean and Chiasson, 1966; Foy et al., 1967), *Lycopersicon esculentum* (Foy et al., 1973), *Phaseolus vulgaris* (Foy et al., 1972), and *Triticum aestivum* (Foy et al., 1967; Foy and Fleming, 1982) did not show differences in the accumulation of Al by aerial tissues, and differences in translocation of Al were not apparent after analysis of stem exudates in Al-tolerant and Al-sensitive cultivars of *Phaseolus vulgaris* (Foy et al., 1972).

It would appear that leaf tissue analyses, which fail to differentiate between Al in the apoplasm and symplasm, may not be a suitable method for investigation of exclusion. Because toxic effects of Al are manifest primarily in root tissue, differences between cultivars in exclusion from the root symplasm is possible without being reflected by differences in the translocation of Al to the shoots. Although tissue analyses are of some value, other approaches to test the exclusion hypothesis will be required. In the following four sections, data supporting immobilization at the cell wall, selective permeability of the plasma membrane, formation of a plant-induced pH barrier in the rhizosphere, and exudation of chelate ligands will be discussed.

## B. Immobilization at the Cell Wall

The cell wall represents a potentially large sink for Al, and the extent to which Al is immobilized in the cell wall could reduce uptake into the symplasm. Nonmetabolic uptake of Al has been demonstrated in roots of *Gleditsia triacanthos, Pinus taeda* (Schaedle et al., 1986), *Brassica oleraceae, Lactuca sativa,* and *Pennisetum clandestinium* (Huett and Menary, 1979). In *Brassica oleraceae* and *Lactuca sativa,* Al uptake by excised roots was biphasic, with a rapid initial phase (60 minutes) followed by a linear or slightly curvilinear phase (Figure 10–6). Aluminum uptake during the first phase, likely representing nonmetabolic accumulation in the apoplasm, accounted for 82% of total uptake and was accompanied by desorption of Ca (Huett and Menary, 1979). Such data suggest that a large portion of Al absorbed by excised roots is absorbed in the apoplasm by exchange with Ca. Clarkson (1967) found 85 to 95% of the total Al accumulated in the roots of *Hordeum vulgare* to be tightly bound to cell wall material and suggested that

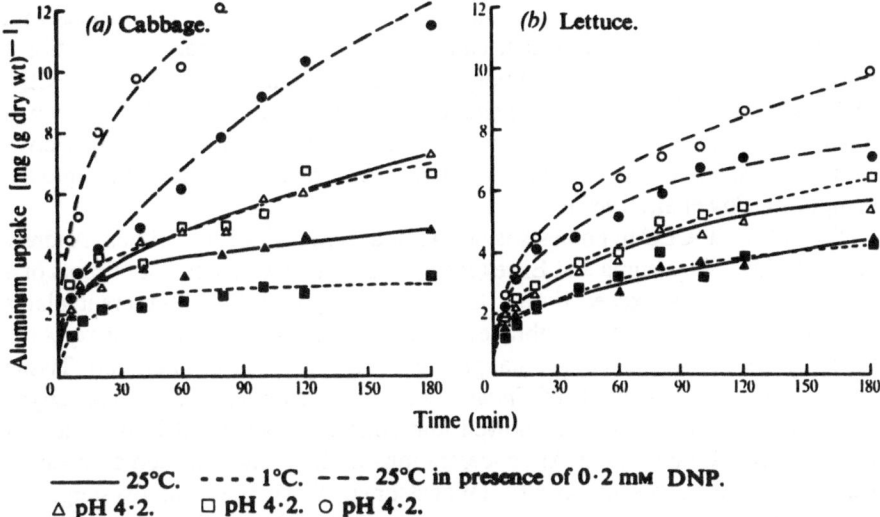

**Figure 10-6.** Time course of Al uptake by excised roots of *Brassica oleracea* (a) and *Lactuca sativa* (b) from solutions of 1.0 mM $Al_2(SO_4)_3$ and 0.5 mM $CaSO_4$. Triangles represent data from 25°C, squares from 1°C, and circles from solutions with DNP. Open symbols represent uptake at pH 4.2; closed symbols represent uptake at pH 4.0. These data demonstrated the primarily nonmetabolic nature of Al uptake. (Reproduced by permission from D.O. Huett and R.C. Menary, 1979.)

interactions between Al and the free carboxyl groups of polygalacturonic acids in the middle lamella of the cell wall could account for such binding. The failure of high concentrations of Ca and Na to desorb Al from cell wall isolates suggested that absorption was more than a simple cation exchange process.

The way in which differences in Al absorption in the apoplasm contribute to differential Al tolerance is still a matter of speculation. In apparent contradiction to the discussion above, cultivar tolerance to Al has been associated with low cation-exchange capacity (CEC) of roots in *Gossypium hirsutum* (Kennedy et al., 1986), *Lolium* sp. (Vose and Randall, 1962), *Hordeum vulgare, Triticum aestivum* (Foy et al., 1967; Mugwira and Elgawhary, 1979), and x *Triticosecale* (Mugwira and Elgawhary, 1979). However, in *Ipomoea batatus*, root CEC did not differ between six cultivars differing in tolerance to Al (Munn and McCollum, 1976). In *Triticum aestivum*, short-term uptake of Al by roots of Al-tolerant (low CEC) cultivars was less than the uptake by roots of Al-sensitive (high CEC) cultivars (Mugwira and Elgawhary, 1979). However, root CEC could not account for differences in the amount of Al absorbed by excised roots of *Brassica oleraceae, Lactuca sativa,* and *Pennisetum clandestinium* (Huett and Menary, 1979). Wagatsuma (1983b) also showed that root Al concentrations of a number of crop species grown in the presence of Al were positively correlated with the CEC of the roots, whereas Al translocation to the leaves was not (Wagatsuma, 1984).

Low root CEC could contribute to Al tolerance in several ways: (1) Species with low CEC accumulate monovalent cations in preference to polyvalent cations, resulting in selective exclusion of polyvalent cations (Takagi, 1976); (2) low CEC reduces binding of Al on exchange sites on the roots, which may be the first step in ion uptake (Mugwira and Elgawhary, 1979); and (3) low uptake of cations relative to anions reduces acidity of the rhizosphere, hence reducing uptake of Al into the symplasm (Foy et al., 1967).

The cell wall–CEC hypothesis is not without flaws (Clarkson, 1966a; Andrew et al., 1973). Wagatsuma and Ezoe (1985) found that the Al content of roots of a number of species exceeded root CEC and suggested that Al-tolerant plants excluded Al at the plasma membrane, leading to polymerization of hydroxy Al in the apoplasm. A second conceptual problem is the question of saturation. Tissue analyses of roots and shoots of *Cucumis sativus, Fagopyrum esculentum, Oryza sativa,* and *Raphanus sativus* showed that concentrations of Al in plant leaves increased rapidly only after Al concentrations in the roots exceeded a threshold value that corresponded to root CEC (Wagatsuma, 1984). These authors suggested that translocation to the shoots increased when root exchange sites were saturated by Al. This raises the question of whether the cell wall could protect plants from metal injury throughout their lifetimes or at Al concentrations sufficient to saturate cell wall binding sites (Kennedy et al., 1986; Taylor, 1987).

The cell wall, however, is not a static structure. Cation exchange sites may become saturated with Al, but if root growth is not inhibited, continued biosynthesis of cell wall material may provide sufficient new sites for immobilization of Al. Because the root tip is the site of cell wall synthesis and perhaps the primary site of Al phytotoxicity, synthesis of new cell wall material could protect the sensitive meristem throughout the lifetime of the plant. In *Vigna unguiculata,* the terminal centimeter of roots was the greatest site of Al accumulation, and Al concentrations were higher in an Al-sensitive cultivar than in an Al-tolerant cultivar (Horst et al., 1983). If immobilization at the cell wall was the primary mechanism of tolerance, then the Al-tolerant cultivar should have shown higher concentrations of Al in the root tip. Although this was not the case, these data do demonstrate that the dynamics of Al uptake and accumulation at the root tip may be a more important factor in Al toxicity and tolerance than uptake and accumulation at the whole root level.

## C. Selective Permeability of the Plasma Membrane

The plasma membrane may also act as a selective barrier to the uptake of Al; however, verification of this hypothesis presents technical difficulties. To demonstrate the potential role of the plasma membrane in exclusion of Al, it would first be necessary to differentiate between Al uptake into the apoplasm and symplasm. If restricted uptake into the symplasm was demonstrated, then the role of the plasma membrane in restricting Al uptake would need to be separated from that of the apoplasm. Nonetheless, limited data do support the hypothesis that the plasma membrane plays a role in exclusion of Al. For example, excised roots of a number of plant species showed increased uptake of Al with exposure to a range of metabolic inhibitors or anaerobiosis (Huett and Menary, 1979; Wagatsuma, 1983a), and species differences in membrane resistance of anaerobiosis were paralleled by differences in the tolerance of the species to Al (Wagatsuma, 1983a). These data suggest that the plasma membrane acts as a barrier to the movement of Al into the cytosol, that the effectiveness of this barrier is reduced under nonmetabolic conditions, and that differential permeability of the plasma membrane to Al may be involved in Al tolerance (Wagatsuma, 1983a). Because the plasma membrane represents the ultimate barrier between the cytosol and its external environment, understanding the role the plasma membrane plays in Al tolerance seems germane to an understanding of the physiology of Al stress.

## D. Plant-Induced pH Barrier in the Rhizosphere

The solubility of Al decreases rapidly in the range of pH 4.0 to 5.0; hence, plants that can maintain a relatively high pH in the root apoplasm or rhizosphere may create a pH barrier at the root-soil interface that could reduce Al toxicity in three ways: (1) hydrolysis or polymerization of Al, resulting in the formation of less toxic, soluble ion species, (2) formation of the sparingly soluble $Al(OH)_3 \cdot 3H_2O$ at near neutral pH, and (3) precipitation of $Al(OH)_2 \cdot H_2PO_4$ (Bartlett and Riego, 1972b; Blamey et al., 1983; Kinraide et al., 1985; Wagatsuma and Ezoe, 1985; Martin, 1986; Taylor, 1988c).

The plant-induced pH hypothesis is supported by studies that have demonstrated a relationship between an ability to maintain a relatively high pH in the growth medium and the Al tolerance of cultivars of *Hordeum vulgare* (Foy et al., 1967), *Pisum sativum* (Klimashevskii and Berezovskii, 1973), *Secale cereale* (Mugwira et al., 1976, 1978), *Triticum aestivum* (Foy et al., 1965, 1967, 1974; Dodge and Hiatt, 1972; Mugwira et al., 1976; 1978; Mugwira and Patel, 1977; Mugwira and Elgawhary, 1979; Foy and Fleming, 1982; Fleming, 1983), and x *Triticosecale* (Mugwira et al., 1976, 1978; Mugwira and Patel, 1977). In many of these studies, however, cultivar tolerance to Al was measured in nutrient solutions with mixed N, where plant-induced pH first declined and then increased rapidly. In mixed N solutions, pH declines as $NH_4^+$ is utilized in preference to $NO_3^-$. The inflection point of the pH rise coincides with depletion of $NH_4^+$ from the growth solution, and the period of pH rise occurs as $NO_3^-$ is more rapidly depleted (Fleming, 1983; Taylor and Foy, 1985d). Most reported correlations between cultivar tolerance to Al and plant-induced pH used pH of the growth solution at the end of the experimental period, after depletion of $NH_4^+$ had occurred.

Taylor and Foy (1985b, 1985c) questioned these data because cultivar differences in tolerance to Al were observable well before depletion of $NH_4^+$, the rapid pH increase, and the observed differences in plant-induced pH. Because inhibition of $NO_3^-$ uptake is one of the observed toxic effects of Al (Tang Van Hai et al., 1983; Taylor and Foy, 1985d; Jarvis and Hatch, 1986; Keltjens and Van Ulden, 1987a, 1987b), an Al-sensitive cultivar should show reduced rates of $NO_3^-$ uptake and a lower plant-induced pH than an Al-tolerant cultivar. This should be especially acute after depletion of $NH_4^+$, when $NO_3^-$ assimilation is most rapid. In addressing this criticism, Taylor and Foy (1985a, 1985b) found that the Al tolerance of cultivars of *Triticum aestivum* was correlated with the ability to resist acidification of the growth solution during the initial period of pH decline, prior to depletion of $NH_4^+$ (Figure 10–7). This pattern occurred in the pH range of 3.8 to 4.5 and was based upon a large number (20) of cultivars. Thus, small differences in plant-induced pH of the growth solution accounted for the variance over a broad range of tolerance ratings. Differences between cultivars in plant-induced pH of the growth solution appear to be due to differences in the relative uptake of cations and anions, especially $NH_4^+$ and $NO_3^-$ (Dodge and Hiatt, 1972; Mugwira and Patel, 1977; Fleming, 1983; Taylor and Foy, 1985d; Keltjens and Van Ulden, 1987a, 1987b).

Despite the mounting evidence supporting the role of N nutrition and plant-induced pH in Al tolerance, the hypothesis as described above cannot account for all current data. Several studies have failed to demonstrate a relationship between plant-induced pH and tolerance to Al in cultivars of *Glycine max, Hordeum vulgare, Phaseolus vulgaris,* x *Triticosecale,* and *Zea mays* (Foy et al., 1972; Clark, 1977, Mugwira and Elgawhary, 1979; Wagatsuma and Yamasaku, 1985). Also, manipulation of N source (hence, plant-induced pH) had little effect on the Al tolerance of cultivars of *Hordeum vulgare* and *Triticum aestivum* (Wagatsuma and Yamasaku, 1985; Taylor, 1988c). The data from *Triticum aestivum* (Taylor, 1988c) are particularly damaging to the plant-induced pH hypothesis because they

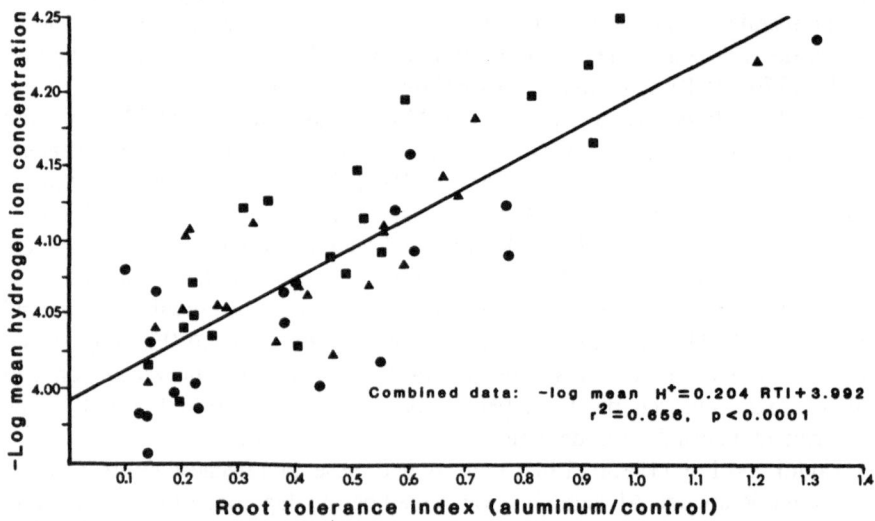

**Figure 10–7.** The relationship between root tolerance index (relative root weight) and the negative log of the mean hydrogen ion concentration induced in nutrient solutions by 20 cultivars of *Triticum aestivum*. Replicate 1 = triangles; replicate 2 = circles; replicate 3 = squares. Plants that resisted acidification of the nutrient solution were more tolerant to Al. (Reproduced by permission from G.J. Taylor and C.D. Foy, 1985a.)

deal with two benchmark cultivars used in earlier studies that supported the hypothesis. Furthermore, plant species appear to respond differently to varying Al concentrations and N source in solutions with constant pH (Kotze et al., 1977; Rorison, 1985). In one study, Rorison (1985) found that *Bromus erectus* (normally restricted to circumneutral soils) was sensitive to both $NH_4^+$ and Al, and *Holcus lanatus* (from soils of widely differing pH) was sensitive to $NH_4^+$, but Al ameliorated this effect, *Deschampsia flexuosa* (restricted to acidic soils) was unaffected by N source, and Al was inhibitory only at high concentrations in $NO_3^-$-based solutions. Although the effect of N on plant-induced pH may have been reduced by maintenance of solution pH at 4.5, it seems unlikely that such control could have fully eliminated pH differences in the rhizosphere or root apoplasm. These results imply that in some species N nutrition and tolerance to Al are not linked as described above, and adaptation to $NH_4^+$ and Al can occur independently.

These data place in question the role of plant-induced pH as a major factor determining tolerance to Al. This is not to suggest that pH does not affect the speciation of Al, the solubility of Al, or plant growth. Indeed, the effect of pH on plant performance under conditions of Al stress has been well established (Kinraide et al., 1985; Alva et al., 1986d; Parker et al., 1988). It would seem, however, that the effect of plant-induced pH in determining cultivar tolerance to Al may be relatively small.

## E. Exudation of Chelate Ligands

Plants could exclude Al from the symplasm if chelate ligands were released into the rhizosphere, where they formed stable complexes with Al, thus reducing activity of the free Al ion. In effect, chelate ligands would compete with root transport sites for complexing with Al, reducing uptake into the cytosol and mitigating toxic effects (Halvorson and Lindsay, 1972, 1977; Lindsay, 1974). This model accounts for the reduced toxicity of a number of Al-chelate complexes when they were supplied to plants in solution culture. For example, Al complexes with citrate, EDTA, humic acids, and soil organic matter extracts were less toxic to *Zea mays* than ionic Al (Bartlett and Riego, 1972a; Tan and Binger, 1986). Similarly, Al-EDTA was less toxic than ionic Al to *Agrostis stolonifera, Medicago sativa, Nicotiana plumbaginifolia, Oryza sativa,* and *Trifolium subterraneum* (Munns, 1965; Clarkson, 1966a; Thawornwong and Van Deist, 1974; Caldwell and Haug, 1980; Conner and Meredith, 1985b), Al-EDDHA reduced the toxicity of Al to *Hordeum vulgare, Fagopyrum esculentum, Glycine max,* and *Gossypium hirsutum* (Foy and Brown, 1964), and citrate reduced the phytotoxicity of Al to *Trifolium pratense* and *Nicotiana plumbaginifolia* (Conner and Meredith, 1985b; Kinraide et al., 1985).

Chelates have also been shown to reduce or eliminate symptoms of Al toxicity other than the growth effects described above. Amelioration of Al toxicity in *Zea mays* by chelation with humic acids was accompanied by a decline in leaf Al concentrations (Tan and Binger, 1986). Aluminum supplied as an EDTA complex prevented accumulation of Al by excised roots of *Zea mays* and prevented an

Al-induced desorption of Ca (Wagatsuma, 1983a). Although ionic Al inhibited the uptake of Ca by excised roots of *Triticum aestivum*, chelated Al was relatively ineffective (Johnson and Jackson, 1964). Also, in plasma membrane isolates from *Pisum sativum*, malic, glutamic, and citric acids all restored $Mg^{2+}$-dependent, $K^+$-stimulated ATPase activity that had been inhibited by Al. Citrate exhibited a similar protective effect on $Mg^{2+}$-dependent, $K^+$-stimulated ATPase activity in *Zea mays* (Suhayda and Haug, 1986; Figure 10–7) and prevented an increase in membrane rigidity induced by Al. The ability of carboxylic acids to detoxify Al may be related to the relative position of OH/COOH groups on the carbon chain. Strong detoxifiers generally have two OH/COOH groups attached to adjacent carbons, or two COOH groups directly connected, permitting incorporation of Al into a stable 5- or 6-bond cyclic structure (Hue et al., 1986).

Evidence that links exudation of chelate ligands with detoxification of Al in the apoplasm is still weak. In a series of experiments, Al-tolerant cell cultures of *Daucus carota* solubilized polymers of Al hydroxide and Al phosphate, and levels of soluble Al in the growth medium paralleled increases in citric acid (Ojima and Ohira, 1983; Ojima et al., 1984). Aluminum-tolerant cell cultures exuded more citric acid into the medium than Al-sensitive cultures (Ojima et al., 1984; Ojima and Ohira, 1985), and addition of malic or citric acid to Al-sensitive cultures mitigated the toxic effects of Al (Ojima and Ohira, 1982, 1985). Exudation of chelate ligands at the root cap in the form of mucilage may also be important. In *Vigna unguiculata*, half the Al in the apical centimeter of the root was bound in mucilage. Mucilage itself represented only 15% of root dry weight, and removal of mucilage before treatment with Al facilitated entry of Al into the root tissue and rendered roots more sensitive to Al (Horst et al., 1982). In a similar study, removal of the root cap of *Zea mays* resulted in rapid entry of Al into the quiescent center and the meristem. Such accumulation was not observed in intact roots treated with Al (Ojima et al., 1984).

Although the mucilage represents only a finite sink for Al, it is important to consider that mucilage is constantly being renewed at the actively growing root tip, provided the Al stress is not too severe (Hecht-Buchholz and Foy, 1981). Regeneration of mucilage would eliminate saturation of the binding capacity as a limiting factor. The biggest conceptual problem with exudation of chelates as an Al-tolerance mechanism is the considerable energetic cost of tolerance to the plant. Chelate ligands exuded into the rhizosphere would be vulnerable to microbial attack; hence, continued renewal of external chelates would be required to maintain a pool of chelate ligands of sufficient size to depress the ion activity of Al to acceptable levels (Taylor, 1987). In an ecological context, however, any adaptation that permits survival could be of adaptive significance.

## F. Internal Tolerance Mechanisms

Despite the presence of a barrier to Al uptake, some Al does enter the symplasm. In *Triticum aestivum*, uptake and distribution of Al in the symplasm varied with external Al concentration and cultivar. In each case, however, the cytosol represented the major pool of Al in the symplasm (63–69%), with smaller pools located in the mitochondria (15–22%) and the nuclei (10–21%) (Niedziela and Aniol, 1983; Aniol, 1984b). Once in the symplasm, the solubility of free $Al^{3+} \cdot 6H_2O$ is limited to less than $10^{-10}$ M at pH 7.0 by the neutral ion $Al(OH)_3 \cdot 3H_2O$. The presence of phosphate in the cytosol further reduces the solubility of $Al^{3+} \cdot 6H_2O$, by precipitation of $Al(OH)_2H_2PO_4$ (Martin, 1986). Even such low concentrations of $Al^{3+} \cdot 6H_2O$ are potentially phytotoxic because of the strong affinity of Al for oxygen donor compounds such as inorganic phosphate, nucleotides, RNA, DNA, proteins, carboxylic acids, phospholipids, polygalacturonic acids, heteropolysaccharides, lipopolysaccharides, flavonoids, anthocyanins, and others (Haug, 1984; Martin, 1986).

Although limited data on uptake of Al into the symplasm are available, such data suggest that internal tolerance mechanisms are required to explain differential tolerance to Al in some species. For example, Niedziela and Aniol (1983) found that an Al-tolerant cultivar of *Triticum aestivum* tolerated five to seven times higher concentrations of Al in mitochondria, nuclei, and the cytosol without lethal effects. Once Al has entered the symplasm, tolerance could be achieved by a number of possible mechanisms, including chelation by ligands in the cytosol, compartmentation in the vacuole, complexation by Al-binding proteins, and the evolution of Al-tolerant enzymes. Each of these possible mechanisms is discussed below.

## G. Chelation in the Cytosol

Chelation of Al by ligands in the cytosol could reduce the activity of monomeric forms of Al and, hence, their phytotoxic effects. Although Al forms stable complexes with a number of oxygen donor ligands, the potential role of carboxylic acids in the detoxification of Al has received the greatest attention. A developing body of literature has documented higher concentrations of carboxylic acids in Al-stressed roots of Al-tolerant crop cultivars when compared with Al-sensitive cultivars. For example, Al decreased concentrations of organic acids in roots of an Al-tolerant and an Al-sensitive cultivar of *Zea mays;* however, the Al-tolerant cultivar maintained higher concentrations (especially malic and transaconitic) than the Al-sensitive cultivar (Suhayda and Haug, 1986). Similarly, Al decreased concentrations of citric, succinic, and levulinic acids in roots of an Al-sensitive cultivar of *Hordeum vulgare,* whereas concentrations of levulinic acid decreased and concentrations of citric and succinic acids were unaffected by Al stress in an Al-tolerant cultivar (Foy et al., 1987a). Unfortunately, these data are inconclusive because reduced concentrations of organic acids are a common symptom of Al stress. Differences in organic acidic concentrations between cultivars grown under conditions of Al stress could be the cause, or simply the result, of differential tolerance. If tolerance were achieved by an exclusion mechanism, for example, then organic acidic concentrations would be lower in the more severely stressed, Al-sensitive cultivars. Aluminum-tolerant cultivars would be protected from stress by the exclusion mechanism, and organic acidic concentrations would be less affected.

In studies with *Hordeum vulgare, Phaseolus vulgaris, Pisum sativum, Triticum aestivum,* and *Zea mays,* Al-tolerant cultivars showed higher concentrations of citrate and malate than roots of Al-sensitive cultivars, both in the presence and absence of Al (Klimashevskii and Chernysheva, 1980; Lee and Foy, 1986). Because differences existed prior to Al stress, they may have played a role in the Al-stress response. In contrast with the findings most other researchers, Cambraia and others (1983) found that concentrations of organic acids, especially transaconitate and malate, increased in the roots of *Sorghum bicolor* under conditions of Al stress. Differences between cultivars were not detected in the absence of Al. It is clear that this aspect of the chelation hypothesis requires further research.

Evidence supporting detoxification of Al by chelation in the cytosol also comes from in vitro studies. Womack and Colowick (1979) and Viola et al., (1980) found that yeast hexokinase activity was inhibited by Al, but this effect was ameliorated by EDTA, EGTA, citrate, malate, 3-phosphoglycerate, phosphate, and catechol. Similar results were reported by Neet and others (1982) for a variety of Al chelates. Formation of an Al-ATP complex may have been responsible for inhibition of hexokinase activity, and amelioration by chelate ligands could have been the result of formation of stable complexes with Al (Womack and Colowick, 1979; Viola et al., 1980; Neet et al., 1982). In another study, malic, glutamic, and citric acids reduced the Al-induced inhibition of a $Mg^{2+}$-dependent, $K^+$-stimulated, plasma membrane ATPase in *Pisum sativum* (Matsumoto and Yamaya, 1986). Once again the inhibitory effect of Al was believed to be due to the formation of an Al-ATP complex and amelioration by carboxylic acids due to the formation of stable complexes with Al. Suhayda and Haug (1986) confirmed the protective effect of citrate on a $Mg^{2+}$-dependent, $K^+$-stimulated, plasma membrane ATPase in *Zea mays* (Figure 10–8) but suggested that citrate prevented a direct effect of Al on the ATPase itself rather than reducing the formation of an Al-ATP complex. Also, Suhayda and Haug (1984, 1985, 1986) demonstrated that citrate (and to a lesser extent oxalate, malate, and tartarate) prevented the binding of Al to calmodulin and partially restored the native structure of calmodulin once an Al-calmodulin complex had been formed. Previously, Siegel and Haug (1983a, 1983b) had demonstrated that Al disrupted the structural integrity of calmodulin and inhibited calcium-calmodulin-dependent phosphodiesterase and calmodulin-stimulated membrane-bound ATPase.

A criticism of the chelation hypothesis is that the stability of Al-chelate complexes in the cytosol has been overestimated because of the failure to recognize that $Al^{3+} \cdot 6H_2O$ is not the major species present in the near neutral pH of the cytosol. The nonhydrolyzed $Al^{3+} \cdot 6H_2O$ has been widely used in calculations of complex stability, but $Al(OH)_3 \cdot 3H_2O$ limits the solubility of free $Al^{3+} \cdot 6H_2O$ to less than $10^{-10}$ M at pH 7.0, and precipitation of $Al(OH)_2H_2PO_4$ even further reduces the solubility of $Al^{3+} \cdot 6H_2O$ (Martin, 1986). Although internal chelation and detoxification of Al is a possible Al-tolerance mechanism, further experimental support is required.

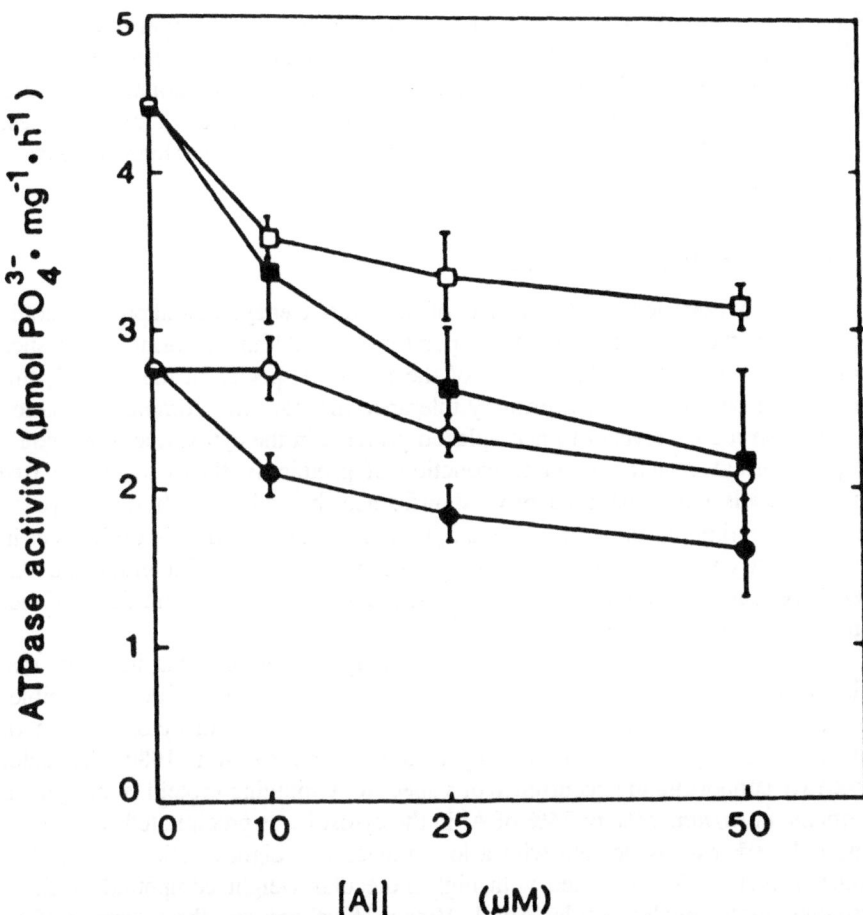

**Figure 10-8.** The effect of Al on a $K^+$-stimulated, $Mg^{2+}$-dependent ATPase in root plasma membranes isolated from an Al-sensitive cultivar of *Zea mays*. Adenosine triphosphatase activity was measured at pH 6.5 at 22°C. Circles represent activity of $Mg^{2+}$-dependent ATPase activity; squares represent activity of $K^+$-stimulated, $Mg^{2+}$-dependent ATPase activity. Open symbols represent data with 100 μM citrate added prior to Al; closed symbols represent data without citrate. Application of citrate reduced inhibition of ATPase activity, possibly a result of formation of an Al-citrate complex. (Reproduced by permission from C.G. Suhayda and A. Haug, 1986.)

## H. Compartmentation in the Vacuole

Aluminum could be immobilized in sites such as the vacuole which may be insensitive to the toxic effects of Al. Although compartmentation in the vacuole has received support as a mechanism of tolerance to other metals (Mathys, 1977;

Kime et al., 1982; Pfeffer et al., 1986), evidence supporting compartmentation of Al is lacking. Furthermore, Clarkson (1969) pointed out that meristematic root cells—those most affected by Al treatment— are not vacuolated in either Al-tolerant or Al-sensitive species. Thus, in order to reduce damage to these tissues, internal detoxification would require a mechanism other than transport to the vacuole.

## I. Aluminum-Binding Proteins

Formation of complexes with a variety of constitutive enzymes and proteins could account for the toxic effect of Al on plant growth. Although formation of such complexes may reduce the activity of the toxic Al species in the cytosol, this should not be regarded as a primary tolerance mechanism. Complex formation would also reduce levels of uncomplexed proteins in the cytosol, and this could impair normal protein function. Induction of protein synthesis to restore free protein levels would have adaptive significance, but this hypothesis has yet to receive experimental support. For example, Larkin (1987) found no differences in calmodulin content of whole roots or root tips between Al-tolerant and Al-sensitive cultivars under both Al-stressed and unstressed conditions. Other enzyme systems have yet to be tested.

Perhaps of greater interest is a growing body of literature that indicates that plants possess relatively low molecular weight (6,000 to 14,000), inducible, metal-binding proteins that may be specifically involved in detoxification of metals in the cytosol (see review by Robinson and Jackson, 1986). To date, evidence supporting the occurrence of a specific Al-binding protein is meager. In *Triticum aestivum*, 71% to 74% of Al in the cytosol was precipitated with TCA, and only 8% was associated with a low molecular weight fraction. These data indicate that Al is associated with high molecular weight compounds such as proteins and/or nucleic acids (Aniol, 1984b). Furthermore, the tolerance of an Al-tolerant and an Al-sensitive cultivar increased upon pretreatment with low concentrations of Al. This effect was more prominent in the Al-tolerant cultivar and was inhibited by cycloheximide, an inhibitor of protein synthesis (Aniol, 1984b). Such results suggest the occurrence of an inducible, Al-binding protein.

The possibility that Al is detoxified by formation of stable metal-protein complexes is interesting, but it is important to note that Al will likely be present primarily as $Al(OH)_3 \cdot 3H_2O$, a neutral species, or $Al(OH)_4^- \cdot 2H_2O$, an anionic species, in the cytosol at near neutral pH. Thus one cannot expect an Al-binding protein to bear resemblance to cation-binding proteins isolated from other species. Until an Al-binding protein has been isolated, Aniol's (1984b) results will remain equivocal.

## J. Evolution of Al-Tolerant Enzymes

If phytotoxic concentrations of Al are present in the cytosol, then conceivably evolution of metal-tolerant enzymes has occurred. Unfortunately, experimental evidence supporting this hypothesis is weak. Woolhouse (1969, 1970) assayed the acid phosphatase activity of excised roots of acid soil–tolerant (Al-tolerant), Pb-tolerant, and calcareous soil ecotypes of *Agrostis tenuis* grown under conditions of Al, Pb, and Ca stress. Acid phosphatase activity declined in each ecotype under each of the stresses imposed. Activity, however, was least inhibited under conditions of Al stress in the Al-tolerant ecotype, under Pb stress in the Pb-tolerant ecotype, and under Ca stress in the calcareous ecotype. Although these data seem to suggest the existence of Al-tolerant enzymes, this is not the only interpretation. If other mechanisms were responsible for Al, Pb, and Ca tolerance, then it would be expected that enzyme activity would be least affected in cases where the tolerance of the plant matched the stress imposed. This criticism was implicit in research with enzymes from Zn-tolerant, Cu-tolerant, Zn- and Cu-tolerant, and Zn- and Cu-sensitive ecotypes of *Silene cucubalis* (Mathys, 1975). In contrast to the experiments of Woolhouse, metals were supplied to both enzyme preparations (in vitro) and to excised roots (in vivo). When metals were supplied in vitro, nitrate reductase, malate dehydrogenase, glucose-6 phosphate dehydrogenase, and isocitrate dehydrogenase were inhibited by metal stress, and there was no evidence for the evolution of tolerant enzymes. When metals were supplied in vivo, nitrate reductase activity from Zn-tolerant plants was stimulated at Zn concentrations that were inhibitory to nitrate reductase activity from sensitive plants. Similar effects were reported for malate dehydrogenase and isocitrate dehydrogenase. Thus, the existence of differential enzyme activity among ecotypes of *Silene cucubalis* was dependent on the presence of intact cells. At present, it would appear that the literature does not support the occurrence of Al-tolerant enzymes. This may reflect a lack of research activity rather than lack of Al-tolerant enzymes. Recent research has indicated the occurrence of Ni-tolerant isozymes of acid phosphatase in *Anthoxanthum odoratum* (Cox and Thurman, 1978) and *Festuca rubra* (Johnston and Proctor, 1984).

## V. Future Research Directions and Concluding Remarks

The intense research effort addressing the acid soil–Al problem has failed to produce a clear understanding of the physiological and biochemical basis of Al phytotoxicity and tolerance. Nonetheless, considerable progress is being made. It appears that Al has several direct toxic effects, possibly in the apoplasm, at the membrane level, and in the cytosol. Perhaps the first effect occurs at the soil-root interface, where alterations in membrane structure and/or function seem likely. Aluminum inhibits the activity of a number of membrane-bound enzymes, especially transport enzymes, and these effects may be the basis of Al-induced mineral deficiencies. Cell elongation may also be reduced because of an effect of Al on cell wall components or cell wall assembly. Once in the cytosol, Al appears to bind to phosphate in DNA, repressing template activity and inhibiting DNA synthesis. Toxic effects on a number of enzyme-mediated reactions, especially in relation to phosphate metabolism, also seem likely. Although research into the effects of Al on the Ca-calmodulin system is still in its infancy, this model of Al toxicity presents an interesting possibility that could explain the toxic effects of Al on many cellular functions.

The concept of a host of adaptations contributing to the Al tolerance of a plant is in agreement with current research findings. Although a number of mechanisms for tolerance have been proposed, none is capable of explaining the range of observations reported. The current literature suggests that Al tolerance is due to a combination of both exclusion and internal tolerance mechanisms. The plasma membrane and the cell wall both seem to play a role in exclusion of Al from the symplasm. The extent to which these structures contribute to differential tolerance to Al, however, is not yet resolved. A plant-induced pH barrier in the rhizosphere would also seem to be a viable mechanism of tolerance, but it is likely of limited importance. The potential role of exudation of chelate ligands deserves further attention. With regard to internal tolerance mechanisms, chelation in the cytosol by carboxylic acids or Al-binding proteins, compartmentation in the vacuole, and evolution of Al-tolerant enzymes are all of potential significance, but assessment of their role in detoxification of Al in the cytosol must await further study.

It would appear that the key to our understanding of both Al phytotoxicity and tolerance lies at the cellular level, as both toxicity and tolerance are expressed in cell culture systems. Currently our knowledge of the cellular basis of Al stress is limited by our inability to accurately identify and quantify monomeric and polymeric forms of Al. Newly developed kinetic techniques and the application of $^{27}$Al NMR spectroscopy may provide the tools necessary to identify the various species of Al that are membrane mobile and/or phytotoxic.

Considering the possible importance of Al injury to the plasma membrane, little attention has focused on the impact of Al on the activity of membrane-bound enzymes and on membrane composition, structure, and function. Evaluation of the role of the plasma membrane, both as a target for Al-induced injury and as a metabolically active barrier to movement of Al into the symplasm, is important. In this area, many interesting questions could be addressed using existing technologies. For example, the toxic effect of Al on membrane function could be evaluated using techniques widely used in the study of ion transport. Also, the role of the plasma membrane in exclusion of Al from the symplasm could be investigated using graphite furnace atomic absorption spectroscopy and protoplast culture to measure the trace quantities of Al that cross the plasma membrane. Such techniques may also play a role in evaluating the role of the cell wall in tolerance.

Another important area of research is the impact of Al on cellular metabolism as it relates to the rapid toxic effects of Al and induced mineral deficiencies. The Ca-calmodulin regulatory system may be a primary site of Al injury and could account for a wide variety of Al-induced injuries, including induced Ca deficiency. Research in this area would not require development of new techniques. Also, $^{31}$P NMR techniques provide an unparalleled view of in vivo metabolic process that could account for a variety of Al-induced injuries, including induced phosphate deficiency. Such information would be particularly valuable in association with more accurate information regarding the distribution of Al among various cellular compartments.

Although cellular processes may be crucial to our understanding of Al stress, this should not lead to neglect of whole-plant processes. The performance of plants under conditions of Al stress will reflect injury to processes such as developmental control, translocation, and carbon partitioning. The impact of Al on the function of growth regulators such as cytokinins has only recently attracted attention and may be of considerable importance. The question of how Al affects resource partitioning, or how plants respond to stress by altering patterns of resource partitioning, has not been addressed. In studies with whole plants, care should be exercised to consider the impact of growth dynamics. For example, the relationship between the toxic effects of Al on growth and the requirement for continued growth to produce new cell wall material or chelate ligands for detoxification of Al requires detailed examination. Of course, in any model of phytotoxicity or tolerance, the unique sensitivity of meristematic tissues must be addressed.

As more research activity is directed toward the biology of Al stress, our understanding of the physiological and biochemical basis of Al phytotoxicity will mature. Judicious use of Al-tolerant and Al-sensitive cultivars or ecotypes, as well as new experimental and analytical techniques will certainly help to further our understanding of the mechanisms of Al toxicity and tolerance. Such progress should bode well for our endeavor to develop plant cultivars or ecotypes adapted for growth on acidic, Al-toxic soils and will certainly aid in our understanding of the role that Al toxicity plays in the vegetation dynamics of ecosystems impacted by acid precipitation.

## References

Alam, S.M., and W.A. Adams. 1979. J Plant Nutr 1:365–375.

Alva, A.K., C.J. Asher, and D.G. Edwards. 1986a. Aust J Agric Res 37:375–382.

Alva, A.K., F.P.C. Blamey, D.G. Edwards, and C.J. Asher. 1986b. Commun Soil Sci Plant Anal 17:1271–1280.

Alva, A.K., D.G. Edwards, C.J. Asher, and F.P.C. Blamey. 1986c. Soil Sci Soc Amer J 50:133–137.

Alva, A.K., D.G. Edwards, C.J. Asher, and F.P.C. Blamey. 1986d. Soil Sci Soc Amer J 50:959–962.

Alva, A.K., D.G. Edwards, C.J. Asher, and S. Suthipradit. 1987. Agron J 79:302–306.

Andrew, C.S., 1976. Aust J Agric Res 27:611–623.

Andrew, C.S., A.D. Johnson, and R.L. Sandland. 1973. Aust J Agric Res 24:325–339.

Andrew, C.S., and R.J. Vanden Berg. 1973. Aust J Agric Res 24:341–351.

Aniol, A. 1983. Biochem Physiol Pflanzen 178:11–20.

Aniol, A. 1984a. Z Pflanzenzuchtg 93:331–339.

Aniol, A. 1984b. Plant Physiol 76:551–555.

Aniol, A., and J.P. Gustafson. 1984. Can J Genet Cytol 26:701–705.

Antonovics, J., A.D. Bradshaw, and R.G. Turner. 1971. Adv Ecol Res 7:1–85.

Armiger, W.H., C.D. Foy, A.L. Fleming, and B.E. Caldwell. 1968. Agron J 60:67–70.

Awad, A.S., D.G. Edwards, P.J. Milham. 1976. Plant Soil 45:531–542.

Bache, B.W. 1980. *In* T.C. Hutchinson, and M. Havas. eds. *Effects of acid precipitation on terrestrial ecosystems*, 183–202. Plenum Press, New York.

Bache, B.W., and W.M. Crooke. 1981. Plant Soil 61:365–375.

Baligar, V.C., R.J. Wright, T.B. Kinraide, C.D. Foy, and J.H. Elgin, Jr. 1987. Agron J 79:1038–1044.

Bartlett, R.J., and D.C. Riego. 1972a. Plant Soil 37:419–423.

Bartlett, R.J., and D.C. Riego. 1972b. Soil Sci 114:194–200.

Bennet, R.J., C.M. Breen, and V. Bandu. 1985b. S Afr J Bot 51:363–370.

Bennet, R.J., C.M. Breen, and M.V. Fey. 1985a. S Afr J Bot 51:355–362.

Bennet, R.J., C.M. Breen, and M.V. Fey. 1985c. S Afr J Plant Soil 2:1–7.

Bennet, R.J., C.M. Breen, and M.V. Fey, 1985d. S Afr J Plant Soil 2:8–17.

Bennet, R.J., C.M. Breen, and M.V. Fey. 1987a. S Afr J Plant Soil 3:11–17.

Bennet, R.J., C.M. Breen, and M.V. Fey. 1987b. Environ Exp Bot 27:91–104.

Bertsch, P.M. 1987. Soil Sci Soc Amer J 51:825–828.

Bertsch, P.M., G.W. Thomas, and R.I. Barnhisel. 1986. Soil Sci Soc Amer J 50:825–830.

Blamey, F.P.C., D.G. Edwards, and C.J. Asher. 1983. Soil Sci 136:197–207.

Boye-Goni, S.R., and V. Marcarian. 1985. Crop Sci 25:749–752.

Brenes, E., and R.W. Pearson. 1973. Soil Sci 116:295–302.

Buss, G.R., J.A. Lutz, Jr., and G.W. Hawkins. 1975. Agron J 67:331–334.

Caldwell, C.R., and A. Haug. 1980. Physiol Plant 50:183–193.

Cambraia, J., F.R. Galvani, M.M. Estevao, and R. Sant'Anna. 1983. J Plant Nutr 6: 313–322.

Carvalho, M.M. De, C.S. Andrew, D.G. Edwards, and C.J. Asher. 1980. Aust J Agric Res 31:61–76.

Carvalho, M.M. De, H.V.A. Bushby, and D.G. Edwards. 1981a. Soil Biol Biochem 13:541–542.

Carvalho, M.M. De, D.G. Edwards, C.S. Andrew, and C.J. Asher. 1981b. Agron J 73:261–265.

Clark, R.B. 1977. Plant Soil 47:653–662.

Clarkson, D.T. 1965. Ann Bot 29:309–315.

Clarkson, D.T. 1966a. J Ecol 54:167–178.

Clarkson, D.T. 1966b. Plant Physiol 41:165–172.

Clarkson, D.T. 1967. Plant Soil 27:347–356.

Clarkson, D.T. 1969. *In* Rorison, I. H. ed. *Ecological aspects of the mineral nutrition of plants:* A symposium of the British Ecological Society, 381–397. Blackwell Scientific Publications, Oxford, England.

Clarkson, D.T., and J. Sanderson. 1971. J Exp Bot 22:837–851.

Conner, A.J., and C.P. Meredith. 1985a. Planta 166:466–473.

Conner, A.J., and C.P. Meredith. 1985b. J Exp Bot 36:870–880.

Cox, R.M., and D.A. Thurman. 1978. New Phytol 80:17–22.

Culvenor, R.A., R.N. Oram, and J.T. Wood. 1986. Aust J Agric Res 37:397–408.

Cumming, J.R., R.T. Eckert, and L.S. Evans. 1985. Can J Bot 63:1099–1103.

Cumming, J.R., R.T. Eckert, and L.S. Evans. 1986 Can J For Res 16:864–867.

Cunningham, S.D., and D.N. Munns. 1984a. Soil Sci Soc Amer J 48:1273–1276.

Cunningham, S.D., and D.N. Munns. 1984b. Soil Sci Soc Amer J 48:1276–1280.

Devine, T.E., C.D. Foy, A.L. Fleming, C.H. Hanson, T.A. Campbell, J.E. McMurtrey III, and J.W. Schwartz. 1976. Plant Soil 44:73–79.

Dodge, C.S., and A.J. Hiatt. 1972. Agron J 64:476–481.

Edwards, J.H., B.D. Horton, and H.C. Kirkpatrick. 1976. J Amer Soc Hort Sci 101: 139–142.

Entry, J.A., K. Cromack, Jr., S.G. Stafford, and M.A. Castellano. 1987. Can J For Res 17:865–871.

Ernst, W.H.O. 1975. *In* T.C. Hutchinson, S. Epstein, A.L. Page, J. Van Loon, and T. Davey, eds. *International Conference on Heavy Metals in the Environment*, Vol. II(1), 121–136.

Ernst, W. 1976. *In* T.A. Mansfield, ed. *Effects of air pollutants on plants*, 115–133. Cambridge University Press, Cambridge.

Fageria, N.K., and J.R.P. Carvalho. 1982. Plant Soil 69:31–44.

Fiskesjo, G. 1983. Physiol Plant 59:508–511.

Fleming, A.L. 1983. Agron J 75:726–730.

Fleming, A.L., and C.D. Foy. 1968. Agron J 60:172–176.

Foy, C.D. 1974. *In* E.W. Carson, ed. The plant root and its environment, 601–642. University Press of Virginia, Charlottesville.

Foy, C.D. 1983a. Iowa State J Res 57:339–354.

Foy, C.D. 1983b. Iowa State J Res 57:355–391.

Foy, C.D. 1984. *In* F. Adams, ed. *Soil acidity and liming*, 2d ed., 57–97. Amer Soc Agron, Crop Sci Soc Amer, and Soil Soc Amer, Madison, Wisc.

Foy, C.D., and J.C. Brown. 1963. Soil Sci Soc Amer Proc 27:403–407.

Foy, C.D., and J.C. Brown. 1964. Soil Sci Soc Amer Proc 28:27–32.

Foy, C.D., G.R. Burns, J.C. Brown, and A.L. Fleming. 1965. Soil Sci Soc Amer Proc 29:64–67.

Foy, C.D., R.L. Chaney, and M.C. White. 1978. Ann Rev Plant Physiol 29:511–566.

Foy, C.D., and A.L. Fleming. 1982. J Plant Nutr 5:1313–1333.

Foy, C.D., A.L. Fleming, and W.H. Armiger. 1969. Agron J 61:505–511.

Foy, C.D., A.L. Fleming, G.R. Burns, and W.H. Armiger. 1967. Soil Sci Soc Amer Proc 31:513–521.

Foy, C.D., A.L. Fleming, and G.C. Gerloff. 1972. Agron J 64:815–818.

Foy, C.D., G.C. Gerloff, and W.H. Gabelman. 1973. J Amer Soc Hort Sci 98:427–432.

Foy, C.D., H.N. Lafever, J.W. Schwartz, and A.L. Fleming. 1974. Agron J 66:751–758.

Foy, C.D., E.H. Lee, and S.B. Wilding. 1987a. J Plant Nutr 10:1089–1101.

Foy, C.D., D.H. Smith, Jr., and L.W. Briggle. 1987b. J Plant Nutr 10:1163–1174.

Franco, A.A., and D.N. Munns. 1982. Soil Sci Soc Amer J 46:296–301.

Furlani, P.R., and R.B. Clark. 1981. Agron J 73:587–594.

Gomes, M.M.S., J. Cambraia, R. Sant'Anna, and M.M. Estevao. 1985. J Plant Nutr 8:457–465.

Grimme, H. 1983. Z Pflanzenernaehr Bodenk 146:666–676.

Halvorson, A.D., and W.L. Lindsay. 1972. Soil Sci Soc Amer Proc 36:755–761.

Halvorson, A.D., and W.L. Lindsay. 1977. Soil Sci Soc Amer J 41:531–534.

Hampp, R., and H. Schnabl. 1975. Z Pflanzenphysiol 76:300–306.

Hanson, W.D., and E.J. Kamprath. 1979. Agron J 71:581–586.

Hartel, P.G., and M. Alexander. 1983. Soil Sci Soc Amer J 47:502–506.

Hartel, P.G., A.M. Whelan, and M. Alexander. 1983. Soil Sci Soc Amer J 47:514–517.

Haug A (1984) CRC Crit Rev Plant Sci 1:345–373.

Haug, A.R., and C.R. Caldwell. 1985. *In* J.B. St. John, E. Berlin, and P.C. Jackson, eds. *Frontiers of membrane research in agriculture* (Beltsville symposium 9), 359–381. Rowman & Allanheld, Totowa, N.J.

Haynes, R.J., and T.E. Ludecke. 1981. Plant Soil 62:241–254.

Hecht-Buchholz, Ch. 1983. Plant Soil 72:151–165.

Hecht-Buchholz, Ch., and C.D. Foy. 1981. Plant Soil 63:93–95.

Hoddinott, J., and C. Richter. 1987. J Plant Nutr 10:443–454.

Hodges, S.C. 1987. Soil Sci Soc Amer J 51:57–64.

Hohenburg, J.S., and D.N. Munns. 1984. Agron J 76:477–481.

Horst, W.J. 1985. Z Pflanzenernaehr Bodenk 148:335–348.

Horst, W.J., A. Wagner, and H. Marschner. 1982. Z Pflanzenphysiol 105:435–444.

Horst, W.J., A. Wagner, and H. Marschner. 1983. Z Pflanzenphysiol 109:95–103.

Huck, M.G. 1972. Plant Cell Physiol 13:7–14.

Hue, N.V., G.R. Craddock, and F. Adams. 1986. Soil Sci Soc Amer J 50:28–34.

Huett, D.O., and R.C. Menary. 1979. Aust J Plant Physiol 6:643–653.

Huett, D.O., and R.C. Menary. 1980. Aust J Plant Physiol 7:101–111.

Jardine, P.M., and L.W. Zelazny. 1986. Soil Sci Soc Amer J 50:895–900.

Jardine, P.M., and L.W. Zelazny. 1987a. Soil Sci Soc Amer J 51:885–889.

Jardine, P.M., and L.W. Zelazny. 1987b. Soil Sci Soc Amer J 51:889–892.

Jarvis, S.C., and D.J. Hatch. 1985. J Exp Bot 36:1075–1086.

Jarvis, S.C., and D.J. Hatch. 1986. Plant Soil 95:43–55.

Jarvis, S.C., and A.D. Robson. 1983a. Aust J Agric Res 34:341–353.

Jarvis, S.C., and A.D. Robson. 1983b. Aust J Agric Res 34:355–365.

Jarvis, S.C., and A.D. Robson. 1983c. Plant Soil 75:235–243.

Johnson, R.E., and W.A. Jackson. 1964. Soil Sci Soc Amer Proc 28:381–386.

Johnston, W.R., and J. Proctor. 1984. New Phytol 96:95–101.

Keltjens, W.G., and P.S.R. Van Ulden. 1987a. Plant Soil 104:227–234.

Keltjens, W.G., and P.S.R. Van Ulden. 1987b. J Plant Nutr 10:841–856.

Kennedy, C.W., W.C. Smith, Jr., and M.T. Ba. 1986. J Plant Nutr 9:1123–1133.

Kerridge, P.C., and W.E. Kronstad. 1968. Agron J 60:710–711.

Keser, M., B.F. Neubauer, and F.E. Hutchinson. 1975. Agron J 67:84–88.

Keser, M., B.F. Neubauer, F.E. Hutchinson, and D.B. Verrill. 1977. Agron J 69:347–350.

Keyser, H.H., and D.N. Munns. 1979a. Soil Sci Soc Amer J 43:500–503.

Keyser, H.H., and D.N. Munns. 1979b. Soil Sci Soc Amer J 43:519–523.

Keyser, H.H., D.N. Munns, and J.S. Hohenberg. 1979. Soil Sci Soc Amer J 43:719–722.

Kime, M.J., R.G. Ratcliffe, and B.C. Loughman. 1982. J Exp Bot 33:670–681.

Kinraide, T.B., R.C. Arnold, and V.C. Baligar. 1985. Physiol Plant 65:245–250.

Kinraide, T.B., and D.R. Parker. 1987a. Physiol Plant 71:207–212.

Kinraide, T.B., and D.R. Parker. 1987b. Plant Physiol 83:546–551.

Kirkpatrick, H.C., J.M. Thompson, and J.H. Edwards. 1975. Hort Sci 10:132–134.

Klimashevskii, E.L., and K.K. Berezovskii. 1973. Soviet Plant Physiol 20:51–54.

Klimashevskii, E.L., and N.F. Chernysheva. 1980. Soviet Agric Sci 1980(2):5–8.

Klimashevskii, E.L., and V.M. Dedov. 1975. Soviet Plant Physiol 22:1040–1046.

Klimashevskii, E.L., Yu. A. Markova, M.L. Seregina, D.M. Grodzinskii, and T.D. Kozarenko. 1970. Soviet Plant Physiol 17:372–378.

Klimashevsky, E.L. 1970. Agrochimica 14:232–241.

Konishi, S., S. Miyamoto, and T. Taki. 1985. Soil Sci Plant Nutr 31:361–368.

Kotze, W.A.G., C.B. Shear, and M. Faust. 1977. J Amer Soc Hort Sci 102:279–282.

Lance, J.C., and R.W. Pearson. 1969. Soil Sci Soc Amer Proc 33:95–98.

Larkin, P.J. 1987. Aust J Plant Physiol 14:377–385.

Lee, C.R. 1971. Agron J 63:604–608.

Lee, E.H., and C.D. Foy. 1986. J Plant Nutr 9:1481–1498.

Lee, J., M.W. Pritchard, J.R. Sedcole, and M.R. Robertson. 1984. J Plant Nutr 7: 1635–1650.

Lindsay, W.L. 1974. *In* E.W. Carson, ed. *The plant root and its environment*, 507–522. University Press of Virginia, Charlottesville.

Long, F.L., and C.D. Foy. 1970. Agron J 62:679–681.

Lowendorf, H.S., A.M. Baya, and M. Alexander. 1981. Appl Environ Microbiol 42:951–957.

Lowther, W.L. 1980. New Zealand J Exp Agric 8:131–138.

McCain, S., and M.S. Davies. 1984. New Phytol 96:589–599.

McCormick, L.H., and F.Y Borden. 1972. Soil Sci Soc Amer Proc 36:799–802.

McCormick, I.H., and F.Y. Borden. 1974. Soil Sci Soc Amer Proc 38:931–934.

MacLean, A.A., and T.C. Chiasson. 1966. Can J Soil Sci 46:147–153.

Malavolta, E., F.D. Nogueira, I.P. Oliveira, L. Nakayama, and I. Eimori. 1981. J Plant Nutr 3:687–694.

Martin, R.B. 1986. Clinical Chem 32:1797–1806.

Mathys, W. 1975. Physiol Plant 33:161–165.

Mathys, W. 1977. Physiol Plant 40:130–136.

Matsumoto, H. 1988. Plant Cell Physiol 29:281–287.

Matsumoto, H., E. Hirasawa, H. Torikai, E. Takahashi. 1976. Plant Cell Physiol 17: 127–137.

Matsumoto, H., and S. Morimura. 1980. Plant Cell Physiol 21:951–959.

Matsumoto, H., S. Morimura, E. Takahashi. 1977a. Plant Cell Physiol 18:325–335.

Matsumoto, H., S. Morimura, and E. Takahashi. 1977b. Plant Cell Physiol 18:987–993.

Matsumoto, H., and T. Yamaya. 1986. Soil Sci Plant Nutr 32:179–188.

Mayz De Manzi, J., and P.M. Cartwright. 1984. Plant Soil 80:423–430.

Mengel, D.B., and E.J. Kamprath. 1978. Agron J 70:959–963.

Meredith, C.P. 1978a. Plant Sci Letters 12:17–24.

Meredith, C.P. 1978b. Plant Sci Letters 12:25–34.

Mesdag, J., I.A.J. Slootmaker, J. Post Jr. 1970. Euphytica 19:163–174.

Morimura, S., and H. Matsumoto. 1978. Plant Cell Physiol 19:429–436.

Morimura, S., E. Takahashi, and H. Matsumoto. 1978. Z Pfanzenphysiol 88:395–401.

Mugwira, I.M. and S.M. Elgawhary. 1979. Soil Sci Soc Amer J 43:736–740.

Mugwira, I.M., S.M. Elgawhary, and K.I. Patel. 1976. Agron J 68:782–786.

Mugwira, I.M., S.M. Elgawhary, and S.U. Patel. 1978. Plant Soil 50:681–690.

Mugwira, I.M., and S.U. Patel. 1977. Agron J 69:719–722.

Munn, D.A., and R.E. McCollum. 1976. Agron J 68:989–991.

Munns, D.N. 1965. Aust J Agric Res 16:743–755.

Munns, D.N., Keyser, H.H., V.W. Fogle, J.S. Hohenberg, T.L. Righetti, D.L. Lauter, M.G. Zaroug, K.L. Clarkin, and K.W. Whitacre. 1979. Agron J 71:256–260.

Naidoo, G., J.McD. Stewart, and R.J. Lewis. 1978. Agron J 70:489–492.

Neet, K.E., T.C. Furman, and W.J. Hueston. 1982. Arch Biochem Biophys 213:14–25.

Niedziela, G., and A. Aniol. 1983. Acta Biochem Pol 30:99–105.

Ohki, K. 1985. Agron J 77:951–956.

Ohki, K. 1986. Crop Sci 26:572–575.

Ojima, K., H. Abe, and K. Ohira. 1984. Plant Cell Physiol 25:855–858.

Ojima, K., and K. Ohira. 1982. *Plant tissue culture*, Proc. 5th Int. Cong. Plant Tissue Cell Culture, 475–476, Japanese Association for Plant Tissue Culture, Tokyo.

Ojima, K., and K. Ohira. 1983. Plant Cell Physiol 24:789–797.

Ojima, K., and K. Ohira. 1985. Plant Cell Physiol 26:281–286.

Osborne, G.J., J.E. Pratley, and W.P. Stewart. 1981. Field Crops Res 3:347–358.

Ouellette, G.J., and L. Dessureaux. 1958. Can J Plant Sci 38:206–214.

Ownby, J.D., and L. Dees. 1985. New Phytol 101:325–332.

Pan, W.L., A.G. Hopkins, and W.A. Jackson. 1988. Commun Soil Sci Plant Anal 19: 1143–1153.

Parker, D.R., T.B. Kinraide, and L.W. Zelazny. 1988. Soil Sci Soc Amer J 52:438–444.

Parker, D.R., L.W. Zelazny, and T.B. Kinraide. 1987. Soil Sci Soc Amer J 51:488–491.

Pavan, M.A., and F.T. Bingham. 1982. Soil Sci Soc Amer J 46:993–997.

Pavan, M.A., F.T. Bingham, and P.F. Pratt. 1982. Soil Sci Soc Amer J 46:1201–1207.

Pfeffer, P.E., S.-I. Tu, W.V. Gerasimowicz, and R.T. Boswell. 1987. Planta 172: 200–208.

Pfeffer, P.E., S.-I. Tu, W.V. Gerasimowicz, and J.R. Cavanaugh. 1986. Plant Physiol 80:77–84.

Pierre, W.H., J.R. Webb, and W.D. Shrader. 1971. Agron J 63:291–297.

Rasmussen, H.P. 1968. Planta 81:28–37.

Rechcigl, J.E., R.B. Reneau, Jr., D.D. Wolf, W. Kroontje, and S.W. Scoyoc. 1986. Commun Soil Sci Plant Anal 17:27–44.

Rhue, R.D. 1979. *In* H. Mussell and R.C. Staples, eds. *Stress physiology in crop plants,* 61–80. John Wiley and Sons, New York.

Rhue, R.D., and C.O. Grogan. 1977. Agron J 69:755–760.

Rhue, R.D., and C.O. Grogan, E.W. Stockmeyer, and H.L. Everett. 1978. Crop Sci 18:1063–1067.

Rice, W.A., D.C. Penney, and M. Nyborg. 1977. Can J Soil Sci 57:197–203.

Robinson, N.J., and P.J. Jackson. 1986. Physiol Plant 67:499–506.

Rorison, I.H. 1965. New Phytol 64:23–27.

Rorison, I.H. 1985. J Ecol 73:83–90.

Rorison, I.H. 1986. Experimentia 42:357–362.

Sampson, M., D. Clarkson, and D.D. Davies. 1965. Science 148:1476–1477.

Santoro, L.G., T.E. Soares, and A.C. Magalhaes. 1984. Phyton (Argentina) 44:75–80.

Sarkunan, V., and C.C. Biddapa. 1982. Oryza 19:188–190.

Sarkunan, V., C.C. Biddapa, and S.K. Nayak. 1984. Curr Sci 53:822–824.

Sartain, J.B., and E.J. Kamprath. 1975. Agron J 67:507–510.

Schaedle, M., F.C. Thornton, and D.J. Raynal. 1986. J Plant Nutr 9:1227–1238.

Schier, G.A. 1985. Can J For Res 15:29–33.

Sheppard, L.J., and M.J.S. Floate. 1984. Plant Soil 80:301–306.

Siegel, N., R. Coughlin, and A. Haug. 1983. Biochem Biophys Res Commun 115:512–517.

Siegel, N., and A. Haug. 1983a. Physiol Plant 59:285–291.

Siegel, N., and A. Haug. 1983b. Biochem Biophys Acta 744:36–45.

Siegel, N., C. Suhayda, and A. Haug. 1982. Physiol Chem Physics 14:165–167.

Smith, R.H., S. Bhaskaran, and K. Shertz. 1983. Plant Cell Reports 2:129–132.

Soileau, J.M., O.P. Engelstad, and J.B. Martin, Jr. 1969. Soil Sci Soc Amer Proc 33:919–924.

Stolen, O., and S. Andersen. 1978. Hereditas 88:101–105.

Suhayda, C.G., and A. Haug. 1984. Biochem Biophys Res Commun 119:376–381.

Suhayda, C.G., and A. Haug. 1985. Can J Biochem Cell Biol 63:1167–1175.

Suhayda, C.G., and A. Haug. 1986. Physiol Plant 68:189–195.

Takagi, S. 1976. Soil Sci Plant Nutr 22:423–433.

Tan, K.H., and A. Binger. 1986. Soil Sci 141:20–25.

Tanaka, A., T. Tadano, K. Yamamoto, and N. Kanamura. 1987. Soil Sci Plant Nutr 33:43–55.

Tang Van Hai, X. De Cuyper, and H. Laudelout. 1983. Agronomie 3:417–421.

Taylor, G.J. 1987. J Plant Nutr 10:1213–1222.

Taylor, G.J. 1988a. In H. Sigel, ed. *Metal ions in biological systems*, Vol. 24, *Aluminum and its role in biology*, 123–163. Marcel Dekker, New York.

Taylor, G.J. 1988b. In H. Sigel, ed. *Metal ions in biological systems*, Vol. 24, *Aluminum and its role in biology*, 165–198. Marcel Dekker, New York.

Taylor, G.J. 1988c. Can J Bot 66:694 699.

Taylor, G.J., and C.D. Foy. 1985a. Amer J Bot 72:695–701.

Taylor, G.J., and C.D. Foy. 1985b. Amer J Bot 72:702–706.

Taylor, G.J., and C.D. Foy. 1985c. J Plant Nutr 8:613–628.

Taylor, G.J., and C.D. Foy. 1985d. Can J Bot 63:2181–2186.

Taylor, G.J., and C.D. Foy. 1985e. J Plant Nutr 8:811–824.

Thawornwong, N., and A. Van Diest. 1974. Plant Soil 41:141–159.

Thornton, F.C., M. Schaedle, and D.J. Raynal. 1986. Can J For Res 16:892–896.

Thornton, F.C., M. Schaedle, and D.J. Raynal. 1987. Environ Exp Bot 27:489–498.

Tracey, A.S., and T.L. Boivin. 1983. J Amer Chem Soc 105:4901–4905.

Unruh, L., and D. Whitney. 1986. Paper presented to the Maximum Wheat Research Systems Yield Workshop, Denver, Colorado, March 5–7.

Van Praag, H.J., F. Weissen, S. Sougnez-Remy, and G. Carletti. 1985. Plant Soil 83: 339–356.

Van Wambeke, A. 1976. In M.J. Wright and S.A. Ferrari, eds. *Plant adaptation to mineral stress in problem soils*, 15–24. Cornell Univ Agric Exp Sta.

Veltrup, W. 1983. J Plant Nutr 6:349–361.

Viola, R.E., J.F. Morrison, W.W. Cleland. 1980. Biochemistry 19:3131–3137.

Vose, P.B., and P.J. Randall. 1962. Nature 196:85–86.

Wagatsuma, T. 1983a. Soil Sci Plant Nutr 29:323–333.

Wagatsuma, T. 1983b. Soil Sci Plant Nutr 29:499–515.

Wagatsuma, T. 1984. Soil Sci Plant Nutr 30:345–358.

Wagatsuma, T., and Y. Ezoe. 1985. Soil Sci Plant Nutr 31:547–561.

Wagatsuma, T., and M. Kaneko. 1987. Soil Sci Plant Nutr 33:57–67.

Wagatsuma, T., M. Kaneko, and Y. Hayasaka. 1987. Soil Sci Plant Nutr 33:161–175.

Wagatsuma, T., and K. Yamasaku. 1985. Soil Sci Plant Nutr 31:521–535.

Wallace, S.U., and I.C. Anderson. 1984. Agron J 76:5–8.

Wallace, S.U., S.J. Henning, I.C. Anderson. 1982. Iowa State J Res 57:97–106.

Weis, C., and A. Haug. 1987. Arch Biochem Biophys 254:304–312.

Whelan, A.M., and M. Alexander. 1986. Plant Soil 92:363–371.

Wissemeier, A.H., F. Klotz, W.F. Horst. 1987. J Plant Physiol 129:487–492.

Womack, F.C., and S.P. Colowick. 1979. Proc Natl Acad Sci USA 76:5080–5084.

Wood, M., J.E. Cooper, and A.J. Holding. 1984a. Plant Soil 78:367–379.

Wood, M., J.E. Cooper, and A.J. Holding. 1984b. Plant Soil 78:381–391.

Woolhouse, H.W. 1969. In I.H. Rorison, ed. *Ecological aspects of the mineral nutrition of plants*. A symposium of the British Ecological Society, 357–380. Blackwell Scientific Publications, Oxford.

Woolhouse, H.W. 1970. In J.B. Harborne, ed. *Phytochemical phylogeny*, 207–231. Academic Press, London.

Wright, R.J., V.C. Baligar, and S.F. Wright. 1987. Soil Sci 144:224–232.

Zhang, G., and G.J. Taylor. 1988. Commun Soil Sci Plant Anal. 19:1195–1205.

Zhao, X.-J., E. Sucoff, and E.J. Stadelmann. 1987. Plant Physiol 83:159–162.

# Index

## T

## Z